WARD LIBRARY

R

HC 79 .E5 G52 1986

Gilpin, Alan.

Environmental planning

CALIFORNIA STATE UNIVERSITY, HAYWARD
LIBRARY

ENVIRONMENTAL PLANNING

BY THE SAME AUTHOR

The Human Environment: The World After Stockholm
Environment Policy in Australia
The Australian Environment: Twelve Controversial Issues
Air Pollution
Dictionary of Energy Technology
Dictionary of Environmental Terms
Dictionary of Economic Terms

"The field of environmental planning embraces all aspects of physical and social problems which have a direct bearing on the quality of our surroundings"

From *The Human Habitat*, A contribution by the Greater London Council towards the United Nations Conference on the Human Environment, Stockholm, 1972

ENVIRONMENTAL PLANNING

A Condensed Encyclopedia

by

Alan Gilpin
Commissioner of Inquiry
Ministry of Planning and Environment
New South Wales, Australia

np NOYES PUBLICATIONS
Park Ridge, New Jersey, U.S.A.

Copyright © 1986 by Alan Gilpin
No part of this book may be reproduced in any form
without permission in writing from the Publisher.
Library of Congress Catalog Card Number: 86-17985
ISBN: 0-8155-1103-5
Printed in the United States

Published in the United States of America by
Noyes Publications
Mill Road, Park Ridge, New Jersey 07656

10 9 8 7 6 5 4 3 2 1

Library of Congress Cataloging-in-Publication Data

Gilpin, Alan.
  Environmental planning.

  1. Environmental policy--Dictionaries. 2. Land use,
Urban--Dictionaries. 3. City planning--Dictionaries.
4. Regional planning--Dictionaries. I. Title.
HC79.E5G52 1986   333.7'03'21   86-17985
ISBN 0-8155-1103-5

Preface

Until recent times, the traditional label for physical or land-use planning over large geographical areas has been, in the United States, "city and regional planning," while in Britain it has been known as "town and country planning."

Within the last few years a new term, "environmental planning," has emerged tending to subvert these older descriptions while ushering in some change of substance and emphasis. Certainly, it embraces planning in the older sense with its strong, though by no means exclusive, emphasis on physical planning; on the other hand, it represents a form of planning which places much greater emphasis on environmental pleasantness and the merits of the natural, if modified, world.

Environmental planning is not, therefore, a new trendy label for an old and unchanging subject; there is a distinct shift of emphasis which enables its practitioners to grapple more effectively with many controversial land-use issues, using new approaches and techniques, while seeking guidance from many specialized state, federal, national and international bodies, and numerous voluntary bodies and individuals. It is planning into which the principles and practice of environmental impact assessment, itself broadly defined, have been integrated as part of the every-day business of planning, public administration, public participation, and political debate.

This encyclopedia seeks to capture the many ingredients that I have found go into environmental planning, with a view to usefulness not only to the student of planning but to a wide range of professions that become involved or entangled in planning processes, and to the much wider spectrum of the participating public exercising a right to involvement in decision-making processes of various kinds, including public hearings and inquiries.

I would like to thank the many planning agencies in North America and Europe that have assisted me in this task.

July 1986   Alan Gilpin

About the Author

Dr. Alan Gilpin is Commissioner of Inquiry with the New South Wales Government, Australia, being responsible for the conduct of public inquiries into controversial planning and environmental issues. He was previously Assistant Chief Planner with that State's Department of Environment and Planning. In the early 1970s, he served with the Department of Urban and Regional Development in Canberra and as Chairman of the Environment Protection Authority in the State of Victoria.

During his earlier career in England, he was involved in major slum clearance, housing improvement and redevelopment programs. Subsequently, he was engaged in locational planning with the Central Electricity Generating Board in London.

Holding degrees in economics from the Universities of London and Queensland, as well as being a chartered engineer, he has traveled extensively in the United States of America, Canada, Europe, and the Middle and Far East.

NOTICE

To the best of our knowledge the information contained in this publication is accurate; however, the publisher does not assume any liability whatsoever for the accuracy or completeness of such information. It is recommended that users intending to follow any of the regulations or procedures mentioned in the book should consult original documents before proceeding.

Contents and Subject Index

Abercrombie Plans for London, UK 1
Accessibility 1
Acid Mine Drainage 2
Acid Rain 2
Activity Pattern 4
Advertisement Control 4
Agency for International Development (AID), US 5
Air Pollution 5
Air Quality Act, 1967, US 5
Alaska National Interest Lands Conservation Act, 1980, US 6
Alaska, US 6
Amazon Basin and Forest 6
Amenity 7
Anadromous Fish and Great Lakes Fisheries Act, 1965, US 7
Antarctica Conventions 7
Anthracite and Bituminous Coal Mine Subsidence Insurance
 Fund, 1961, Pennsylvania, US 8
Anti-Urbanism 8
Aquifer and Aquifer Recharge Areas 8
Asbestos-In-Schools Rule, 1982, US 8
Athens Charter, 1933 8
Australia 8
Australia-Japan Migratory Birds Agreement 8
Australian Environment Council (AEC) 9
Austria 9

x   Contents and Subject Index

Background...10
Backwater Area...10
Back-Zoning..10
Bangladesh...10
Banking, US..10
Barlow Report, 1940, UK......................................10
Belgrade Charter...10
Beneficial Use...10
Benefit-Cost Analysis..11
Benefit/Cost Ratio (BCR).....................................12
Benthos..12
Berm...12
Best Available Control Technology (BACT).....................12
Best Practicable Means.......................................12
Betterment...13
Bhopal Disaster, India, 1984.................................13
Biochemical Oxygen Demand (BOD)..............................13
Biosphere..14
Biostimulants..14
Bituminous Mine Subsidence and Land Conservation Act,
   1966, Pennsylvania, US....................................14
Blanketing...14
Bleve..14
Blight...14
Board of Zoning Appeals, Los Angeles, US.....................14
Brandt Commission..15
Brasilia...15
Brazil...16
British Columbia, Canada.....................................16
Bubble Concept, US...16
Buchanan Report, 1963, UK....................................17
Buffer Zone..17
Building Application, Factors to Be Considered...............17
Building Codes...18
Built Environment..18
Bushfire Prone Areas, New South Wales........................18
Business Improvement District Act, 1981, Kansas, US..........19

California, US...20
Canada...20
Capital Improvement Program..................................20
Car Scrapping Act, Sweden....................................20
Carpool System...20
Central Business District (CBD)..............................20

Central City Community Plan, Los Angeles, US 22
Central Place Theory. 23
Chesapeake Bay Program, US 23
Chicago, Illinois, US 24
Chile 24
China, People's Republic of. 24
Chipko Andolan Movement. 24
Chlorofluorocarbons (CFCs). 24
Circulation Element 25
City and Regional Planning. 25
"City Beautiful". 25
Civic Amenities Act, 1967, UK. 25
Clean Air Acts, 1963 and 1965, US. 26
Clean Air Amendment Act, 1970, US. 26
Clean Air Amendment Act, 1977, US. 26
Clean Lakes Law, 1984, Japan 27
Clean Water Act, 1977, US 27
Clean Water Amendment Act, 1981, US. 28
Clean Water Restoration Act, 1966, US 28
Cluster Housing. 28
Coastal Barrier Resources Act, 1982, US. 28
Coastal Protection. 28
Coastal Protection Act, 1979, New South Wales. 29
Coastal Zone Act, 1971, Delaware, US 29
Coastal Zone Management Act, 1972 29
Coliform Bacteria 29
Colorado, US. 30
Commission for the Environment, New Zealand. 30
Commission on the Height of Buildings, 1913, US 30
Commissioner of Inquiry, New South Wales 31
Committee on Public Participation in Planning, 1969, UK 32
Common Property Resources 32
Commons, The 32
Community Development Block Grant (CDBG), US. 33
Compensation 33
Compensation Depth. 33
Compensatory Open Space 33
Comprehensive Environmental Response, Compensation, and
 Liability Act, 1980 (CERCLA), US. 34
Comprehensive Plan, US 34
Comprehensive Plan, Washington, D.C., US. 36
Compulsory Purchase 36
Conditional Use. 36
Configuration Dredging. 36

Congestion 36
Connecticut, US 36
Conservation 39
Conservation and Development Program, UK 39
Conservation Zones 39
Control of Pollution Act, 1974, UK 40
Conurbation 40
Convention for Cooperation in the Protection and Development of the Marine and Coastal Environment of the West and Central African Region 41
Convention for Cooperation on the Protection of the Marine Environment from Pollution, Kuwait Regional 41
Convention for the Prevention of Marine Pollution by Dumping from Ships and Aircraft 41
Convention for the Prevention of Marine Pollution from Land-Based Sources 41
Convention for the Prevention of Pollution from Ships (MARPOL), 1973 42
Convention for the Prevention of Pollution of the Sea by Oil, 1954 43
Convention for the Protection of the Mediterranean Against Pollution, 1976 43
Convention for the Protection of the Rhine Against Chemical Pollution, 1976 44
Convention for the Protection of the World Cultural and Natural Heritage 44
Convention, Nordic Environmental Protection 45
Convention on Fishing and the Conservation of Living Resources in the Baltic Sea and the Belts, 1973 45
Convention on International Trade in Endangered Species of Wild Fauna and Flora 45
Convention on Long-Range Trans-Boundary Air Pollution 45
Convention on Migratory Species of Wild Animals 45
Convention on Nature Conservation in the South Pacific 46
Convention on the Conservation of Antarctic Marine Living Resources (CCAMLR) 46
Convention on the Protection of the Marine Environment of the Baltic Sea Area 46
Convention on the Protection of the Rhine Against Pollution by Chlorides, 1976 47
Convention on Wetlands of International Importance, Especially as Waterfowl Habitat 47
Conventions for the Conservation of Antarctic Living Resources 47

Conventions for the Protection of the Northeast Atlantic
  Against Pollution. 48
Core 48
Corporate Planning, UK 49
Council for Mutual Economic Assistance (COMECOM) 49
Council for the Environment, Republic of Ireland (EIRE) 50
Council for the Protection of Rural England. 50
Council of Europe. 51
Council of Nature Conservation Ministers (CONCOM),
  Australia 51
Council on Environmental Quality, US .52
Country Parks, UK 52
Countryside Commission, UK. 52
Cubatao Disaster, 1984, Brazil 54
Cumulative Impacts. 54

Decentralization 55
Decibel A [dB(A)] Scale. 56
Delaware, US. 56
Demography 56
Denmark. 57
Department of Environment (Environment Canada). 57
Department of Environment and Planning, New South Wales 57
Department of the Environment, UK 58
Derelict Land Reclamation Programs, UK. 59
Desertification Control 59
Development. 61
Development Application 61
Development Charge, UK 62
Development Control 63
Development Rights 63
Development Standards. 63
Development Zones and Holding Zones .64
Dimensional Variance 64
Distribution of Industry Act, 1945, UK .64
District Structure Plan. 65
"Do The Right Thing" Campaign, New South Wales. 65
Dobry Report, UK 66
Duncan Classification, US. 66

Earthwatch 67
Easements. 67
Ecodevelopment 67
Ecology. 67

xiv   Contents and Subject Index

Economic Commission for Europe (ECE) .67
Economic Commissions for Africa, Latin America, Asia and
  the Far-East. 68
Economic Efficiency. 69
Economic Ends. 69
Economic Growth. 69
Economic System, Functions of 69
Ecosystem. 70
Ecumenopolis 70
Effluent Charge. 70
Egypt 70
Ekistics 70
Eminent Domain. 71
Emission Reduction Credits 71
Emissions Trading Program, US. 71
Enabling Acts, US. 72
Endangered Species Act, 1973, US 72
Enterprise Zones, UK 72
Environment. 73
Environment Agency, Japan 73
Environment Policy Act, 1970, German Democratic Republic 74
Environment Protection Act, 1969, Sweden. 75
Environment Protection (Impact of Proposals) Act, 1974-75,
  Australia 76
Environment Protection (Sea Dumping) Act, 1981, Australia. 76
Environmental Administration, Soviet Union 78
Environmental Assessment, US. 78
Environmental Assessment Act, 1975, Ontario 78
Environmental Assessment Board, Ontario 78
Environmental Conservation Law (ECL), 1977, Korea 81
Environmental Economics. 81
Environmental Education 81
Environmental Effects Statement (EES). .81
Environmental Health. 82
Environmental Health Impact Assessment (EHIA) 82
Environmental Heritage. 82
Environmental Impact Assessment 82
Environmental Impact Statement 84
Environmental Laws 86
Environmental Master Plan, Pennsylvania, US. 86
Environmental Planning 86
Environmental Planning and Assessment Act, 1979, New
  South Wales. 89
Environmental Planning Instrument 90

Contents and Subject Index   xv

Environmental Planning, Objectives of 90
Environmental Planning System 91
Environmental Policy Act, 1971, North Carolina, US..........92
Environmental Precincts.................................93
Environmental Protection Agency (EPA), US................93
Environmental Protection Council, Ghana 96
Environmental Protection Districts........................96
Environmental Protection Law, 1979, China................97
Environmental Protection Service, Israel...................98
Environmental Quality Act, 1970, California, US.............99
Environmental Quality Standards.........................99
Environmental Recovery Areas, UK 99
Environmental Rights Amendment, Pennsylvania, US.........100
Environmental Technical Information System (ETIS), US 100
Environmentally Hazardous Chemicals 100
Euphotic Zone 101
European Economic Community (EEC) 102
European Foundation for the Improvement of Living and
    Working Conditions................................104
Eutrophication 104
Examination in Public, UK 105
Exclusionary Zoning..................................105
Existence Value......................................106
External Effects 106

Facadism...107
Federal Environmental Agency, West Germany..............107
Federal Environmental Assessment Review Office (FEARO),
    Canada 108
Federal Insecticide, Fungicide, and Rodenticide Act, US 109
Federal Land Policy Management Act, 1976, US 109
Federal Water Pollution Control Act, 1956, US 109
Finland 109
Flexible (Average Density of Cluster) Zoning 109
Floating Value.......................................110
Flood Mitigation Works and Measures.....................110
Floodplain..110
Florida, US 111
Focal Point 111
Food and Agriculture Organization (FAO) 111
Food, Drug and Cosmetic Act (FDCA), 1938, US.............112
Footloose Industry 112
Foreground...112

xvi Contents and Subject Index

Forest and Rangeland Renewable Resources Planning Act,
  1974, US...112
Forest Products.......................................112
France...114
Franchise Agreements..................................114
Freeways...114
Frost Commission, Australia...........................114
Functional Pattern....................................115

Garden-City Movement, UK..............................116
Gdansk Convention.....................................116
General Agreement on Tariffs and Trade (GATT).........116
General Improvement Areas (GIAs), UK..................116
General Plan, Los Angeles, US.........................117
Geneva Convention.....................................118
Gentrification..118
Georgia, US...118
German Democratic Republic (East Germany).............118
Ghana..118
Ghetto...118
Glasgow Eastern Area Renewal Project, 1976-87, Scotland.....120
Global Atmospheric Research Program (GARP)............120
Global Environment Monitoring System (GEMS)...........120
Gordon-Below-Franklin Dam, Australia..................121
Grade Separation......................................121
Great Barrier Reef, Australia.........................121
Great Lakes Water Quality Agreement...................121
Great South Western Industrial Estates, US............122
Greater Copenhagen Regional Plan......................122
Greater London Development Plan (GLDP), UK............124
Greater London Plan, 1944.............................124
Greater Stockholm Finger Plan.........................124
Greece...125
Green Alliance..125
Green Ban Movement, Australia.........................125
Green Belt...126
Green Great Wall......................................126
Green Summit, 1984....................................128
Green Tracery Features................................128
Greenhouse Effect.....................................128
Gross National Product (GNP)..........................128
Growth Pole Theory....................................130
Guidelines of State Environment Policy, Indonesia.....130

Habitat Evaluation Procedure (HEP).......................132
Habitat: The United Nations Conference on Human
    Settlements...133
Han River Basin Environmental Master Plan, Korea 133
Hawaii, US..134
Hazard and Problem Areas 134
Hazardous Air Pollutants..................................134
Hazardous Chemicals.......................................134
Hazardous Substance Response Trust Fund, US 134
Hazardous Waste...134
Health..134
Health and Safety at Work Act, 1974, UK 135
Hearing Examiners, Los Angeles, US........................135
Hearings 135
Heavy Metals..135
Helsinki Convention 136
Herbert Report, 1960, UK 136
Heritage Act, 1977, New South Wales.......................137
Heritage Buildings List, UK...............................137
Heritage Coast, UK 138
Heritage Conservation 138
Heritage Criteria 139
Highrise Folly 139
Holford Rules 139
Housing Act, 1949, US 139
Housing Densities 140
Housing Improvement Grants, UK 140
Hungary 141
Hydrological Cycle 141

Incompatible Uses...143
India...143
Indonesia...143
Industrial Air Pollution Inspectorate.....................143
Infant Mortality Rate 144
Infill Development 144
Information and Documentation System for Environmental
    Planning (UMPLIS), West Germany.......................144
Infrastructure 146
Injurious Affection 146
Inner Urban Areas Act, 1978, UK...........................146
Insecticide, Fungicide, and Rodenticide Act, US...........147
Institutional Planning 147
Interaction Matrix..147

xviii Contents and Subject Index

Interim Zoning 147
Internal Rate of Return (IRR) 148
International Atomic Energy Agency (IAEA) 148
International Biological Program (IBP) 149
International Commission for the Protection of the Rhine 149
International Commission on Radiological Protection (ICRP). . . .149
International Commons 150
International Council for the Exploration of the Sea (ICES) 150
International Council of Scientific Unions (ICSU) 150
International Drinking Water Supply and Sanitation Decade
  (IDWSSD) 151
International Energy Agency (IEA) 151
International Labor Organization (ILO) 151
International Maritime Organization (IMO) 152
International Referral System (IRS) (INFOTERRA) 152
International Register of Potentially Toxic Chemicals (IRPTC) . . 153
International Security and Development Act, 1985, US 153
International Union for the Conservation of Nature and
  Natural Resources (IUCN) 153
International Whaling Commission (IWC) 154
Intractable Waste 154
Inversion 154
Iowa, US 156
Iraq 156
Ireland (Eire) 156
Israel 156
Italy 156

Japan 157
Jhumming 157
Joint Committee for Environmental Protection, US-USSR 157
Joint Group of Experts on the Scientific Aspects of Marine
  Pollution (GESAMP) 157
Joint Planning Unit, US 158
Junkyard Control Act, 1967, Tennessee, US 158

Kansas, US 159
Karst Topography 159
Keep America Beautiful 159
Keep Singapore Clean Campaign 159
Kentucky, US 160
Korea 160

Lake Baikal, Soviet Union 161

Contents and Subject Index xix

Land and Environment Court, New South Wales. 161
Land Capability. 161
Land Commission, UK 162
Land Compensation Act, 1973, UK 162
Landfill. 164
Land Use Plan 164
Land Use Planning. 164
Land Use Prediction Models 166
Landscape. 166
Landscape Analysis—Factors to Be Considered 166
Landscape Character 167
Landscape Evaluation, UK 167
Landscape or Amenity Conservation. 167
Landscape Unit. 167
Lapse Rates. 167
Law of the Sea 167
Lead in Gasoline 168
L'Enfant Plan for Washington, D.C., 1974 168
Limnetic Zone. 168
Linkages 168
Litter Control 169
Litter Control, UK 169
Litter Laws, US. 169
Littoral 169
Littoral Drift. 169
Littoral Zone. 170
Living Resource Conservation 170
Local Environmental Plan, New South Wales 170
Local Government Act, 1972, UK. 171
Local Government, Planning and Land Act, 1980, UK 172
Local Plans, UK. 172
Location of Industry, Optimal 173
Location, Theory of 174
London, England. 174
London Convention 174
London Docklands Development Corporation 174
London Smog Incidents 174
Long Shore Drift. 174
Los Angeles, California, US. 175
Los Angeles City Planning Commission, US 175
Los Angeles Smog, US. 175
Love Canal, US 176
Luxembourg 176
L_x Noise Levels. 176

Contents and Subject Index

Macroeconomics 177
Macro Environmental Problems. 177
Maine, US 178
Man and Biosphere Program 178
Mandatory Dedication. 178
Mangrove Parks, Bangladesh 178
Mangroves. 178
Manitoba, Canada 179
Marine Protection, Research, and Sanctuaries Act, 1972,
    (MPRSA), US 179
Marpol Convention 180
Maryland, US 180
Master Plan 180
McMillan Plan, Washington, D.C., US 180
Megalopolis. 180
Merseyside Development Corporation, UK 180
Metropolis. 180
Mexico 181
Mexico City Industrial Disaster, 1984 181
Michigan Episode, US. 181
Microeconomics 182
Micro Environmental Problems 182
Middleground 182
Mine Bonding 183
Mining Wastes 183
Ministry of Environment, Norway. 183
Ministry of the Environment (Environment Ontario) 185
Minneapolis-St. Paul, Minnesota, US 187
Minnesota, US. 187
Missouri, US 187
Mitigating Measures. 187
Model Cities Program, US 187
Mt. Laurel Decision, 1983, New Jersey, US. 188
Multiple Use 188
Multiple-Use Sustained-Yield Act, 1960, US 188
Multiplier 189
Municipalities Planning Code (MPC), Pennsylvania, US. 189

National Ambient Air Quality Standards (NAAQS), US 191
National Capital Development Commission (NCDC), Australia... 192
National Capital Planning Commission, Washington, DC, US 193
National Committee on Environmental Planning and
    Coordination (NCEPC), India 193
National Contingency Plan, US. 194

National Council on Physical Planning and the Environment,
  Greece..194
National Emissions Standards for Hazardous Air Pollutants
  (NESHAP)...195
National Environment Board, Thailand.....................195
National Environment Protection Board, Sweden.............196
National Environmental Policy Act, 1969, (NEPA) US.........197
National Environmental Policy Regulations, 1979, US.........198
National Forest Management Act, 1976, US.................198
National Franchise Board for Environment Protection, Sweden..198
National Heritage Memorial Fund, UK.....................198
National Historic Preservation Act, 1966, US...............198
National Human Monitoring Program, US..................199
National Park..199
National Physical Planning.............................201
National Planning Guidelines, UK........................202
National Pollutant Discharge Elimination System (NPDES), US..206
National Priorities List, US.............................203
National Survey of Air Pollution, UK.....................203
National Stream Quality Accounting Network (NASQAN), US..205
National Trails System Act, 1968, US.....................205
National Urban Policy and New Community Development Act,
  1970, US..206
National Wildlife Refuge System Administration Act, 1966,
  US..206
National Wildlife Refuges, US...........................206
Natural Features.....................................206
Natural Parks, Japan..................................207
Natural Resource Conservation.........................208
Natural Resources Data Bank, Chile......................208
Natural Resources Inventory............................208
Nature and Monuments Conservation Act, 1974, Greenland.....209
Nature Conservancy, UK...............................209
Nature Conservancy Act, 1975, Sweden...................210
Nature Conservation Act, 1976, France...................210
Neighborhood Noise...................................211
Neighborhood Unit....................................211
Neritic Zone...212
Net Present Value (NPV)...............................212
Net Reproduction Rate................................212
Netherlands, The.....................................212
Netting...212
New Brunswick, Canada................................212
New Communities Program, US..........................213

New Hampshire, US 213
New Jersey, US......................................213
New Source Performance Standards (NSPS), US..............213
New South Wales, Australia............................214
New Towns..214
New Towns Act, 1946, UK.............................214
New Towns, France....................................214
New York State, US 215
New York, US..215
New York Zoning Ordinance, 1916, US 215
New Zealand...215
Newfoundland, Canada.................................215
Niagara River Improvement Program, Ontario................215
Noise...215
Noise Abatement Zone, UK 215
Noise Control Act, 1972, US............................216
Noise Exposure Forecast (NEF) 217
Noise Zoning...217
Non-Attainment Areas, US 217
Nonconforming Use 218
Non-Renewable Resources 218
Nordic Environmental Protection Convention.................218
North Carolina, US 219
Norway..219
Nuclear Hazards 219
Nuclear Regulatory Commission (NRC), US..................219

Ocean Dumping..220
Offensive or Hazardous Industry..........................220
Office of Environmental Protection, Switzerland 220
Official Controls, US...................................221
Official Map, US 221
Official Plan, Metropolitan Toronto 221
Official Plan, Toronto 221
Offsetting 222
Ohio, US..222
Oil Spills 222
Oklahoma, US..224
One-Hundred-Year Floodplain 224
Ontario, Canada 224
Ontario Waste Management Corporation....................224
Open Space..224
Open Space Plan 224
Oregon, US..226

Organization for Economic Cooperation and Development
  (OECD)..226
Oslo Convention 228
Overgrazing..228
Overspill..228
Ozone (O_3)...228
Ozonosphere...229

Paris Convention...230
Parker Windscale Inquiry, UK..................................230
Parkway 231
Pedal-Power...231
Pedestrian Precinct 231
Pedway System 231
Pennsylvania, US..231
People Movers...231
Performance Standards 231
Physical Planning..232
Planning 232
Planning Blight 233
Planning Commission, US 233
Planning Gain, UK...234
Planning Inquiry Commission, UK 234
Planning Standards 235
Planning System, UK...235
Planning Trilogy 236
Plat, US...236
Point Source Pollution 236
Poland...237
Policy Plan 237
Pollutant Standards Index (PSI), US...........................237
Polluter Pays Principle..237
Pollution Control Costs..238
Pollution Control Strategy 238
Polychlorinated Biphenyls (PCBs)..............................239
Population Differentiation 239
Population Implosion 239
Post-Closure Liability Trust Fund, US.........................239
Prairie Psychology...239
Prevention of Significant Deterioration (PSD), US 239
Primitive Areas, US...241
Program Impact Statements 241
Project Inquiry, UK...241
Protective Covenant 242

Public Facility 242
Public Hearings and Inquiries 242
Public Hearings and Inquiries, Procedures at 242
Public Local Inquiry, UK 243
Public Participation...................................... 244
Public Policies, Components of 245
Public Trust Doctrine 248

Quality of Life 247
Quebec, Canada... 247

Rain Forest 248
Ranger Uranium Environmental Inquiry, Australia 248
Reasonably Available Control Technology 249
Reclamation Reform Act, 1982, US 249
Recycling 249
Redevelopment of Central Business District Areas Act, 1976,
   Kansas, US 250
Regional Administrative Centers 251
Regional Development Commission, US 251
Regional Environmental Plan, New South Wales 251
Regional Planning Commission, US......................... 251
Regional Seas Program, UNEP............................. 252
Regulatory Impact Analysis (RIA), US..................... 253
Reith Report on New Towns, 1946, UK..................... 253
Residential Development Standards 253
Residential Zones 254
Resource Areas 255
Resource Conservation and Recovery Act, 1976, US 255
Resource Recovery 257
Restriction of Ribbon Development Act, 1935, UK 257
Ribbon Development 257
Right of Common Access, Nordic 257
Risk and Hazard Assessment 258
River Thames Clean-Up, UK 259
River Water Quality Classifications, UK 260
Roads, Hierarchy of 260
Roskill Commission, UK................................... 261
Royal Commission on Environmental Pollution, UK.......... 261
Royal Commission on the Great Barrier Reef, Australia 263
Royal Society for Nature Conservation, UK 263

Saemaul Undong (New Community Movement), Korea 265
Safe Drinking Water Act, 1974, US........................ 265

Safe Growth Plan, Tennessee, US 266
Sanitary Epidemiological Service, Soviet Union 266
Saskatchewan, Canada. 267
Scoping, US. 267
Scott Report, 1942, UK 267
Selective Decentralization 267
Service Activities. 267
Setbacks 268
Severance 269
Sewage Treatment. 269
Shift and Share Analysis 270
Singapore 271
Sinkholes. 271
Site Planning. 271
Site Rehabilitation Objectives 271
Skyways, Skywalks, Pedways 271
Slum Clearance Program, UK 272
Smoke Control Area, UK 272
Social Forestry 273
Social Impact Assessment (SIA) 273
Social Indicator. 274
Socioeconomic Elements. 274
Sociosphere. 274
Soil Erosion. 274
Solid Waste Disposal Act, 1965, US 274
South Africa 275
Southern California Association of Governments, US 275
Spatial Allocation Models 275
Special Environmental Agency (SEMA), Brazil 275
Special Use Corridors 276
Spot Zoning and Rezoning 276
Star Town 277
State and Local Air Monitoring System (SLAMS), US 277
State Development and Redevelopment Plan, New Jersey, US . . .277
State Development Plan, Maryland, US. .277
State Environmental Planning Policies, New South Wales 278
State Implementation Plans (SIP), US. 278
State Plan, Hawaii, US. 278
State Planning Commission, New Jersey, US. 279
State Pollution Control Commission, New South Wales 279
State Regulatory Controls, US 279
Stormwater Runoff. 279
Strategy Plan. 284
Strategy Plan for Southeast England 284

xxvi Contents and Subject Index

Stream Valleys 284
Street Furniture 284
Street Improvement Schemes 284
Streetscape 284
Structure Plan 284
Structure Plan, UK 284
Structure Plan Preparation 285
Structure Plan Principles 286
Subdivision 287
Subdivision Application, Factors to be Considered 287
Sudan, The 289
Sudbury, Ontario 289
Sulfur Dioxide 289
Superfund 290
Surface Mining Control and Reclamation Act, 1977, US 290
Surface or Strip Mining 291
Swales 292
Sweden 292
Switzerland 292
Sydney Metropolitan Waste Disposal Authority 292

Taking Issue, US 295
Tennessee, US 295
Tennessee Valley Authority, US 295
Texas, US 295
Thailand 295
Three Mile Island Incident, Pennsylvania, US 295
Threshold Costs 296
Tiering 297
Times Beach, US 297
Town and Country Amenities Act, 1974, UK 297
Town and Country Planning 297
Town and Country Planning Act, 1932, UK 298
Town and Country Planning Act, 1947, UK 298
Town and Country Planning Act, 1968, UK 299
Town and Country Planning Act, 1971, UK 299
Town Development Act, 1952, UK 300
Townscape 300
Toxic Substances Control Act, 1976, US 300
Trade Waste 301
Traffic Management 301
Traffic Segregation 301
Transfer of Development Rights (TDR), US 301
Transition Areas 302

Contents and Subject Index   xxvii

Transport Modal Split.................................302
Transport Models 302
Transport Plan.......................................302
Transport Planning, Objectives of 303
Transport Policies and Programs (TPPs), UK.................304
Transportation Study 305
Trophic...306
Tropical Cyclone Warning System........................306
Truck Routing.......................................306
Turkey 307
Twilight Zone 307

Union of Soviet Socialist Republics (USSR) 308
United Kingdom (England, Wales, Scotland, and Northern
  Ireland)...308
United Nations Conference on Desertification (UNCOD) 308
United Nations Conference on New and Renewable Sources
  of Energy 309
United Nations Conference on the Human Environment, 1972...309
United Nations Conference on the Law of the Sea (UNCLOS) ...310
United Nations Conference on Trade and Development
  (UNCTAD)...311
United Nations Conferences 311
United Nations Educational, Scientific and Cultural Organiza-
  tion (UNESCO).....................................311
United Nations Environment Program (UNEP)...............312
United Nations Industrial Development Organization (UNIDO) ..314
United Nations Organizations 314
United Nations Water Conference (UNWC)..................314
United Nations World Population Conference, 1974...........315
United States of America 316
Unplanned Growth 317
Urban and Rural Zones Act, 1969-75, Denmark..............318
Urban Areas 318
Urban Conservation Area 318
Urban Consolidation..................................318
Urban Design..318
Urban Development Action Grant (UDAG) Program, US 318
Urban Development Corporations (UDCs), UK................319
Urban Forest Programs 319
Urban Renewal 319
Urban Renewal Act, 1974, Austria 320
Urban Sociology 320
Urban Structure Plan..................................321

Urbanization 321
US-France Memorandum of Understanding, 1984. 322
US-USSR Environmental Protection Agreement 322
Uthwatt Report, 1942, UK 322
Utopian Planners 323
UVCE 323

Vancouver Plan of Action 324
Variance, US 324
Vermont, US 325
Vibration 325
Victor Gruen Foundation for Environmental Planning 325
View Plane Technique 325
Viewer Catchment 325
Virginia, US 325
Visual Elements 325
Visual Impact Measurement 325
Visual Pollution 326
Visual Quality 326

Washington, US 329
Water Act, 1984, Sweden 329
Water Act, 1973, UK 329
Water Classification, New South Wales 329
Water Pollution 331
Water Pollution Control Amendment Act, 1972, US. 332
Water Quality Act, 1965, US 333
Water Quality Improvement Act, 1970, US. 334
West Germany 334
Wetlands 334
Wild and Scenic Rivers 334
Wild and Scenic Rivers Act, 1968, US 335
Wilderness 335
Wilderness Act, 1964, US 336
Williamsburg Restoration, US 336
Wisconsin, US 337
World Bank 337
World Conference on Agrarian Reform and Rural
  Development 339
World Conservation Strategy (WCS) 339
World Data Center on Microorganisms 339
World Food Conference 339
World Health Organization (WHO) 339
World Meteorological Organization (WMO) 340

World Population Conferences, 1974 and 1984. 340
World Population Plan of Action. 340
World Resources Institute 342
World Weather Watch 343
World Wildlife Fund (WWF) 343

Yugoslavia. 345

Zoning. 345
Zoning Administrator, Los Angeles, US .346
Zoning Ordinances, US 346

A

Abercrombie Plans for London, UK: Plans prepared by Sir Patrick Abercrombie (1874-1957) for the County of London in 1943 and the Greater London Area in 1944. The plans were conceived in terms of concentric rings of declining, static, and expanding population in conjunction with a Green Belt (q.v.). The inner ring comprised the densely developed core of London, which was defined as the County of London together with certain adjacent boroughs such as Willesden and Tottenham. The Abercrombie plans suggested an accelerated fall in the population of this area through a dispersal of employment and population; at the same time schemes of redevelopment would be undertaken by the local authorities within which housing, industry, commerce, shopping and social facilities would be introduced on an integrated basis. The environment for the smaller population in this core would be much improved.

The second and third rings consisted of the outer suburbs and the green belt. Very few changes would occur in these more recently developed suburban areas, and the population would remain static. The fourth Abercrombie ring was to be an area of expansion and development beyond the green belt including the development of eight new towns and the expansion of twelve other towns. The Abercrombie plans included quite elaborate proposals for the construction of new highways and railway improvements to match the pattern of development assumed.

The smooth unfolding of the Abercrombie plans was upset in part by an unpredicted level of growth in population which led to incursions into the green belt; this led to critical battles to save the green belt. While centers outside the green belt expanded rapidly, the inner core witnessed a huge growth in city employment. This created a growing army of long-range commuters spending several hours each day traveling to and from work. The problems were accentuated by the quiet shelving of the transport system improvement plans.

In addition, environmental improvement in the inner boroughs was slow in coming. The Greater London Development Plan (q.v.) of 1976 outlined severe housing problems, excessive densities, severe overcrowding, a high incidence of slums and structurally defective dwellings, galloping obsolescence, and a chronic shortage of open space.

Accessibility: The ease by which the public, or a segment of it, may reach places of employment, shopping centers, community facilities,

2   Environmental Planning

and services including higher order services, expressed in terms of time, trouble, effort, or cost.

In respect to a transport system, it is important to distinguish between accessibility from the viewpoint of:

- Workers journeying to work,
- Other journeys made in the course of the day (shopping and recreation), and
- Commercial vehicles (and especially trucks) moving about a metropolitan area delivering goods and services.

Urban structure and city growth may be viewed in terms of accessibility. If transportation is poor, the workplaces, trade centers, and community services tend to assume a pattern of distributed facilities; if it is good, these activities assume more concentrated patterns. Urban spatial structure is intimately connected with the aggregate effort of the community to overcome distance.

Accessibility to nature, open spaces, and coastal facilities and beaches may also present a problem in some cities. Insufficient access has been a problem for open spaces along the Potomac and Anacostia waterfronts in Washington, D.C. Physical barriers like major roads, freeways and institutional lands closed to the public, separate neighborhoods from open spaces. Pedestrian connections and access points in particular to the Anacostia Park and National Arboretum have been unattractive and too few.

One of the objectives of planning is to increase accessibility to places of employment, services and facilities; however, in some circumstances planning may reduce accessibility as, for example, in the establishment of a green belt (q.v.) when the traveling times of those who must live beyond it, but work within it, are increased in consequence. In another instance, the establishment of new radial highways may increase the problems of transverse movement. In the case of wilderness areas, access may be severely restricted. See *Buchanan Report, 1963; Central Place Theory; Environmental Planning; Right of Common Access, Nordic; Transport Planning, Objectives of.*

Acid Mine Drainage: The discharge of mine water which has become contaminated with the sulfur compounds of the coal seams, creating acidity. Acid mine drainage can make groundwater, streams, and lakes unsuitable for recreation, domestic water supply, industry, and agriculture. In Pennsylvania, the Department of Environmental Resources estimated that 1,800 miles of major streams would not meet 1983 national water quality goals because of abandoned mine drainage problems.

Acid Rain: Rain or snow that contains significant amounts of sulfuric acid or nitric acid. Sulfuric or nitric acid is formed when sulfur dioxide (SO_2) (q.v.) or nitrogen oxide (NO_x) emitted by industry or transport undergo a chemical transformation in the atmosphere. The US National Commission on Air Quality, in its 1981 report, found that the process by which man-made pollutants were transformed into acid rain was now "reasonably well known." It described how SO_2 or NO_x gas released into the atmosphere is first oxidized to sulfate or nitrate particles. If water vapor is present, the particles are further transformed into sulfuric or nitric acid which contaminates rain, snow, or fog. Alternatively, sulfate or nitrate particulates may be deposited on the ground in dry form, later combining with surface water or groundwater to produce acid.

In the United States, rain that is 10 to 40 times as acidic as normal rainfall has been occurring frequently in many parts of New England and New York. The average pH of rainfall over substantial portions of the Northeast is 4.3, approximately 10 times normal acidity. Comparable levels of acidity are occurring in many parts of Canada. Highly acidic rainfall has also been observed in a number of southeastern states, particularly Florida, Virginia, Texas and North Carolina, in the Northern Plains states of Minnesota and Wisconsin, and in the Colorado Rockies.

In 1979, the US President established a ten-year comprehensive Federal Acid Rain Assessment Program to be planned and managed by a standing Acid Rain Co-Ordination Committee. The assessment program includes applied and basic research on acid rain effects, monitoring, transport modes, and the study of control measures. Collaboration with the Canadian and Mexican Governments, and with other nations and international organizations, was anticipated.

In August, 1984, New York became the first state to legislate on acid rain. By the end of that year, the areas of the state most at risk had to be identified, and tolerance levels for sulfur deposition determined. Rules governing emissions from the industrial burning of coal and oil were to be produced by 1986, and to come into force by 1988.

In respect to the acid rain issue, the debate in the US Congress has split along regional lines: members from the Northeast have pressed for a tough control program, while members from the Midwest fear a control program that could hamper the region's economy. The West, neither a primary victim of nor contributor to acid rain, does not want to pay to control it.

The Swedish Fisheries Board has now done a national survey of Swedish lakes and has found that 18,000 out of a total of about 90-100,000 are acid. These lakes, poor in calcium, magnesium and nutrient, are situated (mainly in the southwest of Sweden) in areas with coniferous forests and pre-Cambrian bedrock. The pH of the acid lakes varies between 4 and about 5.5. Lakes have a natural defense mechanism or buffering capacity caused by the presence of hydrogen carbonate. However, the majority of Scandinavian lakes and watercourses have low lime content, and thus a low content of hydrogen carbonate, and are therefore particularly vulnerable to acidification. While it was agreed that there were many factors involved in the acidification of lakes and soils in Sweden and other parts of Scandinavia, there was no doubt that sulfur and nitrogen compounds resulting from combustion of fossil fuels in Europe were transported over long distances and contributed to the acidification problem in vulnerable areas.

Acid precipitation over Southern Norway has increased considerably in recent years. Measurements have shown that there is a marked decrease in the pH value in rivers and lakes after rain. In winter, grey snow has sometimes been observed. In the southern-most counties in particular the soil is poor with low calcium levels, so that acid precipitation is only partly neutralized by the soil and thus immediately affects the pH value of rivers and lakes. An observed decline in stocks of fish in a large number of lakes may be related to increased acidity. Acid also has a leaching effect on mineral substances in the soil with possible detrimental effects on forest growth. Norway takes considerable interest in the new OECD research project on the long-range movement of air pollutants.

An October, 1983, report to the West German Bundestag (federal parliament) suggested that more than one-third of West German forests had been damaged by acid rain. In Bavaria and the Black Forest it was claimed that nearly half of the trees had been damaged. New limits on emissions from power stations were announced later in 1983 aiming to reduce sulfur dioxide emissions by over 85 per cent.

In 1984, the Netherlands Environment Minister released a plan to reduce industrial sulfur dioxide emissions by 70 per cent, nitrogen oxide emissions from motor vehicles by 30 per cent and ammonia released from agricultural fertilizers by 50 per cent, by the end of the century. During the same year, the French Government announced its intentions to halve sulfur dioxide emissions by 1990.

However, the United States and British Governments continued to question the relationships between sulfur dioxide emissions and acid deposition. See *Convention on Long-Range Transboundary Air Pollution; European Economic Community; Sulfur Dioxide.*

Activity Pattern: The distribution of various centers of human activity, during daytime and at nighttime, throughout a city or area; this is in many ways similar to functional pattern (q.v.).

Advertisement Control: The planning regulations employed in many countries to restrain the display of advertisements which may offend by reason of location, size, vulgarity, or detract from the environmental quality of the area, district, or street. In the United Kingdom, the approach has been that advertisements constitute development and, as such, require planning permission. However, providing the advertisements fall within the descriptions and measurements set out in the current Advertisement Regulations, they may be "deemed" to have planning permission. For advertisements larger than or different from those defined, specific approval must be sought. There is a right of appeal against refusal of consent or imposed conditions. Procedures exist whereby a local planning authority can keep specific areas virtually free of advertisements.

Agency for International Development (AID), US: A United States agency for overseas aid; the aid program has included projects in the Caribbean region, improving tropical forest management, combating desertification, and crop production in arid zones. Since 1975, AID has required an Initial Environmental Examination (IEE) to be carried out on all projects, and a full environmental impact assessment (q.v.) for projects where the impact is expected to be significant.

Air Pollution: The contamination of the atmosphere with undesirable solids, liquids and gases. In a strict sense, air may be considered polluted when there is added to it any substance foreign or additional to its normal composition. This definition of pollution is much too wide, however, for the purposes of practical air pollution control, and the term "air pollution" is usually restricted to those conditions in which the general atmosphere contains substances in concentrations which are harmful, or likely to be harmful, to man or to his environment. A fuller definition, of American origin, is—"substances present in the atmosphere in concentrations great enough to interfere directly or indirectly with man's comfort, safety or health, or with the full use or enjoyment of his property." In respect to health, this reiterates the dictum that there are no such things as toxic substances, only toxic concentrations. The concept of concentration cannot be divorced, however, from time or duration of exposure; or from the acute or chronic effects likely to arise from high short-term or low long-term exposures respectively.

The degree of air pollution varies according to population density, industrial development, geographical situation, meteorological characteristics, social and economic structure, and local customs. The contaminants arising from human activities (power production, the metallurgical and petrochemical industries, and transport) include sulfur oxides, nitrogen oxides, carbon monoxide, soot, and other compounds; they have clearly defined effects on vegetation, property, and local climate, and are also believed to have adverse effects on human health.

The history of air pollution has been characterized by a number of dramatic incidents, notably the Meuse Valley, 1930; Donora, Pennsylvania, 1948; London, England, 1952 and 1962; Poza Rico, Mexico, 1950; Seveso, Italy, 1967. Yet some of the more spectacular long-term effects have received little publicity. See *London Smog Incidents*.

Air Quality Act, 1967, US: An Act to amend the Clean Air Acts, 1963 and 1965 (q.v.), revamping the whole US clean air program. It provided procedures for the issuance of air quality criteria, the delineation of atmospheric areas and air quality control regions, the setting of standards, and the adoption of implementation plans by the states. The Act also authorized planning grants to air pollution control agencies. If the states failed to act, provision was made for federal intervention.

The Act also extended federal powers to control emissions from new motor vehicles, and promoted a national emission standard study for stationary sources; for the first time, national standards could be set for automotive emissions. In 1970, the National Air Pollution Control Administration established emission criteria for motor vehicle pollutants such as carbon monoxide, hydrocarbons, and nitrogen oxides.

Following the designation of control regions, each state had to formulate ambient air quality standards (as opposed to emission standards) and an enforcement plan. See *Clean Air Amendment Act, 1970, US.*

Alaska National Interest Lands Conservation Act, 1980, US: An Act which expanded the Alaska National Parks System by 45 million acres (18 million hectares), and wildlife refuges under the US Fish and Wildlife Service by 54 million acres (22 million hectares); the US Forest Service also gained about 3 million acres (1.2 million hectares).

Alaska, US: See *National Wildlife Refuges.*

Amazon Basin and Forest: A catchment and forest in the north and west of Brazil occupying more than half the country, but including only four per cent of the population. Three-quarters of the population are concentrated within 200 kilometers of the coast, mainly in the south and the southeast in the four states of Sao Paulo, Minas Gerais, Rio grande do Sul and Rio de Janeiro.

Until very recent years, the aim of Brazilian governments has been to open up the vast Amazon forest for development. Between 1966 and 1976 the official Amazon development agency Sudam approved the setting up of 354 ranches with an average size of 20,000 hectares. Tax rebates were granted to cattle companies. However, the African grasses planted on the cleared forest land grew less well after a few years, and large outlays on fertilizers and on measures against soil leaching and noxious weeds became necessary. Neither the cattle ranches nor the many small farms established have prospered. Between 1976 and 1979 only four additional cattle projects were approved.

In addition, the colonization program that was to have settled about one million families alongside the 4,600 kilometer Transamazon highway, completed in 1970, has been abandoned. Only some 6,000 families moved into the region as official settlers, although a large number of squatters appear to have moved there.

So far about one-tenth of the Amazon forest has been cut down, and destruction is still occurring at the rate of about 1,000 square kilometers a year. Fears have been expressed that the destruction of a large part of this forest could not only produce profound climatic changes in the region, but globally as well. Carbon dioxide absorbed by the forest would then simply augment the concentrations in the atmosphere, promoting the "greenhouse" effect. However, the disappointing results from the cattle projects must relieve the pressures on the forest, at least for these purposes.

In March 1979, President Figueiredo set up a forest policy committee to review the situation and analyze the causes of failure. The committee urged an almost complete reversal of the policy of opening up the 6 million square kilometers of the Amazon forest. It recommended that no more forest land should be leased to companies, and that about 1.2 million square kilometers should be designated as national parks and ecological reserves, and about 900,000 square kilometers as national forests. The failures had been due in part to the belief that the tropical soil was prodigiously fertile; but the topsoil was in fact very thin and ecologically fragile. The exuberant vegetation was mainly the outcome of a closed system of recycling, the nutrients from decaying vegetation being re-absorbed into tree roots without creating a fertile topsoil. In addition, half the rainfall was provided by evaporation from the forest itself.

The committee reaffirmed that the Amazon forest is a valuable resource yielding drugs, fibers, fuel, crops and resins, as well as providing gene pools of rare and valuable species of plants and animals. In 1979, the committee's approach had strong support from the Brazilian President. In that year, three new national parks in the Amazon forest were designated.

Amenity: A word that frequently appears in legislation usually referring to non-marketable environmental benefits such as beauty and tranquillity; public agencies are increasingly required to weigh economic efficiency, narrowly construed, against potential losses in amenity. In Britain, this duty was first incorporated in the Hydro-Electric Power (Scotland) Act, 1943. The "amenity clause" duty was subsequently extended to all public bodies in the Town and Country Planning Act, 1968, and became incorporated in the statutes of the nationalized industries generally.

Anadromous Fish and Great Lakes Fisheries Act, 1965, US: An Act authorizing the US Secretary of the Interior to enter into cooperative agreements with individual States for the purpose of conserving, developing, and enhancing the anadromous fishery resources of the nation that are subject to depletion from water resource developments and other causes. The Act pays particular attention to the fish of the Great Lakes that ascend streams to spawn. The Act established a national policy that wildlife conservation shall receive equal consideration with other features of water resource development programs. See *Great Lakes Water Quality Agreement*.

Antarctica Conventions: See *Conventions for the Conservation of Antarctic Living Resources*.

Anthracite and Bituminous Coal Mine Subsidence Insurance Fund, 1961, Pennsylvania, US: An insurance fund created in 1961 by the Pennsylvania State legislature providing insurance coverage up to $100,000 on residential and commercial structures damaged as a result of mining subsidence. The fund is operated by the Department of Environmental Resources. See *Bituminous Mine Subsidence and Land Conservation Act, 1966.*

Anti-Urbanism: A sense of disaffection with urban development; a genuine belief that urban life is in many ways less desirable than rural life. This may be linked with a view that if cities are inevitable, they have grown too large; urban growth should only be accepted by way of new, relatively small, communities in new locations.

Aquifer and Aquifer Recharge Areas: The underground stratum of earth, gravel or porous stone that contains water and can be used as a water supply. The aquifer recharge areas are those land areas through which surface water percolates and aids in replenishing ground water in the aquifer.

Asbestos-In-Schools Rule, 1982, US: Part of the anti-asbestos program of the Environmental Protection Agency (q.v.), a requirement to identify and remove friable materials containing asbestos from all schools in the United States. A field survey revealed that many schools had friable materials containing asbestos, though many were taking voluntary action to have it removed promptly. Violations of the rule are taken under the Toxic Substances Control Act, 1976 (q.v.).

Athens Charter, 1933: A statement which emerged from a series of studies and discussions conducted during the International Congress for Modern Architecture held in Athens in 1933. The charter addressed a range of contemporary problems relating to building standards, open space, slum clearance, work and industrial location, traffic, and heritage matters. It has had considerable influence on urban design. However, the 95 components of the charter did not deal deeply with economic, sociological, ecological, and legal factors, which are also vital considerations in urban planning. The Athens Charter was largely the work of Le Corbusier (1887-1965) whose visionary city planning schemes had a profound influence on modern architecture and town design.

Australia: See *Australia-Japan Migratory Birds Agreement; Australian Environment Council; Council of Nature Conservation Ministers; Environment Protection (Impact of Proposals) Act, 1974-75; Frost Commission; Gordon-Below-Franklin Dam; Green Ban Movement; National Capital Development Commission; New South Wales, Australia; Ranger Uranium Environmental Inquiry; Royal Commission on The Great Barrier Reef.*

Australia-Japan Migratory Birds Agreement: An agreement between the Australian and Japanese governments for the protection of migratory birds and birds in danger of extinction; it was signed in Tokyo in 1974 and ratified in 1981. Management plans, as required under the terms of the agreement have now been developed. The Towra Point Nature Reserve on the southern side of Botany Bay, Australia, plays an important part in the agreement.

Australian Environment Council (AEC): A Council established in 1971, following agreement between the prime minister of Australia and the state premiers, to provide a forum for consultation between ministers with responsibility for environmental matters. The Council is supported by a standing committee, and a number of working groups set up to study specific issues. The standing committee meets about four times a year, the full Council twice a year.

The Council has examined and made recommendations in respect to such issues as motor vehicle emissions, lead in gasoline, air pollution monitoring and disperants, waste disposal, dumped motor car and tire disposal, fiscal measures to promote pollution control, and environmentally hazardous chemicals. The Council agreed upon the need for a national chemicals notification and assessment scheme to operate from 1985. The scheme provides a mechanism for evaluating new chemicals before they enter the Australian market, and provides for a similar review of those chemicals already in use which may present health and environmental concerns. The implementation of timely and comprehensive control measures may thus be facilitated.

Austria: See *Urban Renewal Act, 1974.*

B

Background: The total landscape which is perceived by an observer from a particular viewpoint, ranging from 4 to 6 kilometers to infinity. Within this range, the observer is unaware of individual details and discerns broader landscape units as patterns of light and dark. See *Foreground*.

Backwater Area: An area subject to flooding that does not experience significant flow velocities. As such it may be filled without adverse effects on flood levels or velocities. In determining areas that may be so classified, it is necessary to consider the total flood plain (q.v.) situation and have regard to the cumulative effects of other such classifications.

Back-Zoning: Descriptive of rezoning when land is deprived of one zoning and given a more adverse zoning. See *Zoning; Zoning Ordinances*.

Bangladesh: See *Mangrove Parks*.

Banking, US: A basis for the emissions trading program (q.v.) in which firms that reduce their pollutant emissions through approved programs can gain Emission Reduction Credits (ERC) which may be held for the firm's own current or future expansion, or sold to another firm in the region. It provides an incentive to reduce pollution below the requirements of law. A number of states have provided banking facilities for ERCs. See *Offsetting*.

Barlow Report, 1940, UK: The report of the Royal Commission on the Geographical Distribution of the Industrial Population; the commission sat between 1937 and 1940 under the chairmanship of Sir Anderson M. Barlow. The report urged the importance of preventing further industrial congestion in the Greater London area, and the need for encouraging a greater diversity of occupation in areas which had become too dependent on specialized industries.

Belgrade Charter: See *United Nations Educational, Scientific and Cultural Organization*.

Beneficial Use: In the context of environmental planning, a use of the environment (q.v.) or any element or segment of the environment that is conducive to public benefit, welfare, safety, or health, and which requires protection from the effects of waste discharges, emissions, deposits, and despoilation. Beneficial uses include:

- Potable water supply for drinking, domestic and municipal purposes;
- Agricultural and industrial water supply;
- Habitats for the support and propagation of fish and other aquatic life;
- Recreational activities such as bathing, fishing, and boating;
- Scenic and aesthetic enjoyment;
- Navigation; and,
- Wildlife habitats.

A residual use is a use other than a beneficial use, e.g. the disposal of liquid effluents after treatment into water.

The term also occurs in British planning law. Questions of beneficial use arise when a landowner, having been refused planning permission or had harsh conditions imposed on a planning consent, or experiencing planning blight (q.v.), feels he/she can no longer obtain "beneficial use" out of his/her land and so serves a purchase notice on the local planning authority asking in effect to be "bought out."

Benefit-Cost Analysis: A technique developed in the United States in response to a legal requirement imposed in 1936 on the water resource projects of the Federal Government; this requirement specified that projects be undertaken only "...if the benefits to whomsoever they may accrue are in excess of the estimated costs..." (US Flood Control Act, 1936). Efforts to implement this requirement led to the preparation of a *Green Book* by a Federal interagency committee composed of the representatives of the major water-resource agencies. This report embodied the general principles of economic analysis as they were to be applied in the formulation and evaluation of Federal water resource projects. The principles of benefit-cost analysis (known as cost-benefit analysis in the United Kingdom) have since been applied to a wide range of projects and to the design of public policies in various areas, including environmental planning. The principles have been applied, despite the inherent difficulties, to pollution control, transportation, urban development, electric power, health, education, welfare, and in social issues such as equity, income distribution, and unemployment. Benefit-cost analyses proceed by adding-up total money costs and benefits (reduced to present worth), generally disregarding who incurs the costs and receives the benefits. If the money aggregate is positive, this is taken to indicate that the gainers could compensate the losers and still be better off after the project is completed. If the net monetary gain is negative, the compensation test fails. The net monetary gains are generally held to measure the efficiency of government projects, or policy changes. See *Regulatory Impact Analysis (RIA)*.

Benefit/Cost Ratio (BCR): A ratio calculated in the following way—

$$\frac{\text{total discounted benefits}}{\text{total discounted costs}}$$

If the ratio is greater than 1, the total benefits exceed the total costs. Theoretically (according to this yardstick of economic appraisal) the project is economically "worthwhile."

Benthos: Plants and animals living in or on the bottom of a sea, attached or unattached, up to the high-water mark.

Berm: A barrier adjacent to a highway to intercept and deflect sound (to create an acoustic shadow); the barrier must be continuous unless pockets of high noise can be accepted. They need not be massive. Effective barriers can reduce noise by up to 25 dB(A) (q.v.), although reductions of 10 dB(A) are in practice more common. A reduction of 10 dB(A) is equivalent to halving the relative loudness.

A full earth berm may be 18 meters wide and at least 3 meters high. Alternatively, a wall may be provided some 3 meters high at a much lower cost in terms of land. Intermediate arrangements are also possible.

Best Available Control Technology (BACT): Or best practical control technology. The most exacting control requirements that may be imposed on a source of pollution after taking energy, environmental and economic costs into account; determined by the control authorities case by case. See *Best Practicable Means; Clean Air Amendment Act, 1977; Reasonably Available Control Technology.*

Best Practicable Means: The most commonly used technical approach to pollution control from industrial premises. The word "practicable" is often taken to mean "reasonably practical" having regard among other things to local conditions and circumstances, to the financial implications, and to the current state of technical knowledge. Administration of the concept is much easier than the air quality standard approach. Once the required equipment is installed, subsequent inspections are confined to ensuring that the equipment is in good working order, and is actually being operated. The concept is also sufficiently flexible to permit improvements over time.

The definition of the words "best practicable means" in British legislation is of considerable importance. It refers not only to the provision and the efficient maintenance of appliances adequate for preventing the escape of noxious or offensive gases (including smoke, grit, and dust) but also to the manner in which such appliances are used, and to the proper supervision by the owner of any operation in which such gases are evolved. The obligation to adopt the best practicable means is continuous and this may entail alterations in plant and methods as new techniques for controlling emissions become available. The adoption of the best practicable means in any instance may include the installation of such equipment as bag filters, wet washers, multi-cell cyclones, and electrostatic precipitators for arresting mists, fumes, and dusts.

Betterment: Also called unearned increment. Profits likely to accrue to the owner of land as a result of an advantageous rezoning of the land, e.g., from agricultural pursuits to housing. In Britain, the Town and Country Planning Act, 1947 (q.v.) laid down the basic principle that simply to own land in an area scheduled for a profitable change of use did not entitle the owner to anything more than the return that would have been obtained from its sale for the original use. If sale occurred, the profit or "betterment" would go to the public authority.

The experiment ended in the early 1950s because of difficulties in administration, rather than a change of view regarding the principle itself. A central difficulty was determining the size of the gain, for the land in its original use may have undergone many improvements. Decisions, therefore, required a large skilled assessment staff. The costs of this, measured against the return, did not prove attractive. There were also the claims of those who had incurred a loss through disadvantageous rezonings. There was also the problem of the distribution of gains between the local and central government.

Bhopal Disaster, India, 1984: A catastrophic gas leak at a pesticide plant in Bhopal, India, in December 1984, as a result of which over 2,000 people, mostly children and older people, died and some 50,000 suffered from various degrees of blindness, temporary and permanent. Methyl isocyanate, a gas used in the manufacture of the pesticide Sevin at a Union Carbide plant, leaked from an underground storage tank and flowed into neighboring slum areas.

It appeared that pressure had built up in the storage tank, rupturing a valve, and allowing the gas to escape through a scrubber system which failed. The poisonous gas appears to have diffused over some 65 square kilometers of the town in an area populated largely by poor families. This air pollution incident was the world's worst industrial disaster. The disaster has promoted intensive reviews of industrial practices in the United States of America (resulting in a toxic air pollutant control bill in 1985), and in the European Economic Community (q.v.) and the Organization for Economic Cooperation and Development (OECD) (q.v.). See *Air Pollution; Mexico City Industrial Disaster, 1984: Risk and Hazard Assessment.*

Biochemical Oxygen Demand (BOD): The weight of oxygen taken up mainly as a result of the oxidation of the constituents of a sample of water by biological action. The result is expressed as the number of parts per million (or grams per liter) of oxygen taken up by the sample from water originally saturated with air, usually over a period of 5 days at a temperature of 20°C. The result gives some measure of the amount of biologically degradable organic material in polluted waters; although when samples contain substances such as sulfites or sulfides which are oxidized by a purely chemical process, the oxygen so absorbed may form part of the BOD result. The BOD test is no longer an adequate criterion for judging the presence or absence of pollution, for many relatively new pollutants must be considered, e.g. pesticides, industrial organic compounds, fertilizing nutrients, dissolved salts, soluble iron and manganese, and heat.

Biosphere: A part of the physical environment, encompassing the sum of all living organisms (plants and animals, including microorganisms); the lower parts of the atmosphere; the hydrosphere (oceans, inland waters, and subterranean water); and the lithosphere or outer crust of the planet. See *Environment*.

Biostimulants: Substances which stimulate the growth of aquatic plants. For example, the addition of large amounts of nitrogen and phosphorus compounds to lakes may stimulate massive growth of microscopic plants such as blue-green algae or the larger waterweeds. The process is called cultural eutrophication.

Sewage is a major source of biostimulants, particularly nitrogen and phosphorus. A large part of the phosphorus found in sewage is derived from domestic and industrial detergents which contain phosphorus compounds to enhance their cleaning properties.

Bituminous Mine Subsidence and Land Conservation Act, 1966, Pennsylvania, US: An Act passed in 1966 by the Pennsylvania state legislature to protect water supplies and property in the bituminous areas of the state against mine subsidence; it gave additional protection to dwellings and buildings in place before 1966. However, new post-1966 property owners could be required to compensate miners to leave coal pillars for subsidence protection. See *Anthracite and Bituminous Coal Mine Subsidence Insurance Fund, 1961*.

Blanketing: Refers to a situation in which there is a reservation of land for a public purpose, but the project is in the distant future and in the meanwhile the authority is not prepared to acquire the land. The owner may be assured of eventual compensation but in the meantime cannot make any significant improvements and may indeed be inhibited in the use of the land. Blanketing may be regarded as a "planning limbo" or a form of "planning blight" (q.v.).

Bleve: A boiling liquid expanding vapor explosion. A potential hazard with the storage of inflammable gases and other combustibles. See *Risk and Hazard Assessment*.

Blight: See *Planning Blight; Urban Renewal*.

Board of Zoning Appeals, Los Angeles, US: A five-member part-time board appointed by the Mayor with the concurrence of the City Council of Los Angeles; members serve five-year staggered terms. The board handles cases involving parcel map procedures and appeals from the zoning administrator. In cases concerning disapproval of zone variances, the board's decision is final. In cases concerning parcel maps, zone variance approvals or conditional uses, the board's decisions may be further appealed to the City Council. See *Hearing Examiners*.

Brandt Commission: An international commission set up in 1977 at the instigation, among others, of the World Bank (q.v.). Its members, twenty in all, came from the rich north, poor south, and the oil-producing nations. Reporting in 1979, it recommended an emergency program followed by longer term reforms.

The emergency program comprised: a large transfer of resources to less developed countries to reach 0.7 per cent of the gross national product of rich countries by 1985, and 1 per cent by 2000; increased lending by the World Bank and other financial measures; a global food program in the form of aid to agriculture, direct food aid, and larger food stocks; and major reforms in the international economic system. The Report proposed a new institution—a World Development Fund—universal taxes flowing into its coffers.

The world economic recession of that period associated with a reduced ability of richer countries to increase aid and for debtor nations to increase borrowing reduced the prospects of the early realization of the recommendations of the Brandt Report, though not the importance of its message.

However, an increased transfer of funds in the future to poorer countries through a variety of channels could have important implications for environmental programs notably anti-vector campaigns, sewerage and water supply schemes. Aid would need to be specifically tied to objectives such as these, to avoid the risk of dissipation on arms or grandeur.

Brasilia: The national capital of Brazil, and perhaps the largest single coordinated city building effort in modern times. The entire city was to have been built according to a plan submitted by the Brazilian architect, Lucio Costa, in 1957; by 1967, however, the development of Brasilia had deviated significantly from the original plan.

The centrally planned area, known as the Pilot Plan, contains the famous superblocks and some residential accommodation; many government ministries are located there. Satellite towns are grouped around this central area, being separated from it by an expanse of open space. However, squatter settlements have sprung up containing many of the original workforce and others who have gravitated to the capital. The area is now a scenario of middle-class apartments and lower-class shacks; it is considered to mirror Brazilian society as a whole reflecting splendor and poverty and the corresponding attitudes of mind.

Brasilia may be considered as a brilliant concept launched with political fervor and urgency, straining national resources to the limit. Some of the achievements are awesome in an architectural or structural sense.

The concept was then strangled by the realities of Brazilian society coupled with a weakening of political resolve and flow of resources. See *National Capital Development Commission, Australia.*

Brazil: See *Amazon Basin and Forest; Brasilia; Cubato Disaster, 1984; Infant Morality Rate; Special Environmental Agency.*

British Columbia, Canada: See *Federal Environmental Assessment Review Office.*

Bubble Concept, US: A basis for emission trading (q.v.); firms are encouraged to combine the emissions from their numerous outlets for the purpose of defining the emission limitations for the entire industrial facility, and then to develop their own strategies for different levels of control at different sources; the entire plant must stay, of course, within the overall emission ceiling set probably by a State Implementation Plan. The bubble concept enables managers to impose severer limitations on emissions which can be restricted relatively inexpensively, in exchange for reduced controls on outlets more expensive to control. There is no incentive within the bubble concept to reduce the overall emissions below the required limits. By the end of 1984, the Environmental Protection Agency (q.v.) had approved over 30 bubbles. See *Figure 1.*

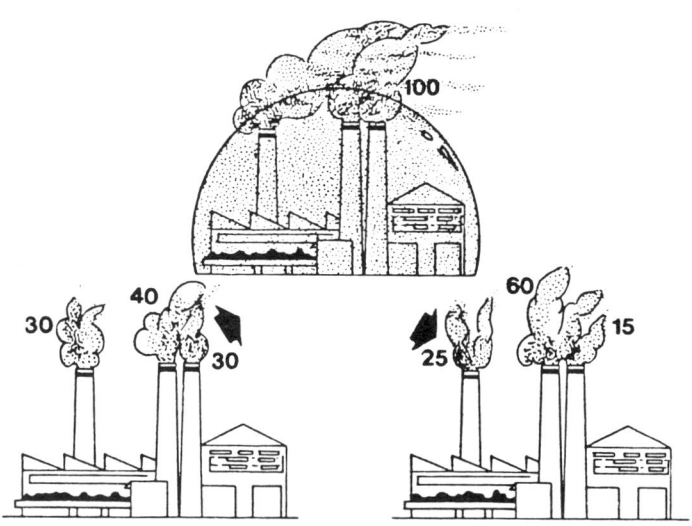

Figure 1: The EPA Bubble Concept. Pollutants of a like kind from each source within a plant can be controlled to minimize costs, providing the total emission from the plant stays within a set limit. Source: U.S. Environmental Protection Agency, *EPA Journal*, Feb. 1980.

Buchanan Report, 1963, UK: A report entitled *Traffic In Towns* which analyzed the growing traffic problem in cities and towns. The report exposed the need for unprecedented investment in road improvements and the rearrangement of the patterns of circulation. It called for the establishment of a hierarchy of urban roads and improved traffic management. It concluded that acceptable environmental standards could in some cases be attained only by imposing a measure of restraint on the free access of vehicles to town centers.

The report stressed the interconnection between land use and traffic, and also that accessibility (q.v.) and environmental considerations often pulled in opposite directions. To secure a satisfactory environment, some sacrifice of accessibility might be involved.

Buffer Zone: An area of land separating land uses which are incompatible with each other, which is (or should be) of sufficient width to prevent any conflict between them. Buffer zones may be established to separate industrial from residential areas; or to separate airports from all other developments. Buffer zones may be planted with various forms of suitable vegetation, shrubs and trees, and may be contoured to form noise bunds. Buffer zones may form also a part of an open space program (q.v.).

Buffer zones are required, for example, around quarries to protect them from encroachment by land uses such as urban and small scale rural residential development which, if permitted too close to quarries, can lead to their premature closure or to the imposition of severe restrictions (including restricted working hours) on their operations.

The amount of land required for a buffer zone around a quarry will vary from deposit to deposit, depending on such factors as size of the resource, quarry design, method and scale of extraction, and topography. The buffer zone has to be of a sufficient width to ensure that noise, vibration, and dust levels at the outer margin are compatible with existing or permissible land uses in the surrounding area. Consideration may also need to be given in the design of the buffer zone to the reduction of visual impact.

The restrictions on land use within buffer zones will, of course, no longer be necessary once the deposit has been worked out and, depending on the nature of the deposit and the manner in which mining is carried out, it may be possible to reduce the size of the buffer zone during the life of the quarry.

Building Application, Factors to Be Considered: Factors to be taken into account by local government, in considering an application for approval of the erection of a building; these factors include:

- Drainage, ventilation, lighting and healthiness of the building;
- Design, materials, stability, building line and height;
- Size, height and lighting of rooms;
- Height of floor levels in relation to level of road;

- Size, height and materials of party walls between buildings;
- The proportion of the site to be covered by the building and the provision of open spaces and light areas;
- The position of the building or any outbuilding or office in relation to other buildings or to the boundaries of the site;
- The provision of storage for water for domestic purposes;
- Means of access generally and particularly the means of access for the purposes of the removal of garbage and other refuse and sewage sludge in appropriate cases;
- Height, materials, stability, design and position of fences (if any) to be erected on or on the boundaries of the allotment on which the building is to be erected;
- Whether the site is subject to flooding or tidal inundation; and
- Whether the site is or probably will be subject to subsidence or slip.
- The provision of parking facilities.

Building Codes: Statutory codes enacted by municipalities for the purpose of safeguarding public safety and health through the regulation of building construction, building use, and maintenance, and through the installation of certain types of services. The codes include building codes, health and sanitary codes, fire codes, plumbing codes, electrical codes, mechanical codes, and housing codes. The codes set forth in great detail the minimum standards to be adopted by the builder, architect and engineer.

As a permit is required before construction can begin, and later for use and occupancy, municipal supervision is insured. Further, the issue of a permit usually involves the filing of a suitable bond.

Built Environment: A reference to buildings, structures, utilities and services which enable people to live, work and play, circulate and communicate, and fulfill a wide range of functions in whole or part; the quality may range from grandeur to blight.

Bushfire Prone Areas, New South Wales: Areas in New South Wales, Australia, which are liable to bushfires (that is, unplanned fires in vegetation) likely to present a hazard to life and property. Such areas must be taken into account by local government in the preparation of a local environmental plan (q.v.). If bushfire hazards may exist on land intended for urban development a draft plan is required to:

- Provide for the creation of a perimeter road or reserve which circumscribes the hazard side of the land intended for that development;

- Provide for the creation of a fire radiation zone managed for hazard reduction and located on the bushland side of the perimeter road;
- Specify minimum residential lot depths for lots adjoining the perimeter road;
- Minimize the perimeter of the area of land which may be developed for urban purposes; and
- Introduce controls which avoid placing inappropriate developments in hazardous areas.

Business Improvement District Act, 1981, Kansas, US: An Act to enable cities in the State of Kansas to establish one or more business improvement districts in order to provide additional services as will restore or promote the economic vitality of the district and the general welfare of the city. The various improvements that may be provided include:

- Beautification of the district,
- Provision of special or additional public services,
- Provision or financial support of public transportation services and vehicle parking facilities,
- Development of plans for the general architectural design of public areas and development of plans for the future development of the district,
- Development, promotion and support of community events,
- Any other services which the city is authorized to perform and which the city does not perform to the same extent on a city-wide basis.

Prior to the establishment of any business improvement district, the Mayor of the city must appoint a district planning committee to review the matter. Following a recommendation from the committee to establish a district, the governing body of the city may adopt a resolution of intent. A public hearing is then required, following which an ordinance may be created. An advisory board for the business improvement district shall be appointed consisting of persons from representative businesses within the district. The advisory body then recommends a program of services and proposed budget to the governing body; on proceeding, the city must establish a special fund for the district. An annual levy may be imposed only on businesses located within the improvement district. See *Redevelopment of Central Business District Areas Act, 1976.*

C

California, US: See *Board of Zoning Appeals, Los Angeles; Central City Community Plan, Los Angeles; Coastal Zone Management Act; Environmental Quality Act, 1970; General Plan, Los Angeles; Hearing Examiners, Los Angeles; Housing Densities; Los Angeles City Planning Commission; Los Angeles Smog; Oil Spills; Southern California Association of Governments; Transport Plan; Transportation Study.*

Canada: See *Acid Rain; Department of Environment (Environment Canada); Federal Environmental Assessment Review Office; Great Lakes Water Quality Agreement; National Park; Ontario, Canada; Sulfur Dioxide; Surface or Strip Mining.*

Capital Improvement Program: A schedule of public improvements such as streets, sewers, parks, water lines, schools, and so on, prepared by the planning commission (q.v.) usually for a five to ten year period and revised annually. It is based on a series of priorities which take account of the need, or desire, for or the importance of the proposed improvements and the community's ability to pay for them. An important determinant of priority is the comprehensive plan (q.v.).

Car Scrapping Act, Sweden: A measure to regulate the disposal of junk cars. To encourage the proper disposal of worn-out cars, the owner is paid a sum of money when he strikes his car from the register; de-registration requires, however, a scrapping certificate made out by an authorized car scrapping firm. See *National Environment Protection Board.*

Carpool System: A system aimed at promoting the more efficient use of the private automobile by increasing occupancy from typically one person per vehicle to several; it has been urged that higher occupancy private automobiles should be permitted to use preferential bus-carpool lanes within the central city as well as those provided on freeways, highways, and other street systems. Parking facilities could also be encouraged to provide reduced rates and preferential locations to higher occupancy private automobiles.

Central Business District (CBD): Consequent upon the scale and intensity of physical development, and the variety of its functions and activities, the center of a metropolitan area, city, or large town. It is often the preferred location for the head offices of major private and public sector organizations; a communications center; a cultural, educational, and recreational center; a commercial and financial center; an entertainment, tourist, and retail center; and contains residential components. Economically, the CBD serves people and companies within and beyond the metropolitan boundary.

In many cities, major suburban shopping and office complexes have developed detracting from the importance of the CBD while easing strains on transport services and parking facilities. Nevertheless, the CBD is prominent and obvious in all cities; it has been likened to the hub of a wheel from which communication and transport "spokes" radiate, or sometimes as a magnet towards which all things tend to gravitate. Its success depends on economic viability in the service as distinct from manufacturing sectors, though travelling to work in the CBD for a better salary can be a long and tedious process for many people.

The CBD always presents fairly intense land use conflicts for land values are high and the retention of land for parks and open space, or for the preservation of historic buildings, becomes correspondingly expensive. The provision of housing for lower-income workers who are needed in the CBD becomes a difficult problem.

Zoning and development control codes have been generally applied by city government to resolve land use conflicts as far as possible. Environmental criteria which can be controlled through planning and building regulations include:

- Bulk and height of buildings,
- Overshadowing,
- Wind tunnel effects,
- On-site parking,
- Continuity of street facades,
- Visual impact of scale of building,
- Pedestrian resting areas,
- Building setbacks and floor-space ratios,
- Preservation of special vistas,
- Proximity to historical features, and
- Sidewalk protection and skyways.

In many North American and European cities the central business districts have undergone profound changes during the post-War years as a result of massive investments in new business developments, dwarfing pre-War structures, coupled with the redesign of freeways and expressways to accommodate the domination of the private car as a means of transport. Large areas of derelict and blighted properties have been cleared.

In Britain and Europe, many city centers had to be reconstructed following devastating damage sustained during the War years; many took the opportunity to restore the historic character of their city centers in meticulous detail. Warsaw is an outstanding example; a cultural victory for the Poles over a barbarous enemy.

In Britain, central London, especially around St. Paul's Cathedral, was extensively reconstructed. Coventry also reconstructed its battered city center in a modern manner; however, the ruins of the bombed Cathedral were retained as a fitting historical approach to the new Cathedral. Other city centers have undergone extensive reconstruction through collaboration between the public and private sectors including Birmingham, Leicester, Liverpool and Newcastle-upon-Tyne.

Central City Community Plan, Los Angeles, US: A key part of the General Plan (q.v.) for the City of Los Angeles which proposes a series of major centers having high density residential and commercial uses at locations throughout the city, connected by a rapid transit system, and separated by low-density residential developments and open spaces.

The central city is intended to be a major center of the Los Angeles metropolitan region, emphasizing the following functions and features: a focal location for business, especially financial institutions and corporate headquarters; one of the region's largest employment concentrations; the major concentration of governmental administrative facilities, providing services not only to the City and the metropolitan area but also much of Southern California; and the location of unique, one-of-a-kind, cultural, recreational and tourist facilities, including the Music Center, the El Pueblo Historical Monument, the Convention Center, Pershing Square Park, and Little Tokyo.

Utilizing an overall design concept, the central city community plan proposes achievement of its objectives and goals by:

- Joining isolated components by new circulation linkages and transitional public and private development;
- Stabilizing the various functional areas by correcting the negative influences of deteriorating development and social problems in adjacent areas;
- Utilizing public investments as catalysts to reverse blight and attract new development;
- Establishing a viable land use mix, including environmental amenities compatible with residential uses;
- Establishing a positive, aesthetic image for Central City which can be enhanced by future growth and change;
- Encouraging excellence in urban design;
- Providing ready access to concentrated development by means of rapid transit stations and people movers.

A major feature of the plan is a proposal for rebuilding a large portion of the central city. Bunker Hill and Little Tokyo are already established redevelopment areas.

Central Place Theory: A body of theory relating to towns and cities, the foundation of which was laid by Walter Christaller in his study of south German towns. Christaller described central place theory as "general deductive theory" designed to "explain the size, number and distribution of towns" in the belief that "some ordering principles govern the distribution".

Christaller regarded the basic function of a city to be a central place providing goods and services to the surrounding tributary area. To perform such a function efficiently, it was argued, a city is found at the center of minimum aggregate travel for its tributary area, i.e., central to the maximum profit area it can command. The greater the centrality of a place, the higher is its order. Higher order places offer more goods, have more establishments and business types, larger populations, tributary areas and tributary populations, do greater volumes of business, and are more widely spaced than lower order places.

Low order goods are generally necessities requiring frequent purchasing but little consumer travel, while high order goods are goods for which the consumer is willing to travel longer distances, although less frequently. Because higher order places offer more shopping opportunities, their trading areas for low order goods are likely to be larger than those of low order places.

Centers of each higher order group perform all the functions of lower order centers plus a group of central functions that differentiate them from, and set them above, the lower order. Central place theory remains one of the most widely applicable of theories relating to systems of cities, particularly in relation to the location of tertiary activity.

It is, however, a simplification of reality holding many factors constant; the theory assumes an even plain and a uniform distribution of natural resources and people. In reality, resource localization distorts the hierarchy. It is best to regard central place theory as providing a valuable partial framework for the understanding of regional structures. See *Growth Pole Theory; Service Activities.*

Chesapeake Bay Program, US: Established in 1976 by the US Congress, a program which created a federal, multiple-state, and multiple-local government partnership to identify water quality baseline conditions, determine trends, and develop related environmental management strategies in relation to Chesapeake Bay. The Bay is one of the world's most productive estuarine systems; fresh water comes to the Bay from a network of over 150 rivers, creeks, and branches. The 12 million people who live in the Chesapeake drainage basin are involved in varied economic activities including agriculture, forestry, coal mining, steelmaking, shipbuilding, leather tanning, plastics and resin manufacturing, chemical production, recreational services, and poultry, seafood, and vegetable processing. The Bay's unique circulation system does not freely exchange substances with the ocean, and some contaminants tend to accumulate in the food chain and sediments.

The dangers of toxic chemicals had been exhibited through the near elimination of osprey due to DDT exposure, and the closure of the James River fishery because of kepone contamination. Nutrient enrichment was apparent in the Potomac River in the 1960s. Submerged rooted grasses began receding at a significant rate in the late 1960s, and by the mid-1970s there was an alarming loss of this important estuarine ecosystem resource. Indeed nutrient enrichment has been identified as the major threat to the productivity and environmental quality of the Bay. The result has been the development of Bay-wide management strategies to decrease nutrient enrichment.

Chicago, Illinois, US: See *"City Beautiful."*

Chile: See *Natural Resources Data Bank*.

China, People's Republic of: See *Desertification Control; Environmental Protection Law, 1979; Green Great Wall*.

Chipko Andolan Movement: A Save-The-Trees campaign which has spread among the Indian public. The Chipko Andolan (tree-hugging) movement started in Ultar Pradesh in 1973, when tree fellers were prevented from cutting down trees by villagers who hugged the trees. The movement has subsequently spread all over the Himalayan region. Some Indian states, attempting to harness this enthusiasm have promoted "social forestry" schemes; cooperating families who engage in tree planting are rewarded at least in one state by a free smokeless stove. The stoves, normally too expensive for families to buy, use only half as much fuel as traditional stoves.

Chlorofluorocarbons (CFCs): Chemical compounds which undergo synergistic effects in the atmosphere. In 1974, a report by the US National Academy of Sciences warned of the possible effects of aerosol sprays in releasing fluorocarbons and halocarbons to the atmosphere. These compounds tend to be carried into the stratosphere which retains them because of the permanent temperature inversion at those altitudes. Under bombardment by ultraviolet radiation the compounds break down, releasing chlorine which reacts with and destroys the atmospheric ozone. The effect is similar therefore to that of high flying supersonic aircraft, the crucial shielding effect of the ozone layer being weakened.

However, the chemistry of the upper atmosphere is complex and still poorly understood. Research is still being conducted to determine the precise history of these substances and the extent to which they contribute chlorine to those parts of the atmosphere where they can catalyze the destruction of ozone.

In May 1977, three US agencies, the Food and Drug Administration, the Environmental Protection Agency, and the Consumer Products Safety Commission, recommended that these substances be phased out within two years as a precautionary measure, and urged other nations to consider similar action. They argued that damage to the shield would increase the risks of skin cancer, upset the growth of vegetation, and cause other adverse effects. The ban became fully effective in April 1979 in relation to the most important substances known as CFC (F-11 and F-12).

The pressure spray (aerosol) products which contained these carbons included deodorants, anti-perspirants, hair sprays, colognes, perfumes, insect repellants, pesticides, household cleaners, shaving creams, and sprays used by asthmatics and hay fever sufferers.

The European Economic Community, for its part, called for a 30 per cent reduction in the use of the two propellants by the end of 1981, using 1976 as the benchmark. It appears that this objective was fully met.

Circulation Element: See *Transport Plan*.

City and Regional Planning: A description which has been generally applied in the United States of America to the physical planning of urban and non-urban areas. It stands in contrast to the description "town and country planning" (q.v.) which has been in vogue in Britain. However, both appellations are tending to give way to the more contemporary description "environmental planning" (q.v.).

"City Beautiful": The theme of Chicago's World Columbian Exposition held in 1893; the exhibition was essentially a protest against city ugliness and an early initiative in favor of civic centers and boulevards, and orderly development. The year 1909 witnessed Daniel Burnham's Plan for Chicago, and the appointment of Chicago's Plan Commission.

Burnham advised future planners: "Make no little plans; they have no magic to stir men's blood and probably will not be realized. Make big plans; aim high in hope and work, remembering that a noble, logical diagram once recorded will never die."

Civic Amenities Act, 1967, UK: An Act "to make further provision for the protection and improvement of buildings of historic interest and of the character of areas of such interest; for the preservation and planting of trees; and for the orderly disposal of disused vehicles and equipment and other rubbish." The Act enabled local authorities, for the first time, to seek the protection of areas of cities, towns and villages, as distinct from the protection of individual buildings. Tree preservation orders became more effective and provided means for planting more trees with new development. The Act made it an offense to abandon vehicles on the highway and required local authorities to provide proper facilities for the collection of old cars and other similar unwanted articles. The Civic Amenities Act was later incorporated in the Town and Country Planning Act, 1971, (q.v.).

Clean Air Acts, 1963 and 1965, US: United States clean air legislation which made available federal technical and financial assistance to state air pollution control agencies, and generally sought to encourage cooperative action by the states and local authorities. The control of air pollution from new motor vehicles was inaugurated.

The Clean Air Act, 1963, designated three specific research areas for special attention: control of motor vehicle exhausts, removal of sulfur from fuels, and the development of air quality criteria for major pollutants. The Act provided an abatement procedure in respect of interstate pollution.

The Clean Air Act, 1965, provided specifically for federal regulation of motor vehicle emissions, authorizing an expanded research program, creating a federal laboratory, and initiating an investigation of new sources of pollution. See *Air Quality Act, 1967, US*.

Clean Air Amendment Act, 1970, US: An amendment to the US Clean Air Acts, 1963 and 1965 (q.v.). Under the 1970 Act, the Environmental Protection Agency (q.v.) was enabled to set national ambient air quality standards for six major pollutants: sulfur dioxide, particulate matter, carbon monoxide, photochemical oxidants (e.g., ozone), oxides of nitrogen, and hydrocarbons. Allowable emission limitations for different kinds of stationary sources of pollution could also be established; motor vehicle pollution was also to be substantially reduced. The Act requires that 1981 and later model cars conform to an emission limit of 1.0 grain per mile (gpm) of nitrogen oxide, and 3.4 gpm of carbon monoxide. The emission of lead has been restrained by restrictions on the lead content of gasoline.

The main administrative mechanism used by State governments to define strategies to achieve national standards has been the State Implementation Plan (SIP). Since 1970, there has been a significant reduction in the emission of some pollutants: particulates by 58 per cent, sulfur oxides by 25 per cent, and carbon monoxide by 27 per cent. The Clean Air Amendment Act, 1970, expired in 1981; it was then extended indefinitely by the US Congress without change pending a thorough review which continued through 1985. See *National Ambient Air Quality Standards (NAAQS), US*.

Clean Air Amendment Act, 1977, US: A further amendment to the US Clean Air Acts, 1963 and 1965 (q.v.). The 1977 Act required the "best available control technology" (BACT) to be installed in new stationary plants. The Act also required the Environmental Protection Agency (q.v.) to establish national emission standards for hazardous air pollutants—pollutants to which national ambient air standards are not applicable because they do not occur on a sufficiently widespread basis. By the end of 1985, standards had been issued in respect of asbestos, beryllium, mercury, vinyl chloride, benzene, arsenic, radionuclides, and coke oven emissions. The Act also paved the way for stricter air quality standards for "clean air areas." The amendments gave the EPA authority to impose economic penalties against non-complying plants. See *National Ambient Air Quality Standards (NAASQ), US; New Source Performance Standards.*

Clean Lakes Law, 1984, Japan: A law enacted by the Japanese Government laying down a basic policy for the conservation of lake water quality; it opened the way for the enforcement of special measures in respect of those lakes which were in great need of action to meet the national environmental water quality standards. It provided for the drafting, in respect of designated lakes, of a Lake Conservation Plan through which pollution loads would be reduced. Ten to twenty lakes, such as Lake Biwa in Shiga Prefecture, Lake Kasumigaura in Ibaraki Prefecture, and Lake Suwa in Nagano Prefecture, are expected to be so designated.

Clean Water Act, 1977, US: United States clean water legislation which amended earlier 1972 legislation. Its aim was to attain and maintain the quality of water necessary for a variety of uses, including public water supplies, propagation of fish and wildlife, recreation, agriculture, industry and navigation. It imposed more stringent controls on toxic pollutants.

Industries discharging to publicly-owned sewage treatment facilities are subject to a separate set of requirements; these stipulate pre-treatment standards for the discharges, prohibit certain discharges, and establish a system of user charges to support the operation and maintenance of publicly-owned facilities.

Stricter requirements in respect of new discharges aim at the adoption of the "best demonstrated control technology, processes, operating methods, or other alternatives, including, where practicable, standards permitting no discharge of pollutants."

The principal enforcement mechanism under the Act is the "national pollutant discharge elimination system." A permit is required within the context of this system for every point source discharge into United States waters; the use of permits ensures compliance with effluent limitations and other requirements under the Act. For example, new power plants must have a permit; such a permit incorporates all applicable effluent and water quality standards promulgated under the Clean Water Act, and the applicant must demonstrate to the permitting agency that these limitations and standards can and will be met. Permits are for five years, and must then be renewed; a case has been put forward for ten year permits to simplify administration and reduce the uncertainties for industry.

By 1985, effluent guidelines for the control of toxic discharges had been issued for such industries as coal mining, coil coating, inorganic chemicals, iron and steel, leather tanning and finishing, ore mining, petroleum refining, porcelain enameling, pulp and paper, steam-electric generation, textile mills, aluminum smelting, battery manufacturing, electrical and electronic components, foundries, metal finishing, pesticides, and pharmaceuticals. See *Insecticide, Fungicide and Rodenticide Act, US; Water Pollution Control Amendment Act, 1972, US.*

Clean Water Amendment Act, 1981, US: A US federal measure to modify the programs under the Water Pollution Control Amendment Act, 1972 (q.v.) to reduce the economic burden on federal taxpayers. The maximum federal share to be paid for the construction of a water pollution control project was reduced from 75 per cent to 55 percent.

The Act also introduced greater flexibility to establish priorities at state and local level; and more "reasonable" treatment requirements were introduced. By 1985, the States were managing over three-quarters of the US construction grants program.

However, between 1972 and 1985, the US federal government had spent over $40 billion to construct wastewater treatment facilities in communities large and small across America. In May, 1985, the Senate Environment and Public Works Committee approved a renewal bill to continue a construction grants program through 1994 at a total cost of $18 billion.

Clean Water Restoration Act, 1966, US: United States clean water legislation which authorized an appropriation of $3.9 billion for construction grants to state water pollution control programs. In addition, the Act, without setting a fixed dollar figure, authorized the appropriation of such additional funds as might be necessary for enforcement, comprehensive planning, and other functions.

Cluster Housing: Groups of houses which share communal open space, thus making more effective use of land.

Coastal Barrier Resources Act, 1982, US: A measure to restrict the expenditure of US federal funds, including federally subsidized flood insurance, in respect of developments on the ecologically fragile system of coastal barrier islands that line the Atlantic and Gulf coasts. The effect was to reinforce measures under the Coastal Zone Management Act, 1972 (q.v.) in this case by discouraging construction projects.

Coastal Protection: Measures to prevent coast erosion including the stabilization of beaches and dunes by mechanical means in the lower parts of the beaches, and by both mechanical and vegetational means on the upper beaches and dunes. Heavy sea walls and revetments may also be used in appropriate cases.

Groins are used extensively to slow beach erosion and to build beaches. Made of wood or other material, and designed to take advantage of the longshore currents that carry the sediments along the beach, groins are usually developed in a series so that their spacing, length and height, form a tapering system.

Natural vegetation has been the most effective measure in stabilizing dunes and upper beaches; the vegetation may consist of herbs, shrubs and trees. Wattles, fences and stakes may be used in conjunction with plantings to inhibit the movement of sand. The planting arrangements are related to the topography, wind, and drift character of the sands. The problem of dune management is complex.

Coastal Protection Act, 1979, New South Wales: The object of this Act, introduced by the New South Wales Government, Australia, is to facilitate the protection of the coastal region of the state. It imposes special controls on development in the coastal zone where such development might adversely affect any beach or dune, floodplain, foreshore, or the shoreline. The term development includes any use, activity, or work within the coastal zone. It specifically includes all forms of residential subdivision.

The coastline of New South Wales holds many of Australia's most attractive and dramatic landscapes: rocky cliffs and headlands, sandy beaches, offshore islands, sand dunes, tidal mudflats, estuarine shallows, salt marshes, wetlands, mangrove areas, and coastal lakes and lagoons. It not only forms an important zone for people but provides highly productive breeding areas for coastal, estuarine, and marine organisms including commercial species, and birdlife.

Coastal Zone Act, 1971, Delaware, US: A measure enacted by the State of Delaware in 1971 designed to protect the coastal zone from deterioration caused by certain industrial uses. Some uses are now prohibited by the Act, while others require a permit. Permit applications are considered by the Department of Natural Resources and Environmental Control; they must be accompanied by written evidence of prior local government zoning approval and an environmental impact statement (q.v.). Appeals from an adverse decision may be filed by the applicant or any aggrieved person with the State Coastal Zone Industrial Control Board.

Coastal Zone Management Act, 1972: Recognizing the need for special protection, an Act passed by the US Congress in 1972 establishing a voluntary federal-state partnership for the conservation and management of coastal resources. Under this partnership, many states made notable progress passing comprehensive management laws; adopting new measures to protect wetlands, mineral resources, historic sites, and other important coastal resources, and developing better management arrangements. By the end of 1984 some 75 per cent of the United States shoreline was covered by federally-approved state coastal zone protection programs. In that same year, a national coastal protection policy was declared and a review undertaken of federal programs significantly affecting coastal resources. The coastline of the United States is remarkably varied, productive and beautiful; over half the population lives within an 80-kilometer wide coastal strip while nearly one-half of the fishing industry depends directly on near-shore waters. Increasing pressures from population growth and developments of all kinds continue to impose strains upon this coastal zone.

Coliform Bacteria: Bacteria, including the bacterium *Escherichia coli*, commonly found in the human large intestine and whose presence in the environment usually indicates contamination by human wastes. Laboratory results are expressed as the number of organisms per 100 ml of sample. While the presence of *E. coli* indicates fecal pollution, the organism is generally considered to be non-pathogenic. To confirm that pathogenic organisms are associated with fecal pollution, samples should be tested for the pathogenic genera Salmonella and Shigella. If Salmonella is detected, further tests are needed in order to differentiate between the organisms of this group, e.g., *Salmonella typhosa* is the causative organism of typhoid fever, whereas *Salmonella typhimurium* is the causative organism only of gastrointestinal upsets.

There are other coliform bacteria more or less similar to *E. coli* that exist in soil and plant materials (particularly *Aerobacter aerogenes* and *Escherichia freundii*); they may be present in water without indicating fecal pollution.

Colorado, US: See *Litter Laws, US*.

Commission for the Environment, New Zealand: A commission created by the New Zealand Government in 1973; its aims are:

- To preserve and enhance the quality of life by creating increasing awareness of the environmental implications of human actions and of natural processes, and by promoting coordinated social, economic, and environmental planning and management towards that end.

- To facilitate understanding of interrelationships of resource use, waste disposal, population patterns and trends, social conditions, people's aspirations, and environmental quality; to identify the implications of these relationships for present and future generations.

- To encourage people to give informed consideration to these relationships and to foster public participation in the formulation of national, regional, and local goals, and in the planning and implementation of projects arising from them.

- To encourage the identification and characterization of resources available to New Zealand, and the establishment of their limits; in the light of these findings to promote policies of wise resource use.

- To stimulate the development of proposals most appropriate to the scale and characteristics of New Zealand's physical and social environment, and to draw attention to the limitations of projects considered environmentally inappropriate.

- To provide information and advice to decision-makers on the environmental implications of their policies and operational proposals.

Commission on the Height of Buildings, 1913, US: A commission set up in 1913 to advise the city of New York on the means of controlling private development. The commission had to decide on which of two quite distinct governmental powers these new controls should be based: eminent domain or the police power. If the former approach were adopted, compensation (q.v.) would have to be paid. If, on the other hand, these controls could be brought under the police power (i.e., the general residual power of government to pass laws in the interests of the general public health, safety, and welfare), then no compensation would be payable; the controls would be analogous to fire or structural regulations.

The whole issue was overshadowed by the Fifth Amendment of the US Constitution which states: "No person. . .shall be deprived of life, liberty, or property, without due process of law; nor shall private property be taken for public use without just compensation."

The commission's report argued against any approach based upon a resort to the power of eminent domain on a variety of grounds. Further, it took the view that the kinds of regulations under consideration were not such as to justify individual compensation. The report continued:

> "While they restrict individual liberty to a certain extent, they do it in such a way as to conserve individual and public interests and rights. They subject the use of urban land to such restrictions as are appropriate and reasonable in the nature and history of this class of property."

This decision determined the objectives and limits of planning controls in the United States; it had become clear that controls could not extend very far beyond the fundamental objectives of separating out grossly incompatible land uses and establishing minimum standards for development. The commissions's report has become a basic document in the history of United States physical planning. It has established a principle of general application that within these bounds no compensation is payable to an owner whose property loses value as the result of a planning decision. See *Eminent Domain; Taking Issue*.

Commissioner of Inquiry, New South Wales: A full-time and permanent appointment under the Environmental Planning and Assessment Act, 1979, of New South Wales, Australia, for the purpose of conducting public inquiries into controversial environmental planning issues. Inquiries have encompassed aluminum smelters, open-cut coal mines, sand and shale extraction, marinas, gas storage installations, recreational lakes schemes, residential development, heliports, regional shopping centers, hotels, off-road vehicles, transmission lines, local and regional environmental plans, and land-use conflicts. Inquiries are also conducted under the Heritage Act, 1977.

The Commissioner presents his report, with findings and recommendations, to the Minister for Planning and Environment; the Minister then proceeds to a decision under the relevant act. The Commissioner's report is then released. The Minister may initiate public inquiries into a wide range of issues; while in prescribed circumstances an objector has the right to require the holding of a public inquiry before the making of a decision.

There is now considerable evidence in New South Wales that public participation procedures have resulted in more carefully considered projects, in environmental terms. While most projects proceed, the range of legal conditions imposed upon construction and operation reflect these contemporary concerns. Developers today increasingly seek to respond to environmental arguments with, in their view, effective environmental responses. The simple theme of "We are going to provide employment," while unmodified adverse environmental consequences are simply to be weighed against this single desirable end, is no longer considered acceptable to most involved. This is the message of the experience of public participation in New South Wales. See *Public Hearings and Inquiries*.

Committee on Public Participation in Planning, 1969, UK: Or Skeffington Committee, a committee asked by the British Government to investigate the best methods including publicity of securing the participation of the public at the formative stages in the making of development plans for their area. The committee's report recommended that the public should be kept informed throughout the preparation of a structure plan or local plan for their area; it proposed the appointment of community development officers and the introduction of community forums for the discussion of local problems to promote this objective. See *Public Hearings and Inquiries*.

Common Property Resources: Those attributes of the natural world which are valued by society but are not in individual ownership and do not enter into the processes of market exchange and the price system. Notable among such resources are the atmosphere, watercourses, ecological systems, and the visual properties of the landscape. These are common property resources of great and increasing "value" presenting society with important and difficult allocational problems, which exchange in private markets cannot resolve.

The key point is that common property can be, and is, used free of charge by any member of society wishing to use it. Perhaps the best example of common property resources are fisheries; fish only become private property on capture. This has led to over-exploitation of fish stocks in many fisheries. Wilderness is another. See *Commons, The*.

Commons, The: Historically, tracts of land owned or used jointly by the members of a community; today the term tends to be used in respect to somewhat wild, waste, and unenclosed land. Prior to the enclosure movements, each community had its commons used for agriculture and sheep and cattle grazing. There was a general tendency for the number of animals to become more numerous than the commons could support, and no one had any responsibility for ensuring the future productivity of this resource. The commons were characterized by low productivity, and exploitation, and were slowly but surely superseded by more efficient forms of agriculture. The characteristic overgrazing and lack of care has been described as "the tragedy of the commons."

Today, on a larger scale, the natural resources of air and water may be regarded as the "commons" of the world; they are exploited by many who have little or no interest in the future productivity of these resources. Hence, for example, intensive fishing and whaling threatens a dramatic loss of productivity. Overfishing is the main threat to marine living resources and a significant threat to freshwater resources; it occurs locally in all regions.

The global commons comprise those parts of the earth's resources beyond national jurisdiction. Apart from the open oceans with their living and other resources, and the general atmosphere, the Antarctic land mass may also be regarded as part of the commons although several countries have claimed parts of it. Outer space, beyond the atmosphere, may also be designated a common likely to become progressively despoiled by space debris such as dead satellites, spent rocket stages and motors.

In his 1981 "state of the environment" report, the United Nations Environment Program (UNEP) executive director Mustafa Tolba recommended to the UNEP governing council a levy on the use of such "international commons" as the oceans, airspace, telecommunication frequencies, and satellite orbits, with the proceeds going to environmental projects in less developed countries. The proposal was not accepted.

Community Development Block Grant (CDBG), US: A program established by the US Housing and Community Development Acts, 1974-77, entitling all cities with populations greater than 50,000 to participate in an annual allocation of federal funds; activities financed must benefit principally families with low or moderate incomes or aid in the prevention of slums or urban blight.

The program consolidated seven earlier grants-in-aid programs administered by the Department of Housing and Urban Development (HUD) into a single program of broad-purpose community development block grants. More specifically the new program covers urban renewal, model cities, neighborhood facilities, rehabilitation, and public facilities including water and sewerage facilities. See *Urban Development Action Grant (UDAG), US.*

Compensation: See *Betterment; Commission on the Height of Buildings, 1913; Taking Issue.*

Compensation Depth: The depth in water at which, because of reduced light penetration, the rate of production of organic material by photosynthesis exactly balances the rate of breakdown of organic material by plant respiration. There is no net production below the compensation depth.

Compensatory Open Space: The replacement of open space taken, for example, for the purposes of a road facility by an equivalent area either adjacent to the road or elsewhere within the same community. If this were done, within the context of a road facility, further property acquisition would be required and the cost should become a road cost to be taken into account in any benefit-cost analysis of a project.

Comprehensive Environmental Response, Compensation, and Liability Act, 1980 (CERCLA), US: Known also as "Superfund", a measure introduced by the US Congress to meet the costs of the clean-up of abandoned hazardous waste sites. The goal of this legislation was to eliminate the most serious threats to public health and the environment posed by hazardous substance spills and uncontrolled hazardous waste sites, and to respond to such hazards in a cost-effective manner. The Act imposed certain "environmental taxes" on the petroleum and chemical industries, re-authorized in 1985, and set up the Hazardous Substance Response trust Fund. Taxes were also imposed on the owners and operators of hazardous waste disposal facilities in order to establish a second fund, known as the Post-Closure Liability Trust Fund. In the event of the release of a hazardous substance, the procedures and methods to be followed are set forth in the National Contingency Plan. To deal with the immediate problems, the Act required the preparation of a national priorities list. See *Resource Conservation and Recovery Act, 1976, US.*

Comprehensive Plan, US: Or general or master plan. A statutory document which sets forth a government's major policies concerning the desirable future physical development of its area; it states the desired ends and not necessarily the means for achieving them. The substantive content of a comprehensive plan varies from state to state according to the enabling legislation, and from local government to local government according to the scope of its authority and responsibility. A plan almost always includes the three physical elements of land use, circulation, and community facilities. A plan may also include an urban design element, a housing element, a renewal element and a historic preservation element. Some comprehensive plans have included non-physical elements on such topics as income, public safety, social services, health care, and employment. The process by which the plan may be linked to the capital improvement program, the budgetary process, and growth management ordinances may be included. It presents a holistic view; the entire planning jurisdiction is covered and the different systems represented in the various elements are all coordinated with each other.

District plans normally follow the comprehensive plan, amplifying its features on a district-by-district basis. Cincinnati became the first United States city to adopt a comprehensive plan in 1925. There appears to be fairly general agreement on the desirable content of a comprehensive plan; it should at least incorporate the following elements:

- Surveys, studies, and data pertaining to existing conditions;

- Factors affecting growth together with projections of such growth;
- Objectives that the community wishes to achieve in respect of the full range of functions: land use, economy, housing, transportation, and community facilities;
- A scheme of future land use, designating the general location and distribution of areas devoted to housing, commerce, and industry;
- A scheme of future public functions and facilities, such as those devoted to recreation, education, institutions, public grounds, and related purposes;
- A scheme of future circulation, designating the general locations of thoroughfares and other transportation routes and necessary terminals and utilities;
- Schemes for strategic locations where important public action is contemplated; this might include the design or redesign of residential neighbourhoods, the central business district (q.v.), or industrial areas;
- Schemes for entire systems of community facilities, i.e., fire protection, sewage and water;
- General suggestions for programming and implementing these actions in terms of time, cost, and other resources including some discussion of the administrative, legal, and fiscal measures that would have to be taken.

A variety of legal and administrative tools may be employed to help the comprehensive plan to succeed:

- Development controls including zoning (q.v.), subdivision regulations, an official map (q.v.), building and housing codes, and the like;
- Coordination measures including the capital improvements program and project reviews by the planning commission (q.v.);
- Incentive measures to stimulate private action consistent with the plan including tax concessions, concessions on the sale of land, and the provision of select public improvements;
- Federal and state aid programs including community development activities, open space acquisition, highway improvement, housing assistance, community facility programs, and revenue sharing.

See *Comprehensive Plan, Washington, D.C.; General Plan, Los Angeles; Policy Plan*.

Comprehensive Plan, Washington, D.C., US: A comprehensive plan (q.v.) for the national capital of the United States of America prepared by the Office of the Mayor in accordance with the provisions of the District of Columbia Self-Government and Governmental Reorganization Act, 1973. The Act required "citizen participation" in the comprehensive plan process; the plan was to be finally submitted to the Council of the District of Columbia which was required to hold public hearings during the adoptive process. The plan eventually became incorporated in the District of Columbia Comprehensive Plan Act, 1983.

The local comprehensive plan elements relate to land use; economic development; housing; environmental protection; transportation; public facilities; urban design; preservation and historic features; downtown development; and human services. The process for preparing the plan is outlined in Figure 2, while plan relationships are shown in Figure 3. See *National Capital Planning Commission*.

Compulsory Purchase: The acquisition of private property and land without the owner's consent for an approved public purpose such as public housing, redevelopment in inner or central city areas, or for new town development on green field sites. The power of compulsory purchase is embodied in British planning and housing legislation; the procedure for arriving at the purchase price is also laid down.

During the late 1950s, Britain resumed her massive national campaign against slums, subsidizing the construction of new dwellings required to accommodate people displaced by slum clearance programs. Quite often the sites of the cleared areas were brought into public ownership for the construction of new publicly-owned dwellings (council houses). See *Eminent Domain*.

Conditional Use: A use which is essential to or would promote the public health, safety, or welfare in one or more zones, but which would impair the integrity and character of the zone in which it is actually located, or in adjoining zones, unless restrictions on location, size, extent, and character of performance are imposed in addition to those imposed in the zoning regulation. A conditional use permit stipulates the conditions which must be met.

Configuration Dredging: Dredging to a pattern calculated to defract wave energy in a certain specified direction, e.g., to minimize wave action in the entrance to a port basin.

Congestion: In respect to a road system, a situation in which traffic flow exceeds 80 per cent of road capacity and drivers are likely to wait at intersections for longer than one cycle of the traffic signals.

Connecticut, US: See *Litter Laws, US*.

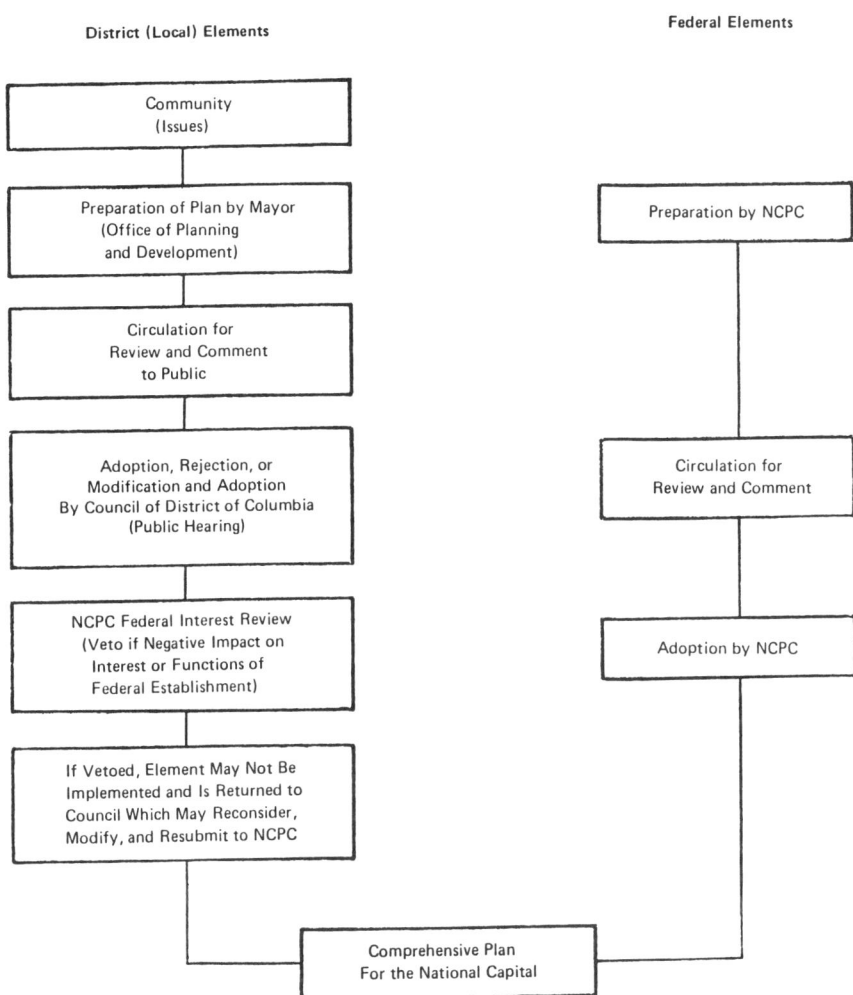

Figure 2: The process for preparing the comprehensive plan for Washington, DC in accordance with the Home Rule Charter.

38   Environmental Planning

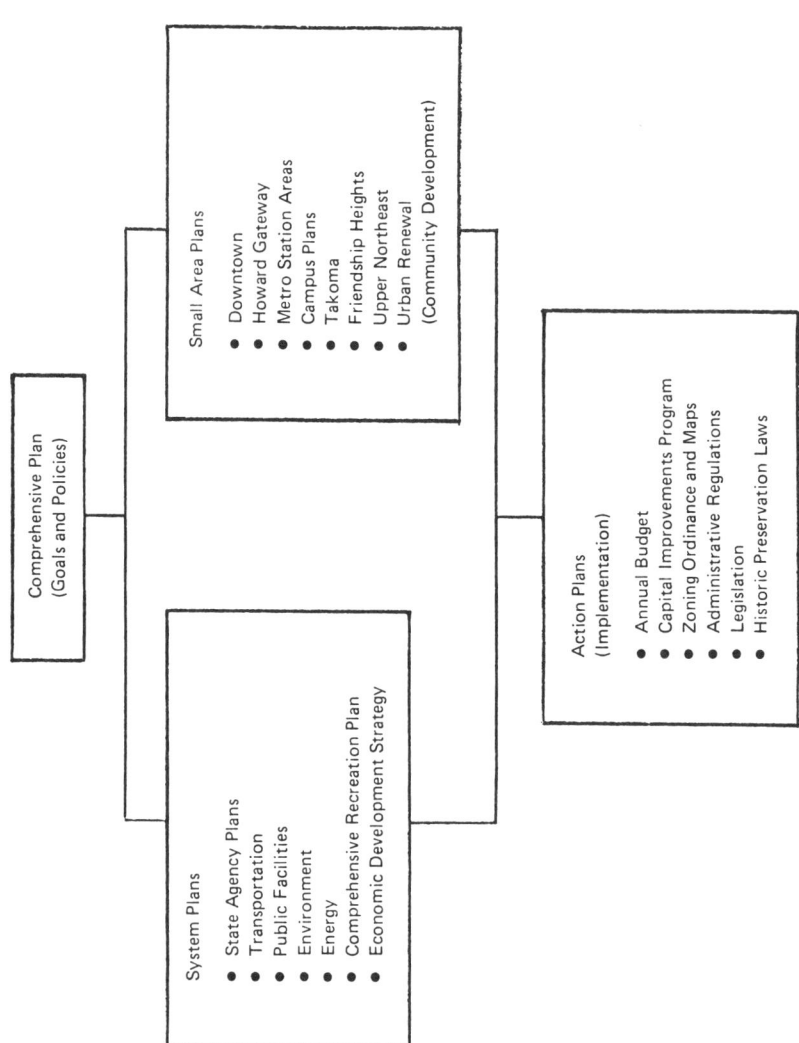

Figure 3: Relationships within the Comprehensive Plan for Washington, DC

Conservation: Defined by the World Conservation Strategy (q.v.) as "the management of human use of the biosphere so that it may yield the greatest sustainable benefit to present generations while maintaining its potential to meet the needs and aspirations of future generations." Conservation is thus something positive, embracing preservation, maintenance, sustainable utilization, restoration, and enhancement of the natural environment.

Living resource conservation is specifically concerned with plants, animals and microorganisms, and with those non-living elements of the environment on which they depend. Living resources have two important properties the combination of which distinguishes them from non-living resources: they are renewable if conserved, and they are destructible if not.

The Strategy defines the three main objectives of living resource conservation as the maintenance of essential ecological processes and life-support systems, the preservation of genetic diversity, and the sustained utilization of species and ecosystems.

Non-renewable resources are natural resources which, once consumed, cannot be replaced, e.g., a ton of coal once consumed is gone forever in that form. Mineral resources generally are regarded as wasting assets of this kind. However, it is difficult to predict what the consequences of exhausting particular resources would be. For any particular mineral the exhaustion process would be gradually accompanied, other things being equal, by a steady rise in its price. A rising relative price intensifies exploration and ensures treatment of lower grades of ore; recycling and reclamation of scrap and rsidues are also encouraged. Meanwhile, developments in substitute materials and processes and in the pattern of demand could mean that a mineral considered "indispensable" at one stage could become redundant in another.

Conservation and Development Program, UK: Britain's national conservation strategy, being a contribution to the development of the World Conservation Strategy (q.v.). The British government was assisted by the World Wildlife Fund (UK), the Nature Conservancy Council, the Countryside Commission, the Council for Environmental Conservation, and the Royal Society of Arts. The program was published in 1983, comprising seven reports and an overview.

Conservation Zones: Zoning controls which have the purpose of securing the preservation of land or buildings for the benefit of the community at large. Land may be set aside by the planning authority for open space or as a place of historical, geographical, or geological importance. Buildings may be marked for preservation as a group because of their architectural or historical significance.

The effect of the controls may be to deny the owner the capacity to use the land or building in the same way as his or her neighbors, and yet at the same time subject him or her to payment of the same rates and txes and to all other liabilities at law in respect of the land.

In many respects, the controls may be regarded as another form of planning blight (q.v.) in that the owner is denied the full use or value of his or her land and will find difficulty selling it. Compensation is not usually payable, though some planning legislation provides for compensation when the land is set aside for particular purposes.

Sometimes the term "urban conservation area" is used to describe an urban area thought to possess special townscape qualities which should be conserved as part of the national estate or local heritage. The term "landscape conservation area" is also used to describe a landscape (natural or rural) which has special scenic, scientific and/or cultural values which should be conserved as part of the national, state, or local heritage. A landscape conservation area may comprise a single topographic feature, such as a mountain or an isolated valley, but more likely will be a collection of topographic features which together form an easily distinguishable landscape unit.

Control of Pollution Act, 1974, UK: An Act passed by the British Parliament which significantly extended the powers of local authorities to deal with noise problems, and was concerned also with the control of the deposit and disposal of waste on land; the prevention of water pollution, including controls over underground water; and the control of pollution of the atmosphere, notably from motor vehicles.

The Act substituted a licensing system for the deposit of waste for the notification system of the Deposit of Poisonous Waste Act 1972. Powers were given to collection and disposal authorities to recycle waste, and to purchase waste for this purpose; and to produce heat and electricity from waste.

The water pollution control provisions replaced almost all the provisions of the Rivers (Prevention of Pollution) Acts 1951 and 1961, the whole of the Clean Rivers (Estuaries and Tidal Waters) Act 1960, and some sections of the Water Resources Act 1963. The approach has been to strengthen existing law, rather than to provide new methods of control. The defenses available to persons causing pollution remained restricted. Furthermore, the controls now apply to virtually all inland and coastal waters. Charges can now be made in respect to pre- and post-1973 discharges of trade effluent to sewers, and also in respect to discharges to rivers and streams.

Conurbation: The English term for megalopolis (q.v.); it may be applied to Greater London, the Midlands, the Manchester-Liverpool area, and Tyneside. The term was coined by Sir Patrick Geddes (1854-1932), the Scottish sociologist, in his book *Cities in Evolution* (1915); he described the waves of population inflow to large cities, followed by overcrowding and the creation of slums, and then the wave of backflow, the whole process resulting in amorphous sprawl, waste and unnecessary obsolescence. As a planner, Geddes laid out some 50 cities in India and Palestine.

Convention for Cooperation in the Protection and Development of the Marine and Coastal Environment of the West and Central

African Region: Signed in Abidjan in 1981, an agreement by the states of West and Central Africa to adopt a comprehensive approach to pollution control and resource management; a related protocol deals with cooperation in pollution emergencies.

Convention for Cooperation on the Protection of the Marine Environment from Pollution, Kuwait Regional: An agreement by the states of the Persian Gulf, signed at Kuwait in 1978, providing for a comprehensive approach to pollution control and resource management; it is supplemented by a protocol concerning cooperation in combating pollution by oil and other harmful substances in cases of emergency.

Convention for the Prevention of Marine Pollution by Dumping from Ships and Aircraft: A convention signed in Oslo, Norway (and known as the Oslo Convention) in February, 1972, by twelve concerned nations. These nations pledged themselves to take all possible steps to prevent the pollution of the sea by dumping, the designated area being the northeast Atlantic. The Convention excluded the Baltic and Mediterranean Seas, as these are subject to separate Conventions. The signatory nations were Belgium, Denmark, Finland, France, Iceland, the Netherlands, Norway, Portugal, Spain, Sweden, the United Kingdom, and West Germany. The convention took effect in 1974.

A Commission was created, composed of representatives from each of the nations concerned, with the following responsibilities: to exercise overall supervision over the implementation of the Convention; to review generally the condition of the seas in the designated area; to review the efficacy of control measures being adopted, and the need for additional or different measures; to receive reports, inventories, and monitoring results.

Dumping was defined as meaning any deliberate disposal of substances and materials into the sea by or from ships or aircraft. However, emergency dumping is excluded from control where the safety of human life, or of a ship or aircraft is threatened.

With this exception, the nations agreed to prohibit the dumping of substances named in a "black list." These include organohalogen compounds, organosilicon compounds, carcinogenic substances, mercury and cadmium compounds, and persistent synthetic materials.

Other substances and materials, named in a "grey list," cannot be dumped without the approval of the appropriate national agency. These include the elements and compounds of arsenic, lead, copper, and zinc; cyanides and fluorides, pesticides and by-products; containers, scrap-metal, tar-like substances, and bulky wastes; and substances which though non-toxic may become harmful due to the quantities in which they are dumped, and which are liable to seriously reduce amenities.

Convention for the Prevention of Marine Pollution from Land-Based Sources: A convention signed in Paris (and known as the Paris Convention) in June, 1974, by fourteen concerned nations. These nations pledged themselves to take all possible steps to prevent pollution of the sea, the designated area being the northeast Atlantic and the Arctic Oceans. The Convention excluded the Baltic and Mediterranean Seas, as these are subject to separate Conventions. The signatory nations were Austria, Belgium, Denmark, France, Iceland, Luxembourg, the Netherlands, Norway, Portugal, Spain, Sweden, Switzerland, the United Kingdom and West Germany. The convention took effect in 1978.

A Commission was created, composed of representatives from each of the nations concerned, with the following responsibilities: to exercise overall supervision over the implementation of the Convention; to review generally the condition of the seas in the designated areas; to draw up programs and measures for the elimination or reduction of pollution from land-based sources; to receive reports, inventories, and monitoring results.

The nations agreed to eliminate, if necessary by stages, pollution of the designated maritime area by certain substances from land-based sources named in a "black list." These include organohalogen compounds, mercury and cadmium and their compounds, persistent synthetic materials, and persistent oils and petroleum hydrocarbons.

They also agreed to reduce and rigorously control the discharge of other substances named in a "grey list." These include the elements and compounds of arsenic, chromium, copper, lead, nickel, phosphorus, silicon, tin and zinc; non-persistent oils and petroleum hydrocarbons; and substances adversely affecting the taste or smell or products derived from the marine environment for human consumption.

Convention for the Prevention of Pollution from Ships (MARPOL), 1973: A convention concluded in London in 1973, at the end of an International Marine Pollution Conference, convened by the then Intergovernmental Maritime Consultative Organization (IMCO) (now International Maritime Organization (q.v.)). The conference, the largest ever held on the subject, was attended by 500 representatives from the leading maritime nations.

The convention was to enter into force twelve months after it had been ratified by 15 countries constituting at least 50 per cent of the world's merchant shipping. The 1973 convention effectively superseded the Convention for the Prevention of the Pollution of the Sea by Oil, 1954, (q.v.) when it came into force in 1983. It is often known as the MARPOL convention.

The new instrument contains provisions aimed at eliminating pollution of the sea by both oil and the noxious substances which may be discharged operationally, and at minimizing the amount of oil which would be accidentally released in such mishaps as collisions or strandings.

The convention applies to any ship of any type whatsoever, including hydrofoil boats, air-cushion vehicles, submersibles, floating craft and fixed or floating platforms, operating in the marine environment. It covers all aspects of intentional and accidental pollution from ships by oil or noxious substances carried in bulk or in packages, sewage and garbage. But it does not deal with dumping which is covered by the Convention for the Prevention of Marine Pollution by Dumping from Ships and Aircraft signed in Oslo in 1972 nor the release of harmful substances directly arising from the exploration, exploitation and associated offshore processing of sea-bed mineral resources.

Convention for the Prevention of Pollution of the Sea by Oil, 1954: A convention which prohibits the deliberate discharge of oil or oily mixtures from all sea-going vessels in specific areas called "prohibited zones". In general, these extend at least 50 miles from all land areas, although zones of 100 miles have been established in areas which include the Mediterranean and Adriatic Seas, the Gulf and Red Sea, the coasts of Australia and Madagascar, and some others. The participating countries undertake to promote the provision of facilities for the reception of oil residues; the Convention also prescribes that every ship which uses oil fuel and every tanker shall be provided with a book in which all the oil transfers and ballasting operations shall be recorded, including entries on accidental or other exceptional discharges and escapes. Although the restrictions imposed by the 1954 Convention have been very effective, the enormous growth in oil movements since then has necessitated more stringent regulations. See *Convention for the Prevention of Pollution from Ships (MARPOL), 1973.*

Convention for the Protection of the Mediterranean Against Pollution, 1976: A Convention adopted by the Mediterranean coastal countries for the purposes of pollution control and resource management. Agreed to in Barcelona in 1976 as an umbrella measure, specific technical protocols elaborate national obligations to control pollution from discrete sources or to cooperate on some aspect of environmental management; protocols so far adopted deal with dumping, cooperation in pollution emergencies, and land-based sources of pollution.

The signatories to the Convention were: Cyprus, Egypt, France, Greece, Israel, Italy, Lebanon, Libya, Malta, Monaco, Morocco, Spain, Tunisia, Turkey, and Yugoslavia. Since the signing, the United Nations Environment Program (q.v.) has been able to develop a Mediterranean Action Plan.

The purpose of the Convention was to achieve a coordinated and comprehensive regional approach for the protection and enhancement of the marine environment in the Mediterranean Sea area. Specifically, it sought to prevent and abate pollution caused by dumping from ships and aircraft; by exploration and exploitation of the continental shelf and the seabed; from discharges from rivers, coastal establishments and outfalls, and from other land-based sources; and by pollution emergencies. The signatory countries undertook to develop programs, conduct monitoring of pollutants, prepare inventories, and pursue relevant research. Under Article 13, the United Nations Environment Program was to receive reports and circulate information, deal with other international bodies, and convene meetings of the signatory nations.

Convention for the Protection of the Rhine Against Chemical Pollution, 1976: This Convention was signed on 3 December 1976 by the five riparian states and the European Community. The parties agreed to eliminate pollution of surface waters by dangerous substances identified in a "black list" (Annex I of the Convention). The list includes such substances as organohalogen, organophosphorus, organotin compounds, carcinogenic compounds, mercury and cadmium and their compounds, persistent mineral oils and petroleum hydrocarbons.

In the meanwhile, any discharge containing one or more of these substances requires prior authorization from a responsible government agency; the setting of maximum permissible concentrations based on toxicity, persistence, and bioaccumulation; strict supervision; and may be allowed for only a limited time period.

The parties also agreed to reduce and rigorously control the discharge to surface waters of hazardous substances included in a "grey list" (Annex II of the Convention). The list contains many elements such as zinc, copper, lead, selenium, arsenic, tin, barium, beryllium, boron and cobalt; biocides; compounds of silicon and phosphorus; cyanides and fluorides; ammonia and nitrites; and non-persistent oils and petroleum hydrocarbons.

National programs are to be established to reduce pollution by these "grey list" substances. The International Commission has a duty to compare the draft programs to ensure that the aims and means coincide. All discharges must be authorized, with conditions imposed to ensure compliance with the quality objectives for Rhine water. Control measures shall take into account the "latest economically feasible, technical advances."

All discharges are to be monitored and inventories drawn up, with the information being supplied to the International Commission. The national inventories are to be updated at intervals of three years. The Commission publishes annual reports in respect of the findings.

In 1984, the North Rhine-Westphalia water authority published a report claiming that the condition of the Rhine River had improved during the previous year. The oxygen content had increased and heavy metals no longer threatened to contaminate drinking water. However, chlorides and hydrochloric acid continued to cause concern. As a whole, the river was still rated as between "massively" and "critically" polluted.

Convention for the Protection of the World Cultural and Natural Heritage: A convention initiated by the United Nations Educational, Scientific and Cultural Organization (UNESCO) (q.v.) and adopted in 1972; by 1979 over 30 UNESCO member countries had become parties to it. The convention established a World Heritage Committee which met for the first time in 1977. The convention created also a World Heritage Fund to allow the financing of preservation projects; the Fund received voluntary contributions from the member nations.

Convention, Nordic Environmental Protection: See *Nordic Environmental Protection Convention*.

Convention on Fishing and the Conservation of Living Resources in the Baltic Sea and the Belts, 1973: Known as the Gdansk Convention, an international convention between the riparian countries of the Baltic embracing the Baltic, the Danish belts and the Sound for the purposes of the rational management of the marine resources of the area. It augmented a concurrent agreement between Sweden and Denmark on the protection of the Sound from pollution. The convention became effective in 1974.

Convention on International Trade in Endangered Species of Wild Fauna and Flora: An international convention which came into force in 1976, providing for the regulation of trade in whole plants and animals, dead or alive, and their readily recognizable parts and derivatives. Excessive international trade has contributed directly to the decline of many wild populations of rare and endangered species. The Convention states that protected species may not be the subject of international trade without a permit granted under conditions specified by the Convention itself.

Convention on Long-Range Trans-Boundary Air Pollution: A convention initiated by the Economic Commission for Europe (q.v.) for the purposes of curbing the incidence of acid rain (q.v.) arising from the long-range transportation of pollutants, notably sulfur dioxide (q.v.). The Convention was signed in 1979 and came into effect in 1983; basically it was a statement of desirable objectives without numerical limits.

In 1983, the Nordic countries proposed a 30 percent reduction in sulfur dioxide emissions in all member countries to be achieved by 1993; the reduction was to be measured from the year 1980. The move was welcomed by most member countries. The United Kingdom argued, however, that it had been achieving significant reductions in sulfur emissions over a period of years so that a further substantial reduction appeared an onerous responsibility; the United States also had reservations because of the lack of conclusive evidence regarding the need. On the other hand, Canada welcomed the proposal and formed what is known as "The 30 per cent Club."

West Germany, at first reluctant, became converted to the new objective. This was expressed in tough 1983 regulations. The new law requires that all coal-burning power plants above 300 megawatts shall be equipped with stack gas scrubbers to remove at least 85 per cent of the sulfur dioxide by 1988, or else be retired by 1993.

Convention on Migratory Species of Wild Animals: An international convention concluded at Bonn, West Germany in 1979. Migratory species as defined by that convention are "the entire population of any species or lower taxon of wild animals, a significant proportion of whose numbers cyclically and predictably cross one or more national jurisdictional boundaries." The Convention obliges parties to protect endangered migratory species and to endeavour to conclude agreements for the conservation of migratory species whose status is "unfavorable."

Convention on Nature Conservation in the South Pacific: An international convention drawn up in June, 1976. This convention imposes general obligations on member countries to safeguard representative samples of natural ecosystems by setting up national parks and reserves.

Convention on the Conservation of Antarctic Marine Living Resources (CCAMLR): An international convention which came into force in 1982; the convention regulates the harvesting of krill, fish, squid, and birds in the southern ocean, an area much larger than that embraced by the Antarctic Treaty. It requires that harvesting decisions shall take into account the possible effects on dependent species. For example, the recovery of endangered whales could be jeopardized by the over-harvesting of krill; the convention requires that this possibility be considered in establishing krill quotas. This is sometimes known as the "ecosystem as a whole" standard. The parties to the convention are Argentina, Australia, Belgium, Brazil, Chile, France, India, Japan, New Zealand, Norway, Poland, South Africa, the Soviet Union, the United Kingdom, the United States of America, and West Germany.

With the completion of the Convention on the Conservation of Antarctic Marine Living Resources, the parties in 1981 agreed to begin special negotiations for a legal regime to govern mineral resource development in the Antarctic. The parties have not accepted the concept of the "common heritage of mankind" as applying to the Antarctic; consequently, the United Nations has not been invited to participate in the negotiations. See *Conventions for the Conservation of Antarctic Living Resources.*

Convention on the Protection of the Marine Environment of the Baltic Sea Area: The outcome of a diplomatic conference on the protection of the marine environment of the Baltic Sea held in the Finlandia Hall, Helsinki, in March 1974. At the close of the conference Denmark, Finland, the German Democratic Republic, the Federal Republic of Germany, the Polish People's Republic, Sweden, and the Union of Soviet Socialist Republics signed the Convention on the Protection of the Marine Environment of the Baltic Sea Area—the "Helsinki Convention". The Convention was to enter into force two months after the seventh instrument of ratification, approval, or accession had been deposited with the Government of Finland. It took effect in 1980.

The convention created a Baltic Marine Environment Commission. One of the duties of the Commission is to define pollution control criteria, objectives for the reduction of pollution, and the objectives in respect of other environment protection measures. Prior to the entry into force of the Convention, the Baltic countries created an Interim Commission to tackle the immediate problems and help in the preparations of each country for the adoption of the terms of the Convention.

Through this new machinery and approach, an effort is being made to bring the present pollution of the Baltic Sea to an end and generally improve conditions. The Helsinki Convention has attracted wide attention as an example of successful regional cooperation towards the protection of marine environments.

Convention on the Protection of the Rhine against Pollution by Chlorides, 1976: An international convention entered into by five riparian countries in 1976 for the purpose of controlling the discharge of chlorides into the Rhine. Under the Convention, the discharge of chlorides into the Rhine must be reduced by at least 60 kilograms per second, considered as an annual average. This objective is to be achieved gradually largely by measures on French territory. The Convention refers specifically to discharges from the Alsace Potassium Mines. The idea is that residual or waste brines, instead of polluting the Rhine, should be pumped or injected into a layer of limestone called "Great Oolite" at a depth of 1,500 to 2,000 meters in an area southwest of Mulhouse.

The French government became responsible for the preparation of an overall plan and the holding of continuing discussions with the International Commission. While the initial costs would be borne by the French Government, major contributions would be made by the Netherlands and West Germany, and a smaller contribution by Switzerland. In the end, only 30 per cent of the cost would be borne by France with 70 per cent being carried by the other three nations.

However, it was only in November 1981 that a tentative agreement was reached by which France would actually take steps to reduce their pollution of the Rhine with salts from the Alsation Potash mines—after years of haggling.

Convention on Wetlands of International Importance, especially as Waterfowl Habitat: An international convention signed at Ramsar which came into force in 1975. Parties to this convention are to take action to create reserves and otherwise protect wetlands that are internationally important for reasons including their habitat value for rare or migratory birds.

Australia has designated the Cobourg Peninsula Flora and Fauna Reserve and Wildlife Sanctuary and the Kakadu National Park, all in the Northern Territory, as wetlands of international significance.

About 80% of the birds in Japan are migratory birds that migrate between Japan and the United States, the Soviet Union, China, Australia, and Southeast Asia. Thus, it is indispensable to promote international cooperation with these countries to protect migratory birds.

For this reason, Japan has concluded agreements for migratory bird protection with the United States, the Soviet Union, and Australia to protect migratory birds and others facing extinction. See *Wetlands*.

Conventions for the Conservation of Antarctic Living Resources: Conventions agreed to by the parties to the Antarctic Treaty, 1959; these included the "Agreed Measures for the Conservation of Antarctic Fauna and Flora," the "Convention for the Conservation of Antarctic Seals" and the "Convention on the Conservation of Antarctice Marine Living Resources." (q.v.)

Australia has accepted the "Agreed Measures for the Conservation of Antarctic Flora and Fauna." The Agreed Measures provide for the protection of wildlife and its habitats in the Antarctic and controls import of exotic species into Antarctica. For Australia, the Agreed Measures are implemented through the Antarctic Treaty (Environment Protection) Act, which received Royal Assent in June 1980.

The "Convention for the Conservation of Antarctic Seals" which came into force in March 1978, applies to seas and ice-flows south of 60°S latitude. It aims to promote and achieve the protection and scientific study of Antarctic seals, while making provision for their rational use as resources. Australia has signed the Convention.

The "Convention on the Conservation of Antarctic Marine Living Resources," concluded at Canberra in May 1980, is the first international agreement based on an ecosystem approach. The Convention establishes clear principles of conservation which are to apply to all harvesting in the seas south of the Antarctic Convergence. It provides for the maintenance of all ecological relationships and aims to prevent irreversible changes to the ecosystem or its component populations. Under the terms of the Convention, an international Commission is to be created with the specific responsibility for drawing up measures to conserve all creatures within the ecosystem. This Commission will be based in Hobart.

Conventions for the Protection of the Northeast Atlantic against Pollution: Three Agreements each dealing with a separate aspect of pollution control in the northeast Atlantic region (including the North Sea). These Agreements are:

- Agreement for Cooperation in Dealing with Pollution of the North Sea by Oil (Bonn, 1969).
- Convention for the Prevention of Marine Pollution by Dumping from Ships and Aircraft (Oslo, 1972), and
- Convention on the Prevention of Marine Pollution from Land-Based Sources (Paris, 1974).

Core: The heart of any large city or town; the core contains the highest land values, the most intense building development per hectare, and the highest concentration of vehicles. In it are to be found offices, government institutions, large retail stores, hotels, restaurants, cultural institutions and theaters, and car parking facilities. Intense economic buoyancy usually ensures a steady stream of development and redevelopment proposals; the problem becomes one of control to ensure balance, efficiency, good architecture, and splendid streetscape, rather than the need for a program of rehabilitation.

However, the core may be surrounded by a frame of commercial, residential uses, and warehousing, in which significant areas of blight are to be found. Here more positive policies are needed to promote positive improvements. See *Central Business District (CBD)*.

Corporate Planning, UK: In Britain, overall planning for a city's physical, economic and social development; an across-the-board approach to solving urban problems resulting in a "community plan." The community plan is a synthesis of two documents, the structure plan (q.v.) and the program plan. Of these only the structure plan becomes a statutory document, while the program plan, being based on the annual budgets, is usually a short-term document. Many local authorities in the United Kingdom have adopted corporate planning structures, headed by a chief executive officer or town manager.

Council for Mutual Economic Assistance (COMECOM): An organization established in 1949 by several eastern European countries for the purpose of centralizing arrangements for trade, credit, and technical assistance among the members. The members currently comprise the Soviet Union, Bulgaria, Cuba, Czechoslovakia, Hungary, Poland, Rumania, the German Democratic Republic, and the Mongolian People's Republic.

At a meeting of the Council for Mutual Economic Assistance in Prague in June 1973, the member countries and the Socialist Federal Republic of Yugoslavia agreed to draw up a detailed overall program of cooperation in environment control. Joint action was agreed:

- To control pollution of inland waters, seas and oceans;
- To eliminate industrial pollution sources by improving technological processes;
- To produce gas and water purifying equipment and related measuring and control instruments;
- In matters relating to the disposal and recycling of wastes.

These decisions strengthened an earlier agreement by the Council in 1971 to draft an environment protection plan. Many research projects have been put in hand. For example, the Hygiene Institutes in Berlin and Moscow investigate the influence of air pollution on human health. Maximum admissible concentrations of chemical substances in human and animal foods, water and soil are the subject of investigations by the Nutrition Institute of the GDR Academy of Sciences and the Kiev Institute of Hygiene and Toxicology. Much of the research is done in cooperation with Soviet partners, but other bilateral agreements exist. Early in 1973, an agreement was signed with Hungary, dealing with scientific and technological cooperation in developing measuring instruments and methods for pollution control.

An environment protection agreement has been concluded between the GDR and Poland. This agreement involves, for example, joint territorial planning. The open-cast mines and power stations situated on both sides of the frontiers present environment problems that are jointly handled.

Protection of the Oder river against pollution concerns both countries and involves common action. When new industrial plants are to be built or existing plants reconstructed near the frontier, environmental hazards are then assessed. Similar cooperation exists between the GDR and Czechoslovakia. Bilateral cooperation in environment and pollution control has developed with Bulgaria, Rumania and Yugoslavia also.

Council for the Environment, Republic of Ireland (EIRE): A Council created by the Irish Government in 1978 to advise the Minister for the Environment on general environmental policy. It is a non-statutory body with a broad membership. In 1979, the Council recommended the adoption of a national environment policy that would "ensure that remedial and preventative action is taken in a planned and rational manner." The Council commented that agreement generally had been reached on what needs to be done to protect the environment, and many environmental protection and improvement measures had already been taken, but the lack of coordination flowing from a central policy had hampered their effectiveness.

The Council continued: "These activities are spread between a large number of government departments, public agencies, and private organizations and, to some extent, have been uncoordinated because of the absence of a national policy. The lack of an overall policy has not been conducive to consideration of the interaction of environmental and other national aims. A policy with clear aims and objectives would provide a better context for considering the implications of decisions which could have a significant impact on the environment. It would also help to ensure that environmental considerations are given adequate weight at an early stage in decision-making."

The Council also highlighted the need for a national policy to deal with developments of national significance. At present these are treated as matters for local planning authorities.

Council for the Protection of Rural England: Formed in 1926, a council bringing together the various national bodies interested in different parts and aspects of the English countryside. The objects of the council are to protect the beauty of the English countryside from disfigurement and injury; to act as a center for obtaining and giving advice and information on matters affecting the protection of rural scenery; and to rouse public opinion to an understanding of the importance of this work and of the need to promote it. There are equivalent bodies for Scotland and Wales.

Council of Europe: A Council created in 1949 comprising a Consultative Assembly and a Committee of Ministers to make decisions and formulate recommendations to governments. The Assembly consists of 140 persons appointed by the national governments. Membership of the Council is limited to European countries; in 1979, there were 21 members.

It has been the task of the Assembly to propose actions to bring European countries close together, to keep under constant review the progress made, and to voice the views of European public opinion on the main political and economic questions of the day. The role of the committee of Ministers is to translate the Assembly's recommendations into action. The recommendations may relate to the lowering of barriers, harmonizing of legislation, or the undertaking of tasks on a joint European basis. Projects have related to economic, legal, social, public health, environmental, education and scientific matters.

Over 70 conventions have been concluded among members of the Council as a result of its deliberations. These include the following nature conservation conventions: Convention on the Conservation of European Wildlife and Natural Habitats; Convention on Underwater Cultural Heritage; Convention on the Protection of International Freshwaters; Convention on the Conservation of Migratory Species of Wild Fauna; Conservation on the Conservation of Antarctic Marine Living Resources.

In 1980, the Council organized the European Campaign for Urban Renaissance. Under the slogan "A Better Life in Towns," the campaign had five main themes: the improvement of urban environmental quality; the rehabilitation of existing and older buildings, housing and areas; the provision of social, cultural and economic opportunities; the achievement of community development and participation; and a developing role for local authorities. Using a similar approach to the European Architectural Heritage Year in 1977, a campaign was organized through national committees, conferences and a program of demonstration projects.

The fourth European Ministerial Conference on the Environment was held in Greece in April, 1984, being attended by 21 member countries together with Finland. The theme was coastal areas, river banks and lake shores, their planning and management in compatibility with ecological balance. A set of guidelines, prepared by Greece, was endorsed and commended to governments and to the Committee of Ministers of the Council of Europe. The guidelines sought to reconcile economic development with environmental protection, and suggested that environmental protection might even advance economic development.

Council of Nature Conservation Ministers (CONCOM), Australia: A council established in 1974, following agreement between the Australian prime minister and the state premiers, with the aim of developing coordinated policies for nature conservation, especially for the reservation and management of adequate areas of land for this purpose and for the conservation and management of wildlife.

Membership consists of the federal and State ministers with responsibility for national parks and wildlife conservation. CONCOM has examined such issues as endangered fauna and flora, nuisance species, law enforcement, the woodchip industry, education and training, and motorized recreational vehicles. It has participated actively in the development of a national management plan for kangaroos.

Council on Environmental Quality, US: A Council created by the National Environmental Policy Act, 1969 (q.v.); the Council comprises three persons appointed by the President of the United States. The responsibilities of the Council include formulating policy recommendations on environmental matters for the President, advice concerning the environmental impact statement procedures and the publication of an annual report on the state of the environment in the United States. The Council functions in a manner convenient to the President. In respect to significant issues, it is normal for the Council to prepare a Presidential Review Memorandum which seeks to shape and influence Administration policy. The President's Environmental Message for 1977 broadened the functions of the Council; by an Executive Order of 24 May 1977, President Carter directed the Council to issue regulations for the preparation of impact statements. The precise relationship of the Chairman of the Council to the President and the White House generally varies, understandably with each Administration, from the cordial to the distant.

The Council has groups employed on various tasks. The energy group seeks to ensure that federal energy policy will evolve with full recognition of national environmental goals; and to identify emerging technologies which hold out the promise of the best results in environmental, social and economic terms. The environmental health group is concerned with a strategy for toxic substances, and the formulation of a program. It has convened a toxic substances committee comprising representatives from all the federal agencies that have responsibilities for monitoring or scientific expertise. Another group is concerned with environmental economics, and the alleviation of any adverse economic effects arising from environmental decisions in specific cases. Another group is interested in recycling, inner city environment, and land use.

Country Parks, UK: Parks established under the Countryside Act, 1968, by the Countryside Commission (q.v.); the principal aim of these relatively small parks is to provide the urban dweller with opportunities to enjoy the countryside through short walks, picnics, or just sitting and enjoying the view. For this reason, they are usually within easy reach of the main centers of population.

Countryside Commission, UK: A commission for England and Wales responsible for the selection of national parks and areas of outstanding natural beauty.

The report of the Hobhouse committee (Cmd. 7121), published in

1947, recommended the setting up of a National Parks Commission and the creation of 12 national parks. This report formed the basis of the legislation of 1949, but the passing of the Town and Country Planning Act in 1947, which transferred planning control from district councils to county councils, had a profound effect on the degree to which the Hobhouse committee's recommendations were accepted.

The Countryside Commission for England and Wales eventually replaced and assumed the functions of the National Parks Commission set up under the National Parks and Access to the Countryside Act to select national parks and "areas of outstanding natural beauty" and to make proposals for the creation of long-distance footpaths and bridleways. Ten national parks have been established: Northumberland, the Lake District, the Yorkshire Dales, the North York Moors and the Peak District in northern England; Snowdonia, the Pembrokenshire Coast and the Brecon Beacons in Wales; Exmoor and Dartmoor in southwest England. They cover 12,618 square kilometers, or 9% of the area of England and Wales. Administration is based on local authority representation, with some members appointed by the Government. The establishment of a new national park has been proposed for mid-Wales.

Within national parks the normal life of the area goes on; designation does not in any way affect the existing ownership of the land. The park planning authorities control development just as county councils do elsewhere throughout England and Wales, but in doing this they pay special regard to the fact that the park is one of Britain's finest landscapes which stands in need of preservation. The land in these designated areas generally remains privately owned, but agreements or orders to secure additional public access may be made by local authorities. Steps are taken to preserve and enhance the landscape's natural beauty by high standards of development control, and by positive measures, for which government grants are available, such as tree planting and preservation, and the removal of eyesores. In the national parks, other measures for the benefit of the public include the provision of car parks, camping and caravan areas, and information centers. All national parks and some other designated areas have warden services. Some 2,414 kilometers of long-distance footpaths and bridleways had been approved by August, 1983. Large stretches of these paths are already public rights of way; the 402 kilometers of the Pennine Way form Britain's first long-distance footpath and the South Downs Way, stretching for some 129 kilometers was the first long-distance bridleway.

Under the Amenity Lands Act (Northern Ireland) 1965 the Ulster Countryside Committee advises on the establishment of national parks and the designation of areas of outstanding natural beauty. Seven areas have been designated. Four amenity areas, acquired by the Ministry of Development, are being managed as country parks. Although there are no national parks as such in Scotland, five large national park direction areas are subject to special planning control under ministerial supervision. The areas are: Loch Lomond-Trossachs; Glen Affric-Glen Cannich-Strath Farrar; Ben Nevis-Glen Coe-Black Mount; the Cairngorms; and Loch Torridon-Loch Maree-Little Loch Broom. Over 90 per cent of the land in Scotland has been designated as countryside within the jurisdiction of the Countryside Commission for Scotland.

The Forestry Commission has formed, and opened to the public, seven forest parks in some of the finest country in Great Britain: Argyll, Glen Trool, Glen More and the Queen Elizabeth Forest Park in Scotland; the Forest of Dean and the Wye valley woods on the borders of England and Wales; Snowdonia in Wales; and the Border Forest Park on the borders of England and Scotland. The parks cover some 243,000 hectares. Camping and other recreational facilities provided in all the parks are used by several million holidaymakers each year. The historic New Forest, in Hampshire, although not a forest park, is also open to the public.

The Commission has also identified some 33 Areas of Outstanding Natural Beauty covering 14,500 square kilometers. They vary considerably in character and cover a wide range of landscape types such as the Bodwin Moor Uplands, the Forest of Bowland and the Deham river valley.

Cubatao Disaster, 1984, Brazil: An incident in the southern Brazilian town of Cubatao in February, 1984 when a leaking gasoline pipeline caught fire incinerating the flimsy huts of hundreds of squatters on the surrounding marshland. About 500 people died, investigators found.

Cumulative Impacts: Impacts and effects on the environment (q.v.) which result from the incremental impacts of individual actions when added to other past, present, and reasonably foreseeable future actions, regardless of the agent or person responsible for such actions. Cumulative impacts can result from individually minor but collectively significant actions taking place over a period of time.

D

Decentralization: A conscious policy of locating or relocating some parts of, or the whole of, an organization in outlying regions away from metropolitan areas with concomitant developments of infrastructure coupled with extensions of existing residential areas or the establishment of new towns (q.v.). The policy may aim at the strengthening of specified regional administrative centers (q.v.). The purpose of such a policy may be to curb the continuing expansion of the large metropolitan areas, providing employment and living opportunities in a wider variety of centers throughout a region, and encouraging the development of regional natural, environmental, and recreational resources.

As an aspect of State regional development policy, it may involve restrictions and limitations on metropolitan location and the provision of incentives for firms to locate in centers which would not otherwise have been chosen.

A general system of incentives has often proved expensive and ineffective in terms of total objective. The tendency in many countries is to offer incentives in relation to selected centers only; this is known as "selective decentralization." It is also often recognized that incentives of a permanent character introduce economic inefficiencies of an equally permanent nature; the tendency, therefore, is to offer incentives of a temporary character, e.g., to overcome the initial disadvantages only of locating or relocating in less preferred places.

A major problem in a complex program of decentralization is coordination. Investment in a new town or regional growth center may be undertaken in the belief that industry and other sources of employment will follow. This may not occur, and the center finishes up serving dormitory purposes only creating new transport problems in relation to a growing body of commuters. Conversely, a center may be successful in attracting industry of a particular kind needing a particular range of skills. If such an industry or corporation wilts, the entire center is affected; hence a measure of diversity is often thought to be essential. Another problem of coordination relates to hospital units, higher educational institutions, and larger shopping complexes; these tend not to materialize until the population has reached a certain threshold level. It is difficult then to induce population growth in the face of glaring gaps in terms of services. The pace may be accelerated in some instances by moving segments of major government departments to new locations as in the case of Canberra, Australia; but such moves to relatively undeveloped areas may result in massive public service resistance which has occurred in some instances.

Decibel A [dB(A)] Scale: An international weighted scale of sound levels which attenuates the upper and lower frequency content and accentuates middle frequencies, thus providing a good correlation in many cases with subjective impressions of loudness and sense of annoyance. Nearly all audible sounds lie between 0 and about 140 dB(A). 0 dB(A) is the "threshold of hearing," while sounds above about 140 dB(A) are not common. An increase of 10 dB(A) means that the noise perceived by a listener has roughly doubled in loudness. A car passing at 70 dB(A) sounds twice as loud as one passing at 60 dB(A). The dB(A) measurement is widely used throughout the world for determining approximately human reaction to noise and is the basis of legislation to control noise in many countries. Examples of typical dB(A) levels are as follows:

	dB(A)
Rustle of leaves	10
Quiet office	50
Busy office	65
Moderate traffic	70
Alarm clock ringing	80
Very noisy factory	90

Delaware, US: See *Coastal Zone Act, 1971; Litter Laws, US*.

Demography: The measurement of certain characteristics of human populations, e.g., the size of a population and its distribution by age, sex, occupation, and location, and trends in these characteristics. Beyond collecting and analyzing statistics, demographers seek to explain patterns of distribution and behavior.

At the macro level, demographers make projections based on assumptions regarding death rate, fertility, life expectancy, and contraception. An important finding at the macro level is that few segments of the world's population are being held in poverty by sheer pressure of numbers. Crowded countries are not necessarily economically sluggish, e.g., Singapore with a population density 250 times that of Argentina has had an economic growth rate three times faster over the last 20 years.

Overcrowding at the family or micro level is of greater concern. Large families are the typical choice of poor people; as there is insufficient money for food let alone health and education, poverty is perpetuated.Hence, poorer nations tend to have the fastest population growth. The problem of population is essentially at the family unit level, and here the solution must lie. Many nations now encourage smaller families, while improvements in social services tend to make smaller families more feasible for poorer people.

Demography has established close causal links between several economic and social variables and declining fertility; very generally, fertility falls as incomes rise, education spreads, health improves, and contraception and the attitudes of parents make this possible. The key is to break the cycle of poverty as has occurred in many countries. See *Population Differentiation; Population Implosion; United Nations World Population Conference, 1974; World Population Plan of Action.*

Denmark: See *Greater Copenhagen Regional Plan; National Physical Planning; Nature and Monuments Conservation Act, 1974, Greenland; Nordic Environmental Protection Convention; Urban and Rural Zones Act 1969-75.*

Department of Environment (Environment Canada): A department created by the Canadian Government in 1971; it has an overall responsibility for protecting the country's environment and natural resources. The department absorbed the functions of several other federal bodies relating to the environment. Environment Canada is organized into seven services—the atmospheric environment service; the environmental protection service; the fisheries service; the lands, forests and wildlife service; the policy planning and research service; and the finance and administration services.

The environmental protection service develops and enforces environmental regulations, guidelines, codes, and protocols introduced under federal legislation in respect of air and water pollution, solid waste management and resource recovery, noise, contaminants, and environmental impact assessment. The service liaises with provincial agencies, industry, other federal agencies, and the general public.

Responsibility for environment protection and nature conservation in Canada is shared by the federal and provincial governments; the municipal governments have powers assigned to them at the discretion of the provincial legislatures. The ten provinces are Newfoundland, Prince Edward Island, Nova Scotia, New Brunswick, Quebec, Ontario, Manitoba, Saskatchewan, Alberta, and British Columbia. The federal government and most of the ten provinces have created departments of the environment; all have created advisory bodies.

Department of Environment and Planning, New South Wales: A department established in New South Wales, Australia, by the State's Environmental Planning and Assessment Act, 1979. The Act gives the Department of Environment and Planning (formerly the Planning and Environment Commission) the main responsibility in New South Wales for ensuring that planning decisions reflect the general aim that all the different uses to which land is put should be based on the wise management of the natural environment, and the social and economic needs of the community.

The Department advises the Minister on key policies that need to be established at the State and regional level to guide local government and ensure an equitable distribution of resources and services. It also ensures that development proposals that are likely to have a major environmental impact, such as mining, give adequate consideration to the nature of those possible environmental impacts and effects.

In developing state and regional policies, the Department consults with local government, public authorities, and community groups. Public authorities are also bound by the Act to ensure that environmental issues are adequately considered in development proposals and activity.

The Act gives local government prime responsibility for making planning decisions about development, ensuring that decisions are consistent with wider regional and state environmental plans and policies. Planning control allows for public consultation on environmental issues at various stages of the planning process. See *Commissioner of Inquiry; Land and Environment Court; Local Environmental Plan; Regional Environmental Plan; State Environmental Planning Policies; State Pollution Control Commission.*

Department of the Environment, UK: In Britain, a Department formed in 1970 bringing into a single unit the three former Ministries of Transport; Housing and Local Government; and Public Building and Works. For the first time, responsibility for planning and land use, for pollution control, for transport, public building and construction were brought together under a single Secretary of State. However, it was neither sensible nor possible to bring together in one Department everything to do with a subject so all-embracing as the environment. Hence other Departments retained some special functions. The Ministry of Agriculture, Fisheries and Food remained responsible for farming and fishing and for the control of pesticides used on the land, the disposal of farm wastes, and the monitoring of contaminants in foodstuffs. The Department of Trade and Industry, responsible for shipping and aircraft, continued to look after oil pollution at sea and noise and pollution around airports. The Department of Employment sought to ensure that factories were safe places in which to work. The Department of Health and Social Security provided medical advice to government as a whole. The Department of Education and Science ensured that the research councils under its care provided a basis of scientific knowledge. In Scotland, Wales and Northern Ireland environmental matters remained the responsibility of the respective departments and offices. The Foreign and Commonwealth Office remained concerned with the international and foreign policy aspects of environmental issues. However, the Secretary of State for the Environment became responsible for coordinating all the pollution control activities in all Departments. Since then, the aim has been to develop a strategic approach to environmental planning and protection, to replace the fragmented approach which previously handicapped the effort.

In November 1976, the Department of the Environment was split into two departments by hiving off those parts concerned with transport. The reason for the split appeared to be purely political, being related to the balance of powers within the Cabinet. The Secretary of State for the Environment remained responsible for the Department of the Environment, encompassing such matters as regional affairs, inner urban and inner city areas, new towns, and devolution issues. Through other ministers assisting the Department, the portfolio extended to housing, construction, planning, development control and land, water and sewerage, pollution, minerals, countryside affairs and sports. However, close collaboration between the Department of Environment and the Department of Transport was established, with joint arrangements on urban transport and planning matters.

Derelict Land Reclamation Programs, UK: Programs for the rehabilitation of land, previously utilized but abandoned and usually disfigured with industrial waste-heaps and ponds, discarded rusting machinery, crumbling buildings, discarded auto bodies and garbage. Derelict land may be reclaimed for a variety of uses. In urban areas it may be restored, where practicable, for housing or industrial use but in practice more of the derelict land in these areas is used to provide public open space for amenity or recreation, since there is a greater need for planned open space in such areas. In other areas the emphasis tends to be on restoration for agricultural uses, mainly rough grazing or woodland; substantial areas of land damaged by mineral working, e.g., colliery spoil heaps, have been successfully reclaimed for such uses. Other examples of after-use are the reclamation of disused gravel pits for use as water recreation areas and as nature reserves, and the treatment of disused railroads to provide public walkways and bridleways in the countryside.

A particular problem, and one to which increasing attention is being paid, is the reclamation of contaminated land in urban areas—for example, the sites of former gas-works, sewage farms and scrap yards, or sites used previously for tipping toxic wastes, e.g., cadmium, lead, phenols, cyanides and tars. A large number of contaminated sites are now being considered for redevelopment for a variety of uses involving various remedial measures to minimize the potential environmental problems. One example of current reclamation projects is the comprehensive development by the Greater London Council at Thamesmead, on the site of the former Woolwich Arsenal, where there is extensive pollution by a wide range of substances including toxic metals, pyrotechnic wastes and gas-works wastes. A national survey of derelict land conducted in 1984 revealed that Britain still had over 45,000 hectares of such land.

Desertification Control: The adoption of measures to combat the encroachment of desert or win back desert areas for useful purposes, on the lines suggested by the United Nations Conference on Desertification (q.v.) held in 1977. The two examples below relate to the immense desert areas in the northern districts of the People's Republic of China, and the relatively small desert areas of Israel:

(1) China—The desertified land in China is widely distributed through the northern districts of the country. Research has indicated that the area of desertified land formed in the 'historical period' covers some 120,000 square kilometers; while the area of desertified land formed in the "recent period" of the last fifty years or so amounts to more than 50,000 square kilometers. The two categories, totalling some 170,000 square kilometers, cover 207 counties, cities, and banners in northern China. Notwithstanding, these northern districts support a population of 35 million people.

In addition to this large area of desert, there are estimated to be some 150,000 square kilometers of "latent desertified land" where desertification will develop if the land is used improperly (UNEP, 1981). Timely measures are needed to protect the environment.

Much of the desertified land is to be found in the western part of the northeastern plain, and in the vast area to the east of the Tarim Basin, extending from the Great Wall and the Kunlun mountain ranges in the south to the Sino-Mongolian and Sino-Soviet borders in the north. It is also found in relatively small areas of eastern Henan and the central and southern regions of the Hebei Plain.

Desertification control measures are being carried out along the following lines:

1. Developing desertified land. Measures include prompt action to re-orient cropping methods and control dry-farming techniques; and the establishment of windbreak networks accompanied by the protection of natural vegetation and a reduction in the pressure of grazing.

2. Protecting desertified land. Measures include building systems of shelter-belts, planting forests in the depressions among the dunes, and growing sand-fixing plants on the dunes.

3. Safeguarding latent desertified land. Measures include careful land use planning based on the land capabilities in the region; and the regulation of grazing in accordance with the carrying capacity of rangeland, thereby ensuring sustained productivity.

See *Green Great Wall.*

(2) Israel—Israel is characterized by a very wide range of physical conditions within a relatively small area; it contains desert, tropical and alpine environments in close proximity to one another. The main topographic formation is the Rift Valley, running north-south along the east of the country for about 400 kilometers. It includes the Sea of Galilee, the Dead Sea (the lowest point on earth), and the Arava desert running south to the shores of the Red Sea. Indeed, the whole southern half of the country can be classified as desert.

The northern half of the country is divided into three regions: the limestone mountains of the north (Galil) and centre (Judean Hills); the alluvial valley of the upper Jordan and the valleys of Jezreel and Bet Shean; and the Mediterranean coastal plain with its sandstone ridges, sand dune areas and highly fertile alluvial soils. The northern part of the country has a Mediterranean climate.

The coastal plain and the alluvial valleys are highly intensive agricultural areas, with citrus and fruit orchards and field crops. The coastal plain is also the location of the country's industry and service activities and is highly urbanized. The mountains have suffered from deforestation and overgrazing, though afforestation programs have brought back a tree cover to some areas.

The early pioneers of the Zionist Movement, arriving a century ago, had found a harsh, desolate country, bare of trees under a pitiless sun. The challenge of restoring trees, forest and shade, early captured their imagination. Progress was at first slow, but between 1949 and 1977, the number of trees planted increased by about 25-fold. By 1977, some 125 million trees had been established covering 500,000 dunams (10 dunams equal 1 hectare) (Jewish National Fund, 1978). The spread of desert areas appears to have been contained.

Development: The application of human, financial, and physical resources to satisfy human needs and improve the quality of life; inevitably development involves modification of the biosphere some aspects of which may tend to detract from the quality of life locally, regionally, nationally, or globally. The benefits and disadvantages of development came more sharply into relief during the 1960s culminating in a series of United Nations conferences relating to the human environment, population, national settlement, water resources and desertification. Development and environmental protection often appeared incompatible; today the view more generally prevails that much-needed development and economic growth are attainable within an acceptable framework of environmental constraints. The Gross National Product (GNP) (q.v.) has ceased to be the sole measure of progress, however, with a new emphasis on overall benefits for the community at large.

Development Application: A formal application by a developer to the local consent authority or planning body for a permit or consent to construct and operate a facility or plant, undertake land subdivision and housing construction, or other major works; depending upon the prospective environmental significance of the project, the application may or may not be accompanied by an environmental impact statement (q.v.). Matters taken into account by consent authorities include:

- The relationship between the proposal and any existing or intended environmental planning instrument or development control plan;

- The likely impact on the environment and the means proposed for mitigating those effects;
- The possible effect on the landscape;
- The social and economic effects of the proposal in the locality;
- The character, scale, density, and height of the development;
- The size and shape of the land in question and the location of the development;
- The nature of adjoining developments;
- The entrance and exit to the site, and provision for the loading, maneuvering and parking of vehicles;
- Traffic generation and its effects on the road system and road safety;
- The availability of public transport;
- The availability of utility services;
- Provision for landscaping and tree preservation;
- The possible effects on the amenities of the area;
- The public interest;
- Representations by any public authority, government department, or member of the public in response to the public exhibition of the details of the project.
- The advice of its own officers;
- The prospects of financial contributions from the developer and/or the state towards the costs of additional public infrastructure such as roads, water and sewage works, and housing for the workforce during construction and later.

See *Development Standards; Zoning.*

Development Charge, UK: A charge introduced in Britain by the Town and Country Planning Act, 1947 (q.v.); the system provided that persons wishing to carry out development should purchase the development value of the land from the government by the payment of a "development charge." This was an attempt to recoup betterment (q.v.); it was abandoned in 1952. However, a similar charge was introduced by the Land Commission Act, 1967; this in turn was abandoned in 1970.

While under these measures charges became payable under certain circumstances, compensation also became payable in adverse circumstances. Compensation became payable where it could be demonstrated that land had development value before the land use controls were introduced; compensation was also payable in respect to the cancellation of "existing use" rights. In specified circumstances, an owner could require a local authority to purchase the land.

Development Control: The process by which environmental plans are implemented insofar as they anticipate and influence the carrying out of development. Plans aim to achieve a logical and efficient distribution of land uses which meet the community's needs in a way that minimizes or avoids adverse environmental consequences.

Development control is concerned with the total amount of development, where and when it occurs, and the way in which it fits into its surroundings.

Most development may require a formal application for consent, permission, approval, or permit, to the planning agency or authority; some development however may be permitted by the plan itself, or permitted subject to development standards (q.v.), without the need for an application for consent.

A development control plan shows in fairly precise detail how it is intended that a particular area of land shall be developed. It shows the location of all roads, public and community land, e.g., for schools and open space. It may also show restrictions on development, e.g., roads to which houses shall not have direct access.

Development Rights: Traditional legal rights associated with the ownership of property, including (but not limited to) the right to exclude trespassers, the right to cut timber, the right to mine or quarry, the right to grow crops or keep animals, the right to build structures for the owner's private use and the right to build structures for profit.

Zoning (q.v.), however, imposes certain restrictions on an individual's use of land in order to protect the public health, safety, and welfare. For most property, these traditional property rights have been modified.

In some instances, "conservation easements" or "scenic easements" are purchased by organizations, individuals or governments for the purpose of limiting development in such a way as to protect the natural environment or preserve scenic views. In some instances, development rights have been purchased from property owners by government in order to restrict the use of their land to agriculture. A recent development is the transfer of development rights (q.v.) to other property owners. See *Development Control; Development Standards, Easements*.

Development Standards: In respect to any development, fixed requirements or standards imposed by regulation or ordinance under environmental planning legislation, including requirements or standards in respect of:

- the area, shape or frontage of any land, the dimensions of any land, buildings or works, or the distance of any land, building or work from any specified point;
- the proportion or percentage of the area of a site which a building or work may occupy;

- the character, location, siting, bulk, scale, shape, size, height, density, design or external appearance of a building or work;
- the cubic content or floor space of a building;
- the intensity or density of the use of any land, building or work;
- the provision of public access, open space, landscaped space, tree planting or other treatment for the conservation, protection or enhancement of the environment;
- the provision of facilities for the standing, movement, parking, servicing, maneuvering, loading or unloading of vehicles;
- the volume, nature and type of traffic generated by the development;
- road patterns;
- drainage;
- the carrying out of earthworks;
- the effects of development on patterns of wind, sunlight, daylight or shadows;
- the provision of services, facilities and amenities demanded by development;
- the emission of pollution and means for its prevention or control or mitigation; and
- such other matters as may be prescribed.

See *Development Application; Development Control; Zoning.*

Development Zones and Holding Zones: A form of zoning (q.v.) particularly applicable where public investments are necessary to overcome natural constraints to development. For example in some communities soils pose a severe limit to on-lot disposal of septic effluent. In such a case, an economical extension of the sewerage system needs to be coordinated with the zoning ordinance.

A development zone is defined where growth can be accommodated and where utilities can be provided without exorbitant expense. Other areas that cannot be efficiently serviced at an early date may be placed in "holding zones" where development is discouraged or prohibited until the necessary services become available.

Dimensional Variance: See *Variance, US.*

Distribution of Industry Act, 1945, UK: An Act enabling the British Government to regulate the location of new factory buildings anywhere in Britain. To establish a factory an industrial development certificate was required; through these certificates, industry was steered away from the more prosperous urban areas such as London and Birmingham towards the more needy "development areas." In 1965, office accommodation became subject to similar procedures, the principal object being to steer new office accommodation away from central and inner London.

District Structure Plan: A plan which shows in a fairly precise way the location of all the principal zones of different uses to which the land is to be put, but not in all cases final detailed boundaries. The purpose of the district plan is to show how the urban structure plan (q.v.) is to be implemented and to provide a sound framework for the preparation of detailed development control plans. See *Development Control*.

"Do The Right Thing" Campaign, New South Wales: The theme of a litter reduction campaign launched in 1978 by the New South Wales Government, Australia, with the support of private industry. The objectives of the campaign were:

- To improve the appearance of the environment;
- To increase public awareness and encourage better community attitudes and behavior towards litter;
- To improve the technology and practice of solid-waste management and litter collection, disposal and recycling.

The campaign was directed towards the seven basic sources of litter: pedestrians, motorists, household garbage, commercial waste, loading and unloading operations, uncovered or inadequately covered trucks, and construction projects. The financing group comprises ten companies involved in the manufacture, packaging and marketing of beverages.

The theme of "Do the Right Thing" was adopted and commercials based on this theme were produced and presented extensively on television and radio. The message appears to have made a noticeable impression on the public. To measure the effectiveness of the campaign, a litter-index study has been undertaken in the Sydney metropolitan area with actual counts of litter deposited over prescribed periods. See *Figure 4*.

Figure 4: The logo selected for the New South Wales government's litter reduction campaign. Source: State Pollution Control Commission, New South Wales, Australia.

Dobry Report, UK; The Report of the Dobry Committee on the British development control system, published in 1975, which supported the idea of environmental impact assessment for selected major developments, with regard to six aspects (namely traffic, roads and public transportation; ground and surface water drainage; publicly provided services; appearance of the surrounding area; employment; and noise and air pollution). Partly in response to this Report, the Secretary of State directed a study through the Department of the Environment (q.v.) into the need for EIA procedures and how best these could be incorporated within the existing development control system. The study recommended that EIA procedures should be undertaken as an experiment, without any amendment of the planning system in the first place. It also included some substantial qualifications about the type of EIA process that it was recommending. See *Environmental Impact Assessment; Environmental Impact Statement.*

Duncan Classification, US: A classification, created by the American sociologist Otis Dudley Duncan and his colleagues, of the interrelations of metropolitan areas and regions of the United States (*Metropolis and Region, 1960*). The classification constitutes an analysis of different levels of regional integration; it distinguishes:

- National metropolises;
- Urban places with diversified manufacturing and metropolitan functions;
- Regional metropolises;
- Regional capitals;
- Urban places with diversified manufacturing and few metropolitan functions;
- Urban places with specialized manufacturing;
- Special cases.

This classification contains elements of hierarchy while suggesting a conclusion that the potential size of individual cities or metropolitan areas depends on the size and integration of the region of which they are a part. See *Central Place Theory.*

E

Earthwatch: A plan for global environmental assessment arising out of the United Nations Conference on the Human Environment, 1972 (q.v.); an internationally financed and coordinated global system of national facilities and services to study the interactions between humanity and the environment, identify trends of significance, provide early warning of environmental hazards, and determine the status of selected natural resources. Earthwatch has four closely-linked components: evaluation and review, research, monitoring, and exchange of information. The responsibility of the United Nations Environment Program (q.v.), Earthwatch contains such important and operational elements as the Global Environmental Monitoring System (GEMS) (q.v.); the International Referral System (INFOTERRA) (q.v.); and the International Register of Potentially Toxic Chemicals (IRPTC) (q.v.).

Easements: Contractual agreements with landowners to protect the public's interest in a parcel of land without purchasing the land outright. An easement can be an agreement to permit public access to, or through, private land. It might also be an agreement to maintain the scenic, open quality of a piece of land. An easement can be an economical way to achieve specific land use objectives, particularly in rural areas. In some cases, landowners may actually donate easements to the municipality to gain federal tax benefits. See *Mandatory Dedication*.

Ecodevelopment: A term to describe ecologically sound development; positive management of the environment for human benefit.

Ecology: The study of ecosystems; the relationships between living organisms and between them and their environment (q.v.). Ecology is concerned frequently with general principles that apply to both animals and plants. Autecology is concerned with single organisms or species, and synecology with communities of species, although these artificial partitions are no longer accepted by all ecologists. Human ecology is the study of the structure and development of human communities and societies in terms of the processes by which human populations adapt to their environments; as a subject it represents an application of the perspectives of the biological sciences to the investigation of topics included in the social sciences. See *Ecosystem*.

Economic Commission for Europe (ECE): A commission set up in 1947 by the Economic and Social Council of the United Nations to initiate action to raise the level of European economic activity, in the wake of the Second World War, and for maintaining and strengthening the economic relations of European countries, both among themselves and with other countries. The members consist of the European members and the United States of America.

Early, the ECE developed a test cycle for the measurement of gaseous emissions from the exhausts of motor vehicles. It was based on European patterns of driving: nevertheless it was adopted in Australia from 1974 as the most appropriate cycle for testing motor vehicle compliance with Australian Design Rule 27, the first serious attempt to reduce hydrocarbon and carbon monoxide emissions to a significant extent in that country.

The environmental activities of the Commission have broadened considerably in recent years. It is now served by a group of Senior Advisers on Environmental Problems. Member countries now submit biennial monographs on their environmental policies and strategies. Working parties address themselves to specific subjects such as air pollution. Seminars have been conducted on a whole range of subjects: environmental impact assessment, economic assessment of environmental damage, desulfurization of fuels and combustion gases, long-range transboundary air pollution, solid wastes and recycling, transportation, non-waste technology and production, long-term planning of water management, and fine particulates.

The Commission also pioneered the Convention on Long-Range Transboundary Air Pollution (q.v.).

Economic Commissions for Africa, Latin America, Asia and the Far-East: Regional economic commissions created by the Economic and Social Council of the United Nations. In 1978, the Economic Commission for Africa established an environment coordination unit for matters covering the various divisions of the regional commission itself, and to provide environmental impact statements for major projects. The Commission started work in 1979 on a study of industrial environments and pollution problems in selected African countries. During the period 1979-85, the program assisted African states in identifying sources of pollution whether industrial or biological, and in identifying major threats to African imports that may arise as a result of environmental concern. The program also aimed at assisting African states in devising pollution control legislation.

In the Economic Commission for Latin America, efforts are being directed towards identifying environmental problems, especially those related to utilizing resources, throughout the region. The food and agriculture program is concerned with developing the productivity of agriculture, while protecting the environment.

The Economic Commission for Asia and the Far-East has studies in hand to: identify major problems of industrial pollution in the region, especially among agro-industries; determine various techniques and alternatives for pollution control in these industries; identify the costs of pollution control and determine the ways and means to minimize costs.

Economic Efficiency: The efficiency with which scarce resources are used and organized to achieve stipulated economic ends. In competitive conditions, the lower the cost per unit of output, without sacrifice of quality, in relation to the value or price of the finished article, the greater the economic efficiency of the productive organization. The social worth of economic efficiency weakens in circumstances in which reduced costs of production are not reflected in lower prices to consumers, in real terms, and in circumstances in which social costs must be borne which are larger than the reduction in the recorded or accountancy costs of the productive organization.

The concept of economic efficiency ignores considerations of equity or fairness, because market prices already imply a particular underlying distribution of wealth which neoclassical writers tended to take as given.

Economic Ends: The objectives of economic activity; ends are both quantitative and qualitative, but with the nature of ends economic science has no direct concern. Economics is concerned only with the number of ends and with their degree of relative intensity. Economic ends form only a part, albeit a very important part, of the full spectrum of public and private objectives. Furthermore, the pursuit of certain specific economic ends in the short term may not be wholly acceptable economic ends to a community in the long term.

Economic Growth: The growth per head of population in the production of goods and services of all kinds to meet final demands, e.g. goods and services for domestic consumption, capital goods for accumulation, export goods to pay for imports. An acceleration of economic growth requires as much emphasis on such elements as better management, better training of labor and improved education, as upon higher capital investment. Furthermore, the quality of investment may count almost as much as the quantity. Capital investment, though important, is but one significant factor in the growth rate of an economy. The "growth rate" is the annual growth, usually expressed as a percentage over the previous year, of productive capacity in a community; in assessing the growth rate per head, an adjustment must be made for changes in population.

Economic System, Functions of: The essential functions to be fulfilled in the economic arrangements of any community; these functions may be defined as follows:

- Generally, to match supply to the effective demand for goods and services in an efficient manner,
- To determine what goods and services are to be produced and in what quantities,

- To distribute scarce resources among the industries producing goods and services,
- To distribute the products of industry among members of the community,
- To provide for maintenance and expansion of fixed capital equipment,
- To fully utilize the resources of society.

In a free enterprise system, the fulfillment of these six economic functions is left to the profit motive and the price mechanism working within a framework of social safeguards. In a socialist society all the operations required are consciously planned by official organizations. Many countries operate a mixed economy, splitting the economy into public and private sectors, the activity of the whole being influenced by direct and indirect planning measures. All systems have tended to neglect the increasing abuse of 'free goods' such as air and water, as these social costs have not fallen within the accountancy systems normally maintained. The increasing concern expressed in recent years in respect of environmental effects has found reflection in the policies of all countries, irrespective of political complexion.

Ecosystem: A natural complex of plant and animal populations and the particular sets of physical conditions under which they exist; the organisms of a locality, together with the functionally-related aspects of environment, considered as a single entity. The word "ecosystem" is derived from two words, "ecology" and "system"; the "eco" part of the word implies environment, while the "system" part of the word implies an interacting, interdependent complex. The word "ecosystem" appears to have been coined by A.G. Tansley in 1935 in an article in the journal *Ecology* entitled "The Use and Abuse of Vegetational Concepts and Terms."

Ecumenopolis: The coming city that will, together with the corresponding open land which is indispensable for anthropos, cover the entire earth as a continuous system forming a universal human settlement. The term was coined by Constantinos A. Doxiadis and first used in the October, 1961, issue of *Ekistics*. See *Ekistics*.

Effluent Charge: A fixed fee levied by a regulating body against a polluter for each unit of waste discharged into public waters. The fee may be uniform for all waste producers in the area, or it may be selective according to the composition of individual wastes or the local absorption capacity. The fee may be charged continuously at all times, or it may be levied only when conditions deteriorate below a specified level.

Egypt: See *International Referral System; Sewage Treatment*.

Ekistics: The science of human settlements. The term was coined by Constantinos A. Doxiados, being first used in his lectures of 1942 at the

Athens Technical University. It envisages the human settlement as a living organism having its own laws and, through the study of the evolution of human settlements from their most primitive phase to megalopolis (q.v.) and ecumenopolis, develops the interdisciplinary approach necessary to resolve its problems.

The goal of ekistics is to develop a system and a methodology: (a) to study all kinds of settlements, irrespective of size, location, etc. in order to draw general conclusions about them; and (b) to study each as a whole, without excluding any of its elements in order to illuminate the knowledge of the field and to solve the specific problems of the settlement being studied. The satisfaction of the inhabitants cannot be ensured unless all needs—economic, social, political, technological, and cultural—are substantially satisfied in a balanced way.

Eminent Domain: The superiority of the sovereign power over all property in the nation, whereby it is entitled to appropriate any part required for the public advantage, compensation being made to the owner. Certain public bodies in the United States are authorized by law to condemn (acquire) property when it is in the public interest, giving fair monetary compensation to the owner. For example, a city may condemn land for a needed highway; or a redevelopment agency may condemn land for an urban renewal project. See *Compensation; Compulsory Purchase; Taking Issue.*

Emission Reduction Credits: See *Banking; Emissions Trading Program.*

Emissions Trading Program, US: An attempt to introduce some free market principles into the use of environmental resources. Trading in discharge permits has represented a natural evolution from some innovative administrative procedures known as the "bubble concept," "offsets," and "banking."

The bubble concept implies the importance of maintaining regional air quality goals. Yet it allows variation, in place of uniformity, in respect to emission controls from point sources within a single industrial process and within the region. A relaxation of discharge at one point requires additional compensatory control at another, within the firm or region.

The "offsets" policy has allowed owners of a new plant to operate provided that they obtained equivalent reductions in the discharge of pollutants from the owners of existing plants. "Banking" has allowed those who achieved discharge reductions below the applicable standard to receive a "credit" for these reductions. These credits could be used later in order to exceed in some instance an otherwise prohibited emission rate.

The introduction in the early 1980s of an "emission trading program" consolidated the bubble and offset procedures and the emission reduction credit. The program enables the states to develop, for federal approval, a system that allows firms to trade permits to discharege pollutants. Participation in the program has been voluntary.

The amount of pollution allowable by individual firms in a region is fixed by permits to operate. Once limitations are set, emissions trading programs allow firms to trade among themselves. The limitation on the total quantity of specific pollutants in the atmosphere of a region remains; regions as a whole must continue to meet ambient air quality goals.

The traditional "command and control" regulation approach does not reward improvements "on current requirements." Under an emission trading program, incentives to achieve reductions in emissions below the legal requirements include the ability to expand the firm involved, or to sell the resultant credits to other firms needing them. New growth can be accommodated while regional air quality requirements are preserved.

The main problem with emissions trading and indeed with the bubble concept, is that regional objectives are not the sole objectives. The protection of people and their activities at a local level is often just as important. The "accountancy" process at a regional level might proceed in a satisfactory way, yet create pockets of intense dissatisfaction as emissions are transferred to less desirable locations, or from high level discharges to low level discharges. The purchaser of a credit or permit in respect to fluorides or fine dust may still encounter difficulties in respect to local planning controls governing location, or permit controls governing heights of discharge.

By the end of 1984, five emission banks were operating in the United States. See *Banking; Bubble Concept; Netting; Offsetting.*

Enabling Acts, US: State laws which permit municipalities and counties to plan for and control land use within their corporate areas; such laws may permit also joint exercises in planning between two or more units of government. As this enabling legislation is permissive only, municipalities or counties that wish to undertake planning must adopt an ordinance to establish a planning commission (q.v.) and stipulate its responsibilities.

In the 1920s, the US Department of Commerce published two documents: Standard State Zoning Enabling Act (1927) and Standard City Planning Enabling Act (1923) which greatly influenced state laws subsequently passed.

Endangered Species Act, 1973, US: Amended in 1978, 1982 and 1985, an Act to protect endangered species in the United States; it requires the maintenance of a list of endangered species and empowers the US Fish and Wildlife Service to approve and review state and agency plans for the recovery of endangered species. It authorizes the prosecution of violators of the wildlife trafficking regulations, attempting to combat the blackmarket in thousands of protected reptiles and birds. A current issue is whether Native Americans should continue to have a right to kill endangered species, such as bald eagles, for religious purposes. The US Lacey Act makes the violation of foreign wildlife laws a violation of US law.

Enterprise Zones, UK: Designated areas in Britain in which businesses pay no rates for ten years and escape most of the red-tape of planning permission procedures. By 1985, the British Government had designated eleven zones and announced the pending designation of many more.

For each enterprise zone, a plan must be prepared by the relevant local authority under the Local Government, Planning and Land Act, 1980, showing the classes of development which will be permitted. The plan must be approved by the Secretary of State for the Environment.

Environment: The physical surroundings or circumstances in which humanity struggles to survive and thrive; it includes the planet Earth and outer space as well as the immediate province of living organisms, the biosphere (q.v.). The environment of the individual includes the abiotic factors of land, water, atmosphere, climate, sound, odors and tastes; and the biotic factors of other humans, animals, plants, bacteria and viruses. The European Commission has defined the environment as "the combination of elements whose complex interrelationships make up the settings, the surroundings and the conditions of life of the individual and of society, as they are or as they are felt." The concept has emerged of the environment as a parcel of things which render a stream of beneficial services and some disservices to people, though largely unpriced, and which take their place alongside the stream of services rendered by real income, commodities, houses, infrastructure, transport, and other people. The idea of a beneficial environment, in these terms, which we may enjoy, seems to relate at least in part to that other concept, the quality of life (q.v.).

Environmental legislation generally defines the term "environment" broadly. For example, the Environmental Assessment Act, 1980, of Ontario, Canada, defines the term as follows:

"(i) air, land or water;

(ii) plant and animal life, including man;

(iii) the social, economic and cultural conditions that influence the life of man or a community;

(iv) any building, structure, machine or other device or thing made by man;

(v) any solid, liquid, gas, odor, heat, sound, vibration or radiation resulting directly or indirectly from the activities of man; or

(vi) any part or combination of the foregoing in the interrelationship between any two or more of them, in or of Ontario."

Thus, Ontario legislation goes far beyond the limits of the natural environment and permits an assessment of almost any aspect of a proposed undertaking including its impact on people and society as a whole.

Environment Agency, Japan: An agency created by the Japanese Government in 1971. It is headed by a Director-General with the rank of State Minister. It has jurisdiction over basic policy planning and promotion,

74 Environmental Planning

general coordination of governmental pollution control activities, coordination of budgetary policy for pollution control expenditures, and responsibility for research and investigation.

The Agency established and enforces national environmental quality standards. It enforces the Air Pollution Control Law, the Water Pollution Control Law, and other laws relating to environmental pollution control. The Agency took over the national parks department, the environmental pollution control department, the environmental protection department, and the divisions of water quality monitoring and water pollution control. The Agency has four bureaus relating to planning and coordination, nature conservation, air quality and water quality.

Other ministries and agencies also have environmental responsibilities. For example, the Ministry of International Trade and Industry handles matters concerning industrial land, industrial location, and prevention of pollution. In June 1979, the Ministry announced that it had established administrative procedures which will require electric power companies to draft preliminary and final environmental impact statements for all large power plant projects. Copies would be available to the public for comment.

In July 1979, the Environment Agency announced that it had reached agreement with the Ministry of International Trade and Industry regarding the implementation of the fourth stage of its control program for oxides of nitrogen from stationary sources. The controls now embrace some 92,000 plants, almost three-quarters of all stationary sources of the pollutant. The aim is to reduce the nationwide emission of oxides of nitrogen by 40 per cent, and effectively reduce the incidence of photochemical smog.

An Agency notification of 1974 sets permissible limits for exhausts from motor vehicles, controlling the emission of carbon monoxide, hydrocarbons, nitrogen oxides, and particulates.

In 1979, the Environment Agency announced plans to begin, in 1980, a ten-year study of the environmental implications of 2,500 chemicals commonly used in Japan.

Environment Policy Act, 1970, German Democratic Republic: An Act passed in 1970 by the People's Chamber of the German Democratic Republic (GDR) giving expression to Article 15 of the GDR Constitution which imposes a duty on public authorities to ensure adequate protection of nature, the rational use and protection of land, purity of water and air, protection of flora and fauna, and the natural beauties of the country.

Section 7 of the Act states that enterprises and organizations must make sure that the land and its resources are sensibly and rationally used; the natural environment must remain as unaffected as possible by their activities. Enterprises are accountable to local popular representative bodies when environment protection is involved. Planned activities must be reported to and defended before the deputies representing the population of a country, district, town or village community as the case may be. The local bodies have the power to impose conditions on enterprises for the purposes of protecting the environment. This power was reinforced by an Act dealing with local popular representative bodies passed by the People's Chamber on 12 July 1973.

The GDR began in the mid 1970s to provide substantial sums of money for air and water quality control, reclamation of land, and efficient and safe waste disposal. More than two-thirds of the budget expenditure spent on environment protection are allocated to the areas of industrial concentration of Cottbus, Halle, Dresden, Leipzig, Karl-Marx-Stadt counties and East Berlin, the capital of the GDR.

Section 27 of the Environment Policy Act stipulates that all enterprises are bound to treat their wastewaters before releasing them to within specified limits for pollutants. For this reason, coupled with increasing resources, progressive improvements have been introduced.

Measurements in the GDR have shown that the pollution load by industrial sulfurdioxide and other exhaust gas emissions is several times higher in densely industrialized areas than in "normal" areas. Section 29 of the Environment Policy Act states that "prevention of air pollution by dust, exhaust gases and unpleasant stenches is a permanent obligation of public authorities, economic executive bodies and the enterprises concerned." The Act goes on to state that the air shall not be loaded with pollutants beyond prescribed limits, and equipment for keeping the air clean has to be operated at optimum efficiency. Limit values have been specified for 113 substances. The major effort is being directed at reducing dust and gas emission from power stations, and from chemical, metallurgical and building material manufacturing plants.

The Environment Policy Act also specifies noise reduction as "a permanent obligation of public authorities, economic executive bodies, the enterprises concerned, and the citizens." Maximum permissible values have been worked out and need to be observed. Noise levels have to be considered in advance when new residential areas, roads, and industrial plants and processes are being planned.

Measures to protect and preserve fauna and flora as well as scenic natural beauty are defined in the First Regulation of the Environment Policy Act. A number of creatures are protected by the Non-huntable Wildlife and Wild Plant Protection Act 1970; these include seals, beavers, and wildcats among the mammals, and several birds including all species of the eagle. Public measures taken in recent years to protect the animals which were threatened with extinction have shown positive results, e.g., the revival of the Elbe beaver.

Environment Protection Act, 1969, Sweden: An Act dealing with water and air pollution, noise and vibrations, unwanted light and other disturbances, and the location of hazardous activities. Under the Act, most types of industry require an operating license which requires an application to the National Franchise Board for Environment Protection (q.v.). Significant changes to the Act became effective in 1981; they were mainly concerned with the pretreating of environmentally dangerous activities, monitoring, and penalties for breaches of the law. The Act now provides for a form of economic sanction that can be imposed parallel with a legal sentence; it consists of a fee high enough to offset the profit made by the company as a result of breaching the regulation. See *National Environment Protection Board; National Franchise Board for Environment Protection.*

Environment Protection (Impact of Proposals) Act, 1974-75, Australia: Legislation passed by the Australian Federal (known as the Commonwealth) Parliament which seeks to ensure that "matters affecting the environment to a significant extent are fully examined and taken into account in and in relation to:

(a) the formulation of proposals;

(b) the carrying out of works and other projects;

(c) the negotiation, operation and enforcement of agreements and arrangements. . . . ;

(d) the making of, or in the participation in the making of, decisions and recommendations and

(e) the incurring of expenditure

by or on behalf of the Australian Government and authorities of Australia, either alone or in association with any other government authority, body or person."

Until 1978 the only legislation of its type in Australia, the Act provides for environmental impact statement (EIS) techniques to be applied to the proposals of the Commonwealth Government, where a significant impact may be expected. The Act is not limited to proposals being developed by the Commonwealth, extending to projects partly financed by the Commonwealth and to private sector activities where these require Commonwealth approval. Such approval is required, for example, where a company needs exchange control approval for the import of foreign capital for a development project, or where export licenses are needed. The breadth of the Act is illustrated also in the definition of the word "environment." The word is defined in the Act as including "all aspects of the surroundings of man, whether affecting him as an individual or in his social groupings." A large number of projects have now been assessed under this legislation, some jointly with the States.

The individual Australian States, as well as the Australian Government have also introduced environmental impact procedures. See *Table 1.*

Environment Protection (Sea Dumping) Act, 1981, Australia: Australian federal legislation which came into effect in March, 1984; under the Act, it is an offense to dump into the sea a wide range of specified substances including mercury, cadmium, polychlorinated biphenyls (PCBs)

Table 1: Environmental Impact Assessment Procedures in Australia

	Jurisdiction	Source	Date
1.	The Commonwealth (including the Australian Capital Territory and Northern Territory)	(a) *Environmental Protection (Impact of Proposals) Act* 1974-75	December 1974
		(b) Order approving administrative procedures under the above Act	June 1975
2.	Victoria	(a) *Environment Effects Act* 1978	May 1978
		(b) Guidelines for Environment Assessment	November 1978 Revised edition, January 1980
3.	New South Wales	(a) Environmental Planning and Assessment Act, 1979, replacing	November 1979
		(b) Environmental Standard, EI-4	October 1974
4.	Tasmania	Guidelines and Procedures for Environmental Impact Studies	undated (published 1975)
5.	Western Australia	Department of Environment and Conservation, Bulletin No. 38: Procedures for Environmental Assessment of Proposals in Western Australia	September 1978
6.	Queensland	Co-ordinator-General's Department, Impact Assessment of Development Projects in Queensland	March 1979
7.	South Australia	(a) No formal guidelines [recommendation of Environmental Protection Council (S.A.) to Cabinet for the introduction of legislation remains in effect, but unimplemented]	December 1973
		(b) South Australian Department for the Environment, Environmental Impact Assessment Handbook (for internal distribution to officers of the Department)	May 1978

and other organohalogen compounds, oil hydraulic fluids, etc. Other substances may not be dumped into the sea except under the conditions of a permit issued by the Minister for Arts, Heritage, and Environment. The Act covers the deliberate dumping of wastes into the sea from ships, aircraft and platforms; wastes arising from normal operations are excluded from control. The Act also covers the incineration of wastes at sea, and the disposal, by sinking, of ships, aircraft or other structures or objects.

Environmental Administration, Soviet Union: A form of environment administration without a single central federal coordinating agency; duties were initially distributed between agencies and ministries by the Decree on the Intensification of Nature Conservation and Improved Utilization of Natural Resources of 1972. See *Sanitary Epidemiological Service and Table 2*.

Environmental Assessment, US: A concise public document prepared by a federal agency to briefly provide sufficient evidence and analysis for determining, in respect to a project, whether an environmental impact statement (q.v.) should be prepared or whether there should be a finding of "no significant impact."

The term "environmental assessment" has replaced earlier comparable terminology used by individual agencies, e.g., "survey" (Corps of Engineers); "environmental analysis" (Forest Service); "normal or special clearance" (Housing and Urban Development); "environmental analysis report" (Interior); and "marginal impact statement" (HEW).

Environmental Assessment Act, 1975, Ontario: In the province of Ontario, Canada, an Act introducing environmental impact assessment procedures for major projects, procedures that would operate alongside an existing set of extensive planning controls administered by local municipalities. However in June, 1977, a government committee review of the Ontario Planning Act noted the possibility of extensive conflict and duplication in application and hearing procedures between the Environmental Assessment Act and the Planning Act. It expressed concern that the existing regulatory system, already bound to take a whole range of factors into account before making a development decision, might be abandoned in favor of one in which the final decision would be made by the Minister of the Environment. In practice, many decisions are made by the municipalities with the more important and significant cases only being determined by the Minister.

Environmental Assessment Board, Ontario: An independent administrative tribunal which reports to the Ontario legislature, Canada, through the Minister of the Environment. It conducts public hearings on environmental issues under the Ontario Water Resources Act, the Environmental Protection Act, the Environmental Assessment Act, and the Consolidated Hearings Act. At the end of 1984, the board had 18 members,

Table 2: The Soviet Union—Allocation of Environmental Responsibilities

Agency	Responsibility
Ministry of:	
Public Health (Sanitary Epidemiological Service)	Public health. Sanitary regulations for waste dumps and industrial wastes. Control over sanitary state of waters.
Land Reclamation and Water Management	Water pollution control. Water resources management and development. Operation of sewage treatment plants.
River Basin Boards (over 100 established)	Regulations concerning water conservation and multipurpose use of water resources. Water consumption. Wastewater discharges. Improvement of wastewater treatment processes. Control of capital investment in wastewater treatment.
Agriculture	Land use. Fertilizers, pesticides and herbicides. Biological control of pests.
Fisheries	Fish resources. Protection of the natural resources of the continental shelf.
Chemicals and Oil	Development and manufacture of pollution control equipment.
Education	Environmental education.
State Committees for:	
Atomic Energy	Nuclear safety.
Forestry	Forest resources. Fire prevention control. Environment control.

(continued)

Table 2: (Continued)

Agency	Responsibility
Safety in Industry and Mining	Mineral resources. Environment control.
Geology	Observation of groundwater regimes. Coordination of plans for water well drilling. Protection of groundwater against pollution and depletion.
Science and Technology	Scientific research on rational use of natural resources and nature protection. Interdepartmental committee on protection of the natural environment and the rational use of natural resources.
USSR Council of Ministers:	Water conservation. Air pollution control. Measures to reduce toxicity of exhaust gases from vehicles. New parks, gardens, boulevards, green belts and forest parks.
Weather Service (under USSR Council of Ministers)	Observing and monitoring the level of pollution of the atmosphere, the soil and bodies of water.
Department of Hydrometeorological Services (under USSR Council of Ministers)	Studies of surface water resources. Account and analysis of natural water quality monitoring.
Republic Ministries of Housing and Utility Services:	Provision and management of water supply and sewage facilities of cities and townships. Monitoring.
Executive Committees of the local Soviets of Working People's Deputies.	Water utilization and conservation. Monitoring.
Councils of Ministers of Union & Autonomous Republics.	Water utilization and conservation.

Source: Derived from *Decree on the Intensification of Nature Conservation and Improved Utilization of Natural Resources*, September, 1972; Ozeranskiy et al. (1977), and later documents.

four of whom, including the chairman, were full-time. The board publishes its own annual report.

Environmental Conservation Law (ECL), 1977, Korea: A comprehensive measure introduced in the Republic of Korea in 1977, replacing the earlier Public Nuisance Prevention Law. The ECL was revised in 1979, transferring its administration from the Minister of Health and Social Affairs to the Administrator of the Office of Environment which was simultaneously created.

The ECL includes many new provisions for environmental quality standards, environmental impact assessment, regular monitoring, designation of special countermeasures zones, surveys, and preservation of ecology. The Law also incorporates provisions to establish an Environmental Conservation Committee under the Prime Minister among others responsible to deliberate environmental master plans, comprehensive policies and coordinate interministrial problems. Furthermore, the establishment of an environmental research institute, self-monitoring of pollutants, and the administration of environmental pollution-prevention industries are provided for.

The ECL also broadens the scope of environmental pollution control. It contains provisions about industrial wastes, motor vehicle emissions and nightsoil of livestock which used to be dealt with in different laws before the promulgation of the ECL. Soil pollution and community noise are also addressed. In addition, liability without fault for compensation of damages to humans was provided in the ECL.

In 1981, the ECL was strengthened mainly through the partial introduction of a pollution charge system, the promotion of an environmental pollution prevention fund and the establishment of the Environmental Pollution Control Service Corporation. See *Han River Basin Environmental Master Plan, Korea; Saemaul Undong (New Community Movement), Korea.*

Environmental Economics: A recognized field of specialization in the economics discipline with its own journals, including the *Journal of Environmental Economics and Management.* It embraces the issues of pollution control and protection of the natural environment; issues in the resolution of which markets have played little or no part, but in which immense common property resources need to be allocated sensibly for the overall public good.

Environmental Education: See *United Nations Educational, Scientific and Cultural Organization.*

Environmental Effects Statement (EES): An alternative and more accurate title for an environmental impact statement (q.v.). The word "impact" while implying strong effects or influences also conveys the impression of sudden events. Such characteristics are usually confined to blasts and explosions. Consequently, the word "effects" is a more precise description of the short- and long-term, direct and indirect, influences of a project or activity. The EES has been consistently used in Victoria, Australia, since its introduction by the author in 1973.

Environmental Health: Those aspects of personal and community health which are influenced by environmental factors (such as contaminated water), in contrast with those aspects of health of an individual nature (such as alcoholism); the term may also be used to describe the application of principles and methods aimed at promoting and safeguarding the health of the community at large. Environmental health science has depended for its success in part upon the study of disease in the group, that is, upon epidemiology. It has sought to protect the public from disease through improved sanitation and hygiene, anti-vector campaigns, and immunization programs.

Environmental Health Impact Assessment (EHIA): An assessment of the impact on the environment and people of elements in a project recognized as having potentially marked health significance; the health component is often a weak point in environmental impact assessments. In 1982, the World Health Organization (q.v.) recommended that EHIA studies should be conducted for all major development projects. Health assessment embraces the following:

- Risks and hazards, direct and indirect, involving explosion, fire, shock, heat, blast, vibration and destruction of property beyond the boundaries of the plant;
- Biological factors such as parasites, helminths, protozoa, bacteria, mycobacteria, rickettsia, and viruses;
- Chemical agents: toxic, carcinogenic or mutagenic chemicals and heavy metals;
- Physical agents: ionizing and non-ionizing radiation, noise, dust and other irritants, excessive temperatures or humidity.

The possible implications for human health may be measured in terms of mortality and morbidity. Many consider that the adverse health effects of the Aswan High Dam, Egypt, was a neglected item in the environmental impact assessment. See *Tables 3 and 4*.

Environmental Heritage: Those buildings, works, relics or places of historic, scientific, cultural, social, archaeological, architectural, natural or aesthetic significance for the city, county, province, state, or nation.

Environmental Impact Assessment: The critical appraisal of the likely effects of a proposed project, activity, or policy on the environment, both positive and negative. Assessments are carried out independently of the proponent, who may have submitted an environmental impact statement (q.v.), for the purpose of assisting the consent or permit authority. The

Table 3: Checklist of Environmental Health Factors That May Be Affected by Development Projects

Effects on inhabitants of project area

    Communicable disease
    Housing and sanitary facilities
    Dietary change
    Effects on groundwater
    Changes in ecological balance
    Changes in agriculture
    Increased risk of road accidents
    Risks to community health from certain industrial processes

Effects on workers

    Work accidents
    Exposure to chemical and physical hazards
    Nutritional status of workers

Indirect effects

    Introduction of new disease vectors
    New infection or reinfection of existing vectors
    Increased propagation and spread of existing vectors

Effects on existing health services

Source: Based on the World Bank Report, 1982.

Table 4: Environmental Health Impact Assessment—Main Steps

1. To assess direct impacts on environmental parameters
2. To assess indirect impacts on environmental parameters
3. To screen environmental parameters which have a health significance (EH factors)
4. To assess increase of exposure
5. To assess increase in risk-group populations
6. To assess health impacts (mortality and morbidity) (ref. risk assessment studies)

Source: Eric Giroult, 1984.

consent authority may be a local authority, or a state or federal government or one of its agencies. Factors to be taken into account in making assessments include:

- Any environmental effects on a community;
- Any environmental impact on the ecosystems of the locality;
- Any diminution of the aesthetic, recreational, scientific, or other environmental quality of value of a locality;
- Any effect upon a locality, place or building having aesthetic, anthropological, archaeological, cultural, historical, scientific, or social significance or other special value for present or future generations;
- The endangering of any species of fauna or flora;
- Any long-term effects on the environment;
- Any risks or hazards which might endanger the safety of the environment on some future occasion;
- Any curtailing of the range of beneficial uses of the environment;
- The effects of any pollution on the environment;
- Environmental problems associated with the disposal of waste;
- Any cumulative environmental effects with other existing or likely future activities;
- Any implications for resources, natural or otherwise, which are, or are likely to become, in short supply.

Environmental Impact Statement: A considered report, following careful studies, disclosing the likely, possible, or certain effects of a proposed project, activity, or policy on the environment, thus alerting the decision-maker, the public, and the government to the environmental risks involved; the findings, particularly after independent environmental impact assessment (q.v.), often enable better informed decisions to be made. A decision may involve the rejection or deferment of the proposal, though more usually the approval of the project subject to a range of conditions which are attached to the development consent or permit. Environmental impact statements are prepared in written form, frequently by consultants, and signed by the person undertaking or coordinating the study. The contents of an EIS customarily include the following:

- A full description of the proposed project or activity;
- A statement of the objectives of the proposed activity;
- A full description of the existing environment likely to be affected by the proposal;

- The identification and analysis of the likely environmental interactions between the proposed activity and the environment;
- The justification of the proposal in terms of environmental, economic and social considerations;
- The measures to be taken in conjunction with the proposal for the protection of the environment and an assessment of the likely effectiveness of those measures;
- Any feasible alternatives to the proposal;
- The consequences of not carrying out the proposal for the proponent, the region, and the state.

In addition to an independent environmental impact assessment (q.v.), a controversial and highly sensitive issue may involve an independent public inquiry (q.v.).

The principle of the environmental impact statement was introduced to the United States of America through the National Environmental Policy Act, 1969 (q.v.). Its use is widely established in other countries, including Canada and Australia. It has made limited progress only in the European Economic Community (q.v.).

The Environmental Assessment Act, 1980, for Ontario, Canada, outlines the contents of an environmental impact statement as follows:

5(3)"An environmental assessment submitted to the Minister pursuant to subsection (1) shall consist of

(a) a description of the purpose of the undertaking;

(b) a description of and a statement of the rationale for

  (i) the undertaking,

  (ii) the alternative methods of carrying out the undertaking, and

  (iii) the alternatives to the undertaking;

(c) a description of

  (i) the environment that will be affected or that might reasonably be expected to be affected directly or indirectly,

  (ii) the effects that will be caused or that might reasonably be expected to be caused to the environment, and

  (iii) the actions necessary or that may reasonably be expected to be necessary to prevent, change, mitigate or remedy the effects upon or the effects that might reasonably be expected upon the environment, by the undertaking, the alternative methods of carrying out the undertaking and the alternatives to the undertaking; and

(d) an evaluation of the advantages and disadvantages to the

86   Environmental Planning

environment of the undertaking, the alternative methods of carrying out the undertaking and the alternatives to the undertaking."

It was not until the 1980s that Britain adopted the principle of separate environmental impact statements and assessments for major projects. The official view had been that the system of planning and development control already embodied, by and large, a consideration of the relevant environmental factors. It was recognized, however, that advantages might derive from some kind of environmental assessment as a separate exercise in respect to a limited number of major developments and this has gradually been introduced. The principle is not yet in general application.

Environmental Laws: Laws generally relating to the control of pollution, nature conservation, and environmental impact statements, as distinct from those relating to city and regional planning. Table 5 indicates the years of introduction of these more specialized pieces of legislation in some fourteen countries.

Environmental Master Plan, Pennsylvania, US: A statement of environmental quality goals and an "environmental agenda" adopted by the Commonwealth of Pennsylvania in July, 1982. The plan was prepared by the state's Environmental Quality Board, through a process of review and public consultation. The Plan has no statutory standing; it is a series of guidelines oriented toward protecting environmental values and natural resources. It is a broad policy plan designed to guide State, regional and local planning programs and individual decisions which determine Pennsylvania's environmental future. More specifically, the Plan deals with:

- Protection of the Heritage (including clean air resources, special protection watersheds, wetlands, natural areas, primitive areas, and scenic areas).

- Conservation of Natural Resources (including farmlands, metropolitan open space, forests and woodlands, fish and wildlife, coal resources, other mineral resources, landslide prone areas, areas with carbonate geology, floodplains, areas with limited water supply, groundwater, and lakes and reservoirs).

- Restoration of Degraded Areas and Protection of Environmental Health (including mined lands, solid and hazardous waste, air pollution, water pollution, drinking water, and radiation).

Environmental Planning: The identification by the community of desirable objectives in respect to the physical environment, including social and economic considerations, and the creation of administrative procedures

Table 5: Major National Environmental Laws in Selected Countries

Country	General	Water	Air	Impact Statements	Nature Conservation	Other
United States of America		1972 1977	1963 1970 1977	1970		1972[a] 1976[b,c]
New York City	1971					
New York State	1970			1975		
Pennsylvania			1946			
United Kingdom	1974	1973	1956 1968		1949	1972[b] 1974[a]
Sweden	1969	1969	1969	1969	1964	1970[b]
West Germany	1974	1957 1976	1974 1981	1975	1976	1977[d] 1975[e]
Canada		1970	1971	1973		1975[g]
Ontario	1971	1970		1975		1973[e]
Australia				1974		
New South Wales	1971	1970	1961	1979		1977[f], 1975[a]
Victoria	1970	1970	1958	1977		
Queensland		1971	1963			
Israel	1965					
South Africa		1971			1971	
Soviet Union	1972	1970				
China	1979	1979	1979	1979		
Japan	1967 1970	1958 1970	1962 1968		1972	1973[g]
Malaysia	1974	1974	1974			
Philippines	1977	1977	1977		1977	1977[d]
Brazil	1973	1976	1976			1979[d]

a noise
b toxic substances
c resource recovery
d wastes
e pesticides
f heritage
g chemicals

88   Environmental Planning

and programs to achieve those objectives. Matters embraced by the subject include city and regional planning (or town and country planning); land use; transportation; employment; health; growth centers and new towns; population and national settlement policies; locational problems; industrial and urban development; state environmental planning policies; regional and local comprehensive plans; planning permit procedures; development control codes; zoning ordinances; subdivision regulations; building codes; housing standards; urban renewal; community development programs; welfare policies; human resource development; growth management concepts; living resource conservation; landscape conservation; heritage conservation; wilderness, national and marine parks; pollution control strategies; environmental impact statements and assessments; public hearings and inquiries; appeal mechanisms and procedures; and international conventions relating to environment protection and pollution control.

Planning having been traditionally associated with the use of land, some writers continue to present it in such terms. For example, it has been defined by Lincoln Allison in his *Environmental Planning* (London, 1975) as "the processes and patterns of action through which the use of land is controlled in a nation-state." However, with the broadening of concerns it may be now argued that land use is simply the principal means by which desirable objectives are attained, and should not be elevated even by inference to the status of an end in itself. A planning concept should reflect a growing concern for the social and economic consequences of physical development.

The Royal Town Planning Institute of the United Kingdom has avoided a preoccupation with land use and land use planning by describing "environmental planning" as a "unique concern with the physical environment of town and country" offering, through the consideration of wider concerns, an "increasingly complex task." In 1972, the Greater London Council, as a major metropolitan planning authority, expressed it even more broadly: "The field of environmental planning embraces all aspects of physical and social problems which have a direct bearing on the quality of our surroundings."

Most writers today recognize the concept of "environmental planning," though not always in the all-embracing sense used here. In such instances, environmental planning is often regarded as an element in a planning trilogy (q.v.) comprising land use planning, transportation planning, and environmental planning. Environmental planning tends to be regarded as embracing such issues only as pollution control, protection of the natural heritage, creation of parks and urban open space, flood control, risk and hazard management, and urban blight.

However, environmental planning should not be viewed as a new and separate sector of traditional planning. Rather, concern for the environment should permeate all facets of the planning process. It is a superior term to *comprehensive planning* in confining the area of concern to the outside, physical world; yet it is also superior to the term *physical planning* in embracing some regard for social and economic considerations. The framework of environmental planning is comprehensive allowing the maximum practicable attainment of economic, social, and physical objectives. See *Comprehensive Plan; Corporate Planning; Development Control; Development Standards; Environmental Impact Statement; Environmental Planning, Objectives of; Heritage Conservation; Land Use Planning; Living Resource Conservation; Local Plans; National Park; New Towns; Planning Commission, US; Planning System, UK; Policy Plan; Pollution Control Strategy; Regional Planning Commission; Structure Plan; Subdivision; Town and Country Planning; Transport Planning, Objective of; Urban Renewal; Zoning; Zoning Ordinance.*

Environmental Planning and Assessment Act, 1979, New South Wales: Comprehensive planning and environmental assessment legislation introduced by the New South Wales Government, Australia, in 1979. It created a Department of Environment and planning (q.v.) and the position of Commissioner of Inquiry (q.v.). The objects of the Act are:

(a) to encourage—
   (i) the proper management, development and conservation of natural and man-made resources, including agricultural land, natural areas, forests, minerals, water, cities, towns and villages for the purpose of promoting the social and economic welfare of the community and a better environment;
   (ii) the promotion and coordination of the orderly and economic use and development of land;
   (iii) the protection, provision and coordination of communication and utility services;
   (iv) the provision of land for public purposes;
   (v) the provision and coordination of community services and facilities; and
   (vi) the protection of the environment;
(b) to promote the sharing of the responsibility for environmental planning between the different levels of government in the State; and
(c) to provide increased opportunity for public involvement and participation in environmental planning and assessment.

The Act defines the planning system as comprising local environmental plans, regional environmental plans, and state environmental planning policies; it sets out development assessment procedures and the powers to impose conditions on developments that proceed. It requires the preparation of environmental impact statements in respect to a wide range of prospective developments and provides for public participation and public inquiries. The Act replaced earlier town planning schemes and interim development orders with "environmental planning instruments." See *Land and Environment Court; Local Environmental Plans; Regional Environmental Plans; State Environmental Planning Policies.*

Environmental Planning Instrument: A statutory document or regulation implementing, e.g., a state environmental planning policy, a regional environmental or structure plan, or a local environmental plan.

Environmental Planning, Objectives of: The aims of environmental planning as they have evolved during the twentieth century. They may be defined as follows:

- The pursuit of social improvement through the medium of ambient physical change maximizing opportunities for individual choice and protecting the individual from the adverse effects of the actions of others.
- The orderly arrangement of various parts of the city and region (residential, commercial, industrial, recreational) to enable each part to perform its functions at least cost and with minimum conflict.
- The provision of an efficient system of transportation and communication within the city and to other centers.
- The achievement of optimal standards in respect of infrastructure, lot sizes, building spacing, sunlight, open space, parking facilities, and aesthetics.
- The provision of comfortable housing in a variety of types to meet the needs of all families and other categories.
- The provision of schools and colleges, recreational and other community services of a high standard in terms of location, size and quality with regard to special needs.
- The provision of a safe and adequate water supply, sewerage system, energy system and other public services.
- The achievement of desirable objectives in respect to the control of air pollution, water pollution, noise and vibration control, and the transport and disposal of hazardous wastes.
- The progressive development of landscaping, tree planting, anti-litter measures, bicycle paths, nature walks, national parks, community access to natural features, and recreational schemes.
- Where relevant, the adoption of special measures to protect the coast and beaches and ensure easy access by the public.

- The identification, preservation and restoration of items of the environmental heritage, either individually or as parts of conservation areas.
- The identification and clearance of substandard dwellings incapable of repair or improvement at reasonable expense, and the rehabilitation of blighted areas.
- The coordination of the proposals of single-purpose agencies concerned with the provision of services such as water, sewerage, and electricity; and the provision of transport by road, rail, sea and air.
- The identification of natural resources and the prospective future demands for those resources with a view to reasonable precautions against the sterilization and alienation of the most valuable of them, so that at some future time those resources are accessible for mining, fishing, extraction, harvesting, or felling.
- Consistency between local environmental plans and any regional, provincial, state, federal or national plans and policies, and relevant international conventions.
- The provision of ample opportunities for public participation (q.v.) at all stages of the planning process, including public hearings and inquiries into "policy and need" as well as individual projects.
- The retention of flexibility within the planning system to cope readily with changing economic and social circumstances.
- The promotion of a balance of population and industry throughout the region, having regard to employment prospects and characteristics, and economic viability.
- The achievement of a fair and reasonable distribution of infrastructure and site development costs between the public and the developer.
- Where appropriate, the adoption of special measures for the protection of wildlife and wilderness.

See *Environmental Planning*.

Environmental Planning System: An organizational and legislative structure within which the environmental decision-making process takes place. In a generalized way, the system's objective is to provide the best framework for making planning decisions. Specifically, the following features are desirable in a planning system:

- It should be based on social, economic and environmental aims.

- It should avoid an over-centralized decision-making process, separating as far as possible state-wide, regional and local issues.
- It should provide for public involvement.
- It should be easily understood, contained in as few acts and regulations as possible and allow for speedy plan preparation and development decisions.
- It should encourage the guiding rather than the restrictive aspect of planning.

The overall planning body, agency, commission, or department, should be responsible for:

- Coordinating the state's policies for environmental planning and land resource management in the state and regional context;
- Coordinating information and activities of other government departments at the state and regional level;
- Integrating land use planning with population distribution and policies on transport and other services;
- Preparing regional environmental plans and sub-regional structure plans in cooperation with the regional development councils (if they have planning functions);
- Exercising all regional environmental planning responsibilities (if the regional development council does not have planning functions);
- Contributing, with others, to research, resource assessment, technical and administrative advice;
- Ensuring that local plans are amended by the council concerned where any inconsistency with State and regional environmental plans is evident;
- Certifying that local plans comply with State, regional or subregional plans;
- Preparing quickly interim guidelines for local plans until more sophisticated regional environmental plans are prepared;
- Preparing guidelines for development control by local councils.

See *Comprehensive Planning; Planning System, UK.*

Environmental Policy Act, 1971, North Carolina, US: An Act to encourage the wise productive, and beneficial use of the natural resources of North Carolina without damage to the environment while preserving the natural beauty of the state. Stage agencies were to prepare environmental impact statements in respect to projects likely to have significant adverse environmental effects, outlining measures to minimize those effects and discussing the alternatives to the project. Copies of such statements were to be made available to the public and to counties, municipalities, and institutions. In 1981, the General Assembly extended the measure through 1990.

Environmental Precincts: Or environmental areas, being convenient parts of a local government jurisdiction which either have a uniform character or are identified by some outstanding feature. The boundaries of such precincts may be natural (such as parks, rivers, cliffs, or ridge lines) or artificial (roads, expressways, shopping centers, and large institutions). Social characteristics and links, important community facilities such as schools, parks, or hotels, may also be reflected in the division of the larger area into parts. The division of a local government jurisdiction (or municipality) into environmental precincts or areas may then form a convenient basis for planning.

The nature of environmental areas will vary. The task of the planning authority is to use judgement, based on local knowledge and experience in deciding which areas should retain their existing character, which should change slightly, which substantially and which completely. In new areas, where any development means change, the question becomes not so much the extent of change as the intensity and kind of development appropriate to the area. Certain planning criteria will help in this judgement. The local government area's location within the wider region and its proper role in the region's growth will determine the extent of pressures to change. There is little point in making large areas available for high density residential development in places where there is no demand for this housing form. Conversely, in areas in which strong pressures for redevelopment exist, plans should make some attempt to accommodate them in the most suitable areas; for if this is not done, the pressures will persist and may in time become so strong as to force accommodation in less suitable areas than would have been the case, had the plan acknowledged their existence.

See Figure 5 for a hypothetical division of a local government jurisdiction into environmental precincts.

Environmental Protection Agency (EPA), US: An agency created by the United States Government in 1970; its purpose was to protect and enhance the environment to the fullest extent possible under the laws enacted by Congress. Its mandate was to mount an integrated, coordinated attack on environmental pollution in collaboration with state and local governments. The Agency began with a staff of 6,000 people located in 48 of the 50 States, following the amalgamation of some fifteen separate agencies or parts of agencies.

The new agency became responsible for the management of the federal

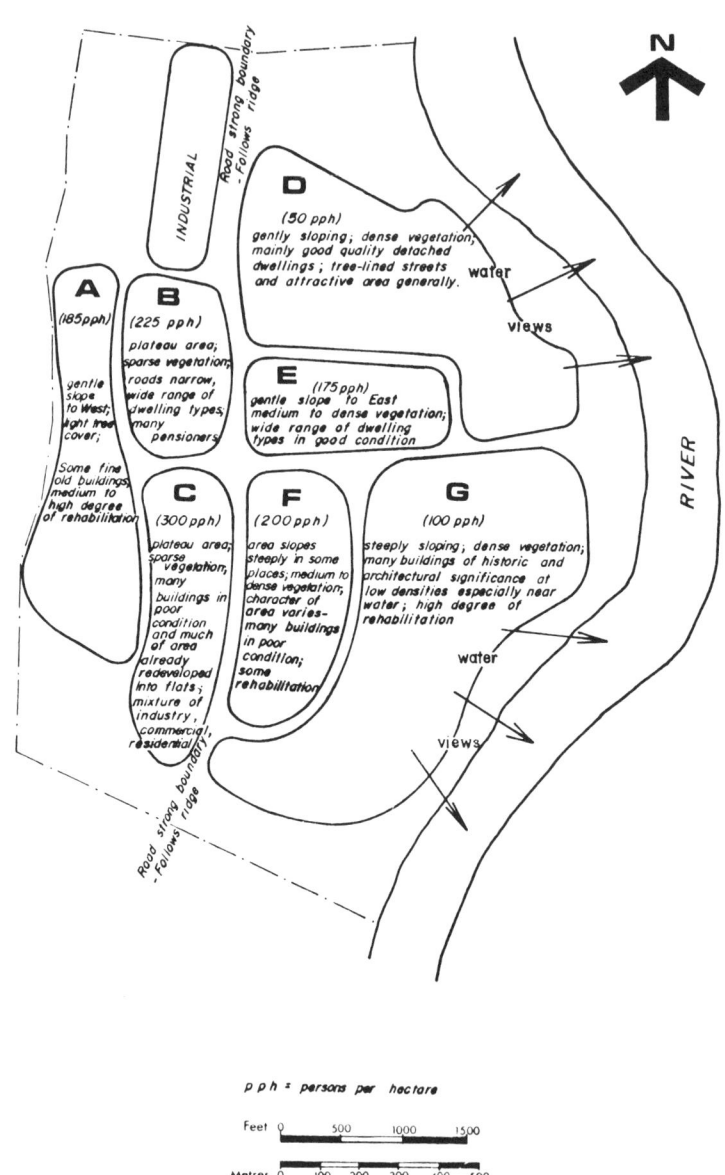

Figure 5: Hypothetical division of a local government jurisdiction into environmental precincts.

programs for air and water pollution, solid waste disposal, pesticide registration, toxic substances, the setting of radiation standards in the environment, and noise control. The agency also accepted responsibility for environmental impact assessment policy, although the public has been given an important say in environmental matters and the courts have been heavily involved.

The activities of the agency today include the development of national programs, technical policies, and regulations for air pollution control; development of national standards for air quality, emission standards for new stationary sources, and emission standards for hazardous pollutants; technical direction, support, and evaluation of regional air activities; and the provision of training in the field for air pollution control personnel. Related activities include the study, identification, and regulation of noise sources and control methods; technical assistance to states and agencies conducting radiation protection programs; and national surveillance and inspection programs for measuring radiation levels in the environment.

In respect to water quality, the agency's functions include the development of national programs, technical policies, and regulations for water pollution control and water supply; water supply quality standards and effluent guidelines; provision of technical direction, support, and evaluation of regional water activities; provision of training in the field of water quality; analyses, guidelines, and standards for the land disposal of hazardous wastes; assistance in the operation of waste management facilities; and analyses on the recovery of useful energy from solid waste.

The agency is responsible for the development of national strategies for the control of toxic substances; criteria for assessing chemical substances, standards for testing, and the rules and procedures for reporting and regulating substances; evaluating and assessing the impact of new chemicals and chemicals with new uses to determine the hazard. It coordinates activities under the Toxic Substances Control Act. Additional activities include the control and regulation of pesticides to protect environmental quality; establishment of the tolerance levels for pesticides in relation to food; monitoring of pesticide residues in food, humans, and non-target fish and wildlife and their environments; and the investigation of pesticide accidents. The agency is required to compile a list or inventory of chemicals manufactured, imported or processed in the United States. Manufacturers and importers are required to give notice to the Environmental Protection Agency of an intention to manufacture or import a substance not on the list.

The agency undertakes enforcement procedures in respect to control programs in air, water (including groundwater), toxic substances, solid waste management, radiation, and noise. It conducts conferences, hearings, and other legal proceedings. It also conducts a national research program to further the technological control of all forms of pollution, using the facilities of its own laboratories as well as those of others through contract procedures.

A 1985 EPA report estimated that the cost of meeting federal air and water pollution standards over the 10 years from 1981 to 1990 would be about $526 billion.

Environmental Protection Council, Ghana: A Council created by the government of Ghana in 1974, bringing together for the first time under one national body all activities and efforts aimed at protecting and improving the quality of the environment.

The details of the Council's functions as spelt out by decree are as follows:

- To advise the Government generally on all environmental matters relating to the social and economic life of Ghana;

- To coordinate the activities of all bodies concerned with environmental matters, and to serve as a channel of communication between those bodies and the government.

- To conduct and promote investigations, studies, surveys, research and analysis (including the training of personnel), relating to the improvement of Ghana's environment and the maintenance of sound ecological systems;

- To serve as the official national body for cooperating and liaising with other national bodies and international organizations on environmental matters;

- To undertake such studies and submit such reports and recommendations with respect to environmental matters as government may request;

- To embark upon general environmental educational programs for the purpose of creating an enlightened public opinion regarding the environment and an awareness of the public's individual and collective role in its protection and improvement;

- Without prejudice to the economic and social advancement of Ghana, to ensure the observance of proper safeguards in the planning and execution of all development projects, including those already in existence, that are likely to interfere with the quality of the environment.

The Council operates through expert committees drawn from its own membership as well as specialists from outside, e.g. the universities, research institutes, ministries, government departments and private bodies to deal with specific subjects within the Council's program of activities.

Environmental Protection Districts: Areas of a community specially designated primarily on the basis of their environmental characteristics;

these characteristics may be natural hazards such as floodplains, areas of potential mineral extraction, or areas with a limited groundwater supply.

Environmental Protection Law, 1979, China: A measure passed and promulgated in September, 1979, by the Standing Committee of the National People's Congress; the law is based on Article 11 of the Constitution of the People's Republic of China.

The law uses the term *environment* to encompass the air, water, land, mineral resources, forests, grasslands, wild plants and animals, aquatic life, places of historical interest, scenic spots, hot springs, resorts and natural areas under special protection as well as inhabited areas of the country.

The law stipulates that, while developing the economy according to a national plan, it is necessary for the state to adopt measures to protect and improve the environment, to entrust the implementation of these measures to certain units, and to control and eliminate existing pollution and other public hazards step by step in a planned way.

More efforts, the law stresses, must be made to control and prevent pollution from liquid and gaseous wastes, slag, dust, sewage, radioactive material and other harmful matter as well as pollution from noise, vibration and toxic odors.

The law stipulates that no enterprise or institution which might pollute the environment can be built near residential areas in cities and towns or beside protected water areas, places of historical interest, scenic spots, hot springs, resorts or natural areas under protection. Those already existing should adopt measures to control their pollution or be moved to other places within a specified period.

As pollutants from industrial enterprises are the main cause of public hazards, it is therefore necessary strictly to keep this source of pollution under control. The law requires the adoption of a series of measures to control and prevent pollution, including developing new technology and methods of processing as well as new products which will be totally or practically pollution-free. It calls for multi-purpose use of gaseous wastes, liquid and slag. Effective measures are to be taken to deal with smoke and dust, devices are to be installed to reduce noise and vibration. Cleaner energy sources such as coal gas, liquefied petroleum gas, natural gas, methane gas and solar energy are to be developed; the dumping of rubbish and slag into rivers, lakes or the sea will be strictly forbidden; more highly effective insecticides with low residual toxicity are to be produced and industrial dust and toxic gases in working areas are to be reduced to standards set by the nation.

Since China has suffered serious ecological damage, the law provides that land must be used rationally according to local conditions, soil erosion prevented and land kept from being turned into deserts. It also stipulates that rivers, lakes, seas and reservoirs must be protected and a good quality of water maintained. Forest and grassland resources and certain wild animals and plants are to be protected and developed, and great care is to be taken in exploring and utilizing mineral resources.

Article 6 of the Environmental Protection Law addresses the necessity of environmental impact reports. Specifically:

Article 6
All enterprises and institutions shall pay adequate attention to the prevention of pollution and damage to the environment when selecting their sites, designing, constructing and planning production. In planning new construction, reconstruction and extension projects, a report on the potential environmental effects shall be submitted to the environmental protection department and other relevant departments for examination and approval before designing can be started. The installations for the prevention of pollution and other hazards to the public should be designed, built and put into operation at the same time as the main project. Discharge of all kinds of harmful substances shall be in compliance with the criteria set down by the State.

The units which have caused pollution and other hazards to the environment shall, according to the principle of 'whoever causes pollution shall be responsible for its elimination', make plans to actively eliminate such, or alternatively submit an application to the competent authorities for approval to transfer the property or move to some other place.

Article 7 also addresses environmental impact studies, as follows:

Article 7
In rebuilding old cities or building new ones, assessments shall be made of the potential environmental effects in industrial and residential areas, public utility facilities, and green belts by reference to the meteorological, geographical, hydrological and ecological conditions, and overall planning and rational layout be made to prevent pollution and other hazards to the public so as to build a clean modern city in a planned way.

Environmental Protection Service, Israel: An agency established in 1972 by the Israeli Government. Initially, its function was that of an advisory body to Government and decision-makers, aiding in the development of a comprehensive national environmental policy. It became increasingly clear, however, that legal authority, enforcement powers and concentration of responsibility would be necessary to implement any national policy. In 1976, the Service was transferred from the Prime Minister's Office to the Ministry of the Interior with its important planning functions. The move had considerable merit, for the Ministry is responsible for:

- A comprehensive physical planning system, integrating local, district, and national levels of planning, administered under the Planning and Building Law of 1965. Environmental considerations have now become an essential part of land-use planning.
- The administration of the local authorities in Israel. Environmental units have been established in 16 of Israel's major towns. They have a positive impact on environmental programs at a grass roots level.
- The administration of several environmental laws.

What has been accomplished is the creation of a comprehensive program for environmental protection involving environmental assessment procedures, emission and ambient standards, and environmental planning and research. Environmental education has been introduced from the preschool level through to tertiary levels in the education system. See *Desertification Control.*

Environmental Quality Act, 1970, California, US: A Californian measure patterned after the National Environmental Policy Act, 1969 (q.v.) requiring the preparation of environmental impact reports (EIRs) in respect to projects requiring government approval which had a potential of "significant impact" on the environment. EIR guidelines are provided by the California Planning Department. In September, 1972, the California State Supreme Court (Friends of Mammoth v. Mano County) ruled that the Act applied to private projects as well as public projects.

Environmental Quality Standards: Levels of exposure to pollutants which should not be exceeded; standards may be statutory or presumptive. Two levels have been adopted by the US Environmental Protection Agency (q.v.):

(a) Primary. Levels judged necessary to protect health with an adequate margin of safety.

(b) Secondary. Levels judged necessary to protect public welfare from any known or anticipated adverse effects.

These are essentially ambient environmental quality standards. Standards may also prescribe the contents of products, e.g., the amount of phosphates in detergents or of pesticide residues in foodstuffs. They may also take the form of emission standards, e.g., the upper limits of what may be emitted from the exhausts of motor vehicles or from the chimneys of industrial plants.

Environmental Recovery Areas, UK: A British term for urban renewal (q.v.). It involves the selection of areas of housing and other mixed uses still structurally sound but lacking certain amenities; an assessment is made of which properties may be retained and which must be demolished. The properties are then rehabilitated, often with some new extensions. Streets may be closed to traffic, parking facilities improved, and tree planting undertaken. The improvement of houses is actively encouraged by central government in the form of grants; the local authority may declare a General Improvement Area. The Housing Act, 1969, is directed almost entirely at enabling this sort of rehabilitation to take place. See *General Improvement Areas; Housing Improvement Grants.*

Environmental Rights Amendment, Pennsylvania, US: An amendment to the Pennsylvania Constitution in 1971, by way of a new section in Article 1. The new section, Article 1, Section 27, is often referred to as the "Environmental Rights Amendment." It states:

> The people have a right to clean air, pure water, and to the preservation of the natural, scenic, historic and aesthetic values of the environment. Pennsylvania's public natural resources are the common property of all people, including generations to come. As trustee of these resources, the Commonwealth shall conserve and maintain them for the benefit of all people.

In addition to creating a constitutional responsibility for preserving environmental values throughout all levels of government, the amendment created a greater obligation to orchestrate programs related to environmental and historical resources at all levels of government. See *Environmental Master Plan, Pennsylvania.*

Environmental Technical Information System (ETIS), US: A system developed in the 1970s by the US Army Corps of Engineers, located in Champaign, Illinois. The ETIS is a computer-based information center designed to assist the Department of Defense in developing environmental impact assessments and statements, as required by the National Environmental Policy Act (q.v.). Most users operate through remote terminals connected to the University of Illinois computer system.

Environmentally Hazardous Chemicals: Chemicals, either as raw materials or final products, which escaping to the environment may prove dangerous to health and welfare, having either immediate or cumulative effects. The term "chemicals" in this context embraces many thousands of substances used: (1) in the home, by way of pharmaceuticals, drugs, cosmetics, and cleaners; (b) in industry in hundreds of processing and manufacturing processes including the preservation and storage of food products; and (c) in agriculture and forestry by way of fertilizers, pesticides, insecticides, and herbicides. Many of these uses have brought beneficial effects to man and his environment; some have resulted in adverse effects in a variety of ways while a few have brought harmful effects of an unprecedented kind. For example, the use of pesticides has prevented untold misery in many tropical and semi-tropical countries through the control of mosquitos and other insects. On the other hand, the thalidomide tragedy in the early 1960s caused great concern about the possibility of some drugs inducing congenital malformation.

The term also embraces a variety of waste products resulting from the use of chemicals in all these areas: e.g., trace metals such as cadmium, lead, mercury, arsenic, beryllium, zinc, copper, antimony, selenium; polycylic aromatic hydrocarbons, polychlorinated biphenyls; and dioxins.

Not all chemical products produced by industry reach the general public, many being used in processing or manufacturing systems; some of these may however present a hazard in the work environment.

Some of the wastes from industry finish in water or air; others are deposited in a variety of landfill arrangements. Land disposal, though often satisfactory for many substances in the right circumstances, may prove hazardous in others. In April 1976, the 57th meeting of the Governing Council of the United Nations Environment Program (UNEP) acknowledged the potentially serious impact of many chemicals on the environment and made a strong appeal to all governments "to take adequate measures to ensure that new chemical substances and physical agents are properly evaluated before they are used and discharged into the environment...." UNEP subsequently established an international register of potentially toxic chemicals, providing information on about 45,000 known chemicals in use in the world.

In 1981, the United States government established a "Superfund" to permit hazardous dumps to be dealt with directly and immediately, without awaiting court litigation. Several hundred dumps were identified as requiring urgent attention.

Administered by the Environmental Protection Agency, it is implemented through regulations that amount to a "cradle to grave" formula for the management of hazardous wastes. Those who make or move wastes, treat or store them, are required to keep track of all the chemicals they handle and make annual reports to the federal government. The regulations also require full records to be kept of the compounds involved.

More than 60,000 chemicals are now in common use in America, and those that are hazardous give rise to some 40 to 60 million tons of waste each year. This waste is created by as many as 750,000 different firms relying on some 500,000 active dump sites most of which do not reach EPA standards. Some areas have suitable sites, while others do not. See *Comprehensive Environmental Response, Compensation, and Liability Act, 1980 (Superfund); Environmental Protection Agency; Land Fill; Love Canal; Michigan Episode; Times Beach.*

Euphotic Zone: The open water zone of the ocean, corresponding to the limnetic zone (q.v.) of a lake. The zone has sufficient sunlight to support photosynthesis and a considerable population of phytoplankton. Usually, sunlight cannot penetrate deeper than 200 meters in most marine habitats;

this depth is frequently considered the lower border of the euphotic zone. Below the euphotic zone lies the abyssmai zone.

European Economic Community (EEC): Popularly known as the European Common Market, a free trade organization created by the Treaty of Rome, 1957. The Community comprises twelve nations: France, West Germany, Italy, Belgium, the Netherlands, Luxembourg (the last three being known as the Benelux countries), together with the United Kingdom, Ireland and Denmark who joined the Community in 1973, Greece which joined the Community in January 1981, and Spain and Portugal which joined in 1985. The population of the Community exceeds 300 million.

The need for a Community environment policy was recognized at a Community summit conference held in Paris in October 1972. Just over a year later, on 22 November 1973, the Council of Ministers adopted the Community's first environmental action program and set out the objectives and general principles of Community environmental policy. This program involved three broad categories of action: action to reduce and prevent pollution and nuisances; action to improve the environment and the quality of life; Community action or, where applicable, common action by the member states in international organizations dealing with the environment.

The second environmental action program, agreed by EEC Environment Ministers on 9 December 1976, provided for a continuation of measures set out and initiated under the first program and introduced a new emphasis on preventive action, particularly as regards pollution, land use and the production of waste. Special attention was to be given also to the protection and rational management of space, the environment and natural resources. An important aspect of the program was a study of how appropriate environmental impact assessment procedures might be introduced into the Community. The polluter-pays principle was endorsed.

Directives, decisions and regulations are intended to be adopted by member countries and embodied in national legislation. Community laws affecting the environment follow this same process. Any member country failing to implement Community law may be brought before the European Court of Justice.

More important, more recent changes to clean air policy have given a new impetus and direction to air pollution control A European Community Directive on standards for sulfur dioxide and suspended particulates became Community law in June, 1981.

The core of the Directive comprised the concentration values in its Annex I. The Directive imposed an obligation (through Article 2) on member countries to ensure that these levels were not exceeded throughout their territory by 1 April, 1982. But where this proved impossible, up to ten years may be allowed for the completion of a program of control (article 3) in limited areas. Implementation of the Directive in the UK was achieved mainly by the extension of smoke control by local authorities in the main conurbations.

Table 6: Directives Issued by the European Economic Community (EEC)

1970 Directive against air pollution by exhaust fumes from motor vehicle engines.

Directive on motor vehicle noise standards.

1973 Directive on polychlorinated biphenyls (PCBs), polychlorinated terphenyls (PCTs) and vinyl chloride monomer.

Directive on detergents.

Directive on the testing of the biodegradability of anionic surfactants.

1975 Directive on drinking water standards.

Directive on bathing water quality standards.

Directive on the sulfur content of gas oils.

Directive relating to the preparation of inventories of harmful wastes and residues.

1976 Directive on the discharge of dangerous substances.

1977 Directive on biological standards for lead and on screening of the population for lead.

1978 Directive on the discharge of titanium dioxide.

Directive on measures to combat oil pollution.

Directives controlling pollution from the wood and pulp mill and other industries.

Directive setting the maximum permitted lead compound in petrol (0.4 g/l).

1979 Directive on the classification of packaging and labelling of dangerous substances (sixth modification to 1967 Directive).

1980 Draft Directive on environmental impact assessments.

Directive on air quality limit values and guideline values for sulfur dioxide and suspended particulates.

1981 Directive for an air quality standard for lead setting an annual mean level (not more than 2 $\mu g/m^3$).

1982 Draft Directive on the assessment of the environmental effects of proposed development projects in the public and private sectors.

1984 Directive in respect of emissions from industrial plant.

Directive in respect of lead in gasoline (petrol).

1985 Draft Directive in respect of motor vehicle exhausts.

Annex II contained a lower set of guide values; unlike the Annex I limit values they were discretionary and intended to serve as reference points for the longer term improvement of air quality. They were also designed for the setting of targets in special zones.

The Directive required air pollution monitoring to be carried out where the Annex I limit values were likely to be approached or exceeded; it also required reports to be made to the Commission of the European Community on monitoring and control measures in areas exceeding the limit values. The coverage and methods of the UK National Survey of Air Pollution proved adequate to meet these requirements.

In 1984, it was estimated by the EEC that air pollution by sulfur dioxide and nitrogen oxides cost the Community between $1.4 billion and $4.2 billion a year. In that year, the European Commission proposed a 25-year abatement program. It was proposed that the sulfur dioxide emissions from large industrial installations should be reduced by 60 per cent; while oxides of nitrogen and dust from these and other sources should be reduced by 40 per cent respectively. It was decided in 1984 that by 1989 lead-free gasoline (petrol) must be on sale throughout the Community; by 1991 all new cars would be obliged to run on it. Leaded gasoline would not be sold anywhere by the end of the century. See *Table 6*.

European Foundation for the Improvement of Living and Working Conditions: A Foundation created in May, 1975, by the European Economic Community (q.v.). The object of the Foundation is "to record all the factors which combine to influence the living and working environment and to make a long-term projection of the factors which might threaten conditions of existence as well as those which might improve those conditions." The Foundation is concerned with the working as well as the living environment, and with short-term and long-term questions. It is based in Dublin, Ireland.

Eutrophication: The ageing of a body of water by the growth of vegetation; these plants flourish and then die, their decay using up the dissolved oxygen of the water with serious impairment of water quality. In lakes, rivers, harbors and estuaries, the accumulation of nutrients is a natural process. These nutrients include carbon, hydrogen, oxygen, sulfur, potassium, calcium, magnesium, nitrogen and phosphorus. Algal "blooms," thriving on these nutrients, may be blue-green, red or brown in color; in small numbers these plants are beneficial as they contribute to the oxygen balance in lakes and streams, and also serve as food for fish. Industrial and domestic activity may greatly accelerate an otherwise slow ageing process by increasing the amount of nutrients entering the water; effluents from these sources contain phosphorus and nitrogen. In addition, fertilizers containing phosphorus and nitrogen may be carried by rainwater runoff or groundwater seepage into lakes or streams.

The effects of advanced eutrophication are:

- Lush algae create problems of water color, taste and odor, resulting in increased costs of water treatment;
- The water is less attractive for boating, swimming and fishing;
- The more desirable types of fish may be eliminated, irrigation canals may become clogged.

Possible remedies include:

- Removal of nutrients from wastewaters, a very costly procedure;
- Bypassing of lakes, diverting of wastewaters to streams below the lake by pipeline;
- Removal of excessive weeds and debris, and dredging of lake sediments.
- Application of chemicals to destroy algal growths, copper sulfate and chlorine being commonly used for this purpose.

Examination in Public, UK: A public inquiry by the Secretary of State for the Environment into selected issues affecting his/her consideration of a structure plan (q.v.). The examination is normally conducted by a panel, appointed by the Secretary of State, consisting of an independent chairman and two other members. It takes the form of a discussion, led by a panel, with selected participants, including the county planning authority responsible for the plan. The examination is confined to those matters that need to be further investigated before a decision can be made. Under the Local Government, Planning and Law Act, 1980, the Secretary of State may dispense with an examination in public on an alteration to, or replacement of, a structure plan. See *Public Hearings and Inquiries; Public Local Inquiry, UK.*

Exclusionary Zoning: The use of zoning (q.v.) in a manner that discriminates against social and economic groups, e.g., seriously impedes or absolutely prevents the construction of low-cost housing. This may not be done directly but by means of local zoning ordinances that render suburban housing costs so prohibitively high that low and moderate income families cannot afford to buy.

Exclusionary techniques hinder the migration of low-income and minority groups from central cities to suburban areas. Exclusionary zoning has often been found illegal on the grounds that zoning ordinances must provide for a wide range of housing types suitable for all income groups; residential zones must also permit construction of multi-family housing such as townhouses or apartments. The zoning ordinances of a municipality must accommodate a fair share of the expected growth of the region. Each municipality has an obligation to assume its share of the costs associated with low-income and middle-income growth, as well as the benefits of high-income residential, commercial and industrial growth. See *Mt. Laurel Decision, 1983, New Jersey.*

Existence Value: The benefits from simply knowing that something is there through reading, films, and pictures; existence value is obviously difficult to measure but it is argued that it should be added to any list of economic resources. Many Americans may find real value in knowing that grizzly bears still roam the Yellowstone Plateau as they did in the days of the mountain man, and that in Alaska a hiker can still wander through land reminiscent of the late pleistocene glaciation.

External Effects: Or externalities. Social costs and benefits, caused by the activities of an industry, which are not reflected in the price at which the product is sold, or do not influence the quantities purchased; costs not borne by those who occasion them, and benefits not paid for by the recipients. In the generation and supply of electricity, two externalities are:

- Those costs to the community of effluent from power stations;
- Those benefits to the community of replacing dirty fuels with clean electricity, thus reducing air pollution in cities at the point of use.

The cost or benefit is external to the industry and does not find its way into the price charged the ultimate consumer; thus the electrical energy market, among others, is imperfect in giving the consumer incorrect information about the cost of resources used to produce power. The economist, R. Turvey, has suggested that externalities should be "internalized", if they are known to have a significant effect on the demand or cost structure of a product, i.e. corrections should be made to allow for them when calculating marginal cost. Marginal cost thus becomes a social opportunity cost, or true cost.

F

Facadism: The practice and philosophy of requiring developers to retain the frontages or facades of important heritage buildings, while redeveloping behind them. The retention of facades is standard planning practice in West Germany, the Netherlands, Italy, the United Kingdom and the United States of America. In Australia, it is practiced in Sydney.

Many noble buildings of heritage quality have splendidly designed and tooled frontages, and entrance lobbies; however, quality often falls off rapidly to the rear (the Queen Anne front and the Mary Anne back). Frequently, nothing but the frontage is worth retaining in heritage terms while the remainder is totally inefficient and unsuitable in functional terms. The solution is often to retain the front wherever this is practicable and feasible; it is not always so.

Some building facades are preserved solely as an incidental to the recycling of buildings in which the interiors alone are refurbished. Thus, a warehouse may become a shopping precinct or a block of home units, with the whole of the exterior preserved; in these instances the motivation may not be primarily heritage preservation but economy, for substantial savings may sometimes be made compared with the cost of the complete clearing of a site coupled with new construction.

Federal Environmental Agency, West Germany: An agency established under the Federal Environmental Agency Establishment Act, 1974; an independent federal authority with headquarters in Berlin. According to Section 2 of the Establishment Act the Federal Environmental Agency has the following principal functions:

- Scientific assistance to the Federal Minister of the Interior in all matters relating to emission (air quality) control and waste management, in particular in the preparation of legal and administrative provisions, in basic research and development to devise suitable measures and in the testing and investigation of processes and installations.

- Establishment and management of an Information System for Environmental Planning and of a central record of environment documentation, education of public opinion on questions relating to the environment, provision of central aids and services for department research and for coordination of research at a federal level, assistance in the investigation of the environmental compatibility of federal measures.

Other functions can be assigned to the agency by law or by the federal ministries.

The Federal Environmental Agency has no executive competence. It can neither issue prohibitions nor permits, nor control plants which influence the environmental situation, nor impose fines. Most federal states have established their own agencies or institutes to provide their administrative authorities with technical advice. At the same time the Federal Environmental Agency offers advice and support to the environment authorities of the federal states, if the latter so desire. See *Information and Documentation System for Environmental Planning (UMPLIS)*.

Federal Environmental Assessment Review Office (FEARO), Canada: Located in Ottawa, an agency responsible for the Canadian environmental assessment and review process (EARP). The process was established by Cabinet decision in December, 1973, being later revised in 1977. It applies to federal departments and agencies and seeks to take environmental matters into account throughout the planning and implementation stages of new projects, programs and activities. It requires an environmental impact assessment to be carried out for all projects which may have adverse effects on the environment before commitments and irrevocable decisions are made.

In addition to projects initiated by federal departments or agencies, the process applies also to projects requiring federal funds and those involving federal funds. The departments and agencies are responsible for assessing the environmental consequences of their own projects and activities or those for which they assume the role of initiator.

The Canadian provinces derive their authority to carry out environmental impact assessment from a variety of legal bases. Manitoba established an environmental assessment and review process by Cabinet Directive in 1975; New Brunswick adopted a similar policy in the same year. Ontario passed specific environmental impact assessment legislation in 1975; Saskatchewan and Newfoundland followed suit in 1980. In 1978, Quebec amended its Environmental Quality Act to provide for environmental impact assessment. British Columbia supplemented EIA procedures within existing statutes through the passage of the Environmental Management Act in 1981.

Generally speaking, the provincial procedures and the federal procedures are similar, containing three stages. First, proposals are screened for their potential impacts; those projects with minor impacts, or with major impacts which can be mitigated, proceed without further review. Projects with major adverse effects are referred to an agency within the government for a public review by an independent board or panel. That body directs the proponent to prepare an environmental impact statement (q.v.) which it studies and makes available to the public. The review body obtains the views of the public and then writes a report containing recommendations to the Cabinet minister responsible for the environment. The minister, in consultation with Cabinet colleagues, decides whether to accept the recommendations or not.

Federal Insecticide, Fungicide, and Rodenticide Act, US: See *Insecticide, Fungicide, and Rodenticide Act, US*.

Federal Land Policy Management Act, 1976, US: Part of US federal law clearly establishing a policy that public lands shall generally be retained in federal ownership, and managed for all Americans under the principles of multiple-use and sustained yield. The Bureau of Land Management (BLM) has developed and implemented a planning and management system aiming to efficiently, effectively, and fairly resolve many of the conflicting demands placed upon public lands. It aims also at the development of needed resources, while protecting other resources, including the environment, from inadvertent damage or destruction. The Act requires that decisions on public land use shall be based on full resource inventories and an analysis of alternative possible uses; such decisions are to be made with the full participation of the public, with a careful balance of competing uses. The Act also directed that lands administered by the Bureau of Land Management be inventoried for wilderness characteristics. The BLM's *Wilderness Management Policy* was published in 1981.

Federal Water Pollution Control Act, 1956, US: The basic US law relating to water pollution control, subsequently amended in 1961, 1965, 1966, and 1972. The Act provided for pollution abatement procedures and water quality standard enforcement procedures. Action could now be taken in respect to pollution of interstate and navigable waters which endangered health and welfare. The Act provided for conferences between federal and state agencies, in respect to interstate pollution; the federal government could in the ultimate bring an enforcement suit. The Act also authorized, for the first time, the making of construction grants to municipalities for sewage treatment plant construction.

An earlier measure, the Water Pollution Control Act 1948, had restricted the US federal role to investigation, research, and surveys, with primary responsibility for water pollution control being left entirely with the states. See *Water Pollution Control Amendment Act, 1972, US*.

Finland: See *Nordic Environmental Protection Convention; Right of Common Access*.

Flexible (Average Density of Cluster) Zoning: A variation of traditional zoning (q.v.); rigid specifications for equal lot sizes and dimensions have often prevented the most efficient and environmentally sensitive use of a development site, whereas averaging and clustering allow smaller lots and higher densities in those parts of the site most favorable to development. The overall density on the entire site may remain at the specified level, although in some cases density bonuses may be offered.

This technique eliminates the need to develop every part of a site to obtain the maximum density of dwellings. It encourages the conservation of open spaces where slope, tree cover, erodibility or other factors restrict building.

Floating Value: A phenomenon whereby potential development value "floats" over a number of competing sites all of which are, to all intents and purposes, identical. Not all will be developed, but all the individual owners have an equal right to believe that theirs is the plot which will attract the development and the development value.

Flood Mitigation Works and Measures: Levees, floodway schemes, drains, flood gates, river banks stabilization, pumping facilities, flood free mounds, diversions, dams, dredging and the like; additional measures include zoning, land acquisition, land exchange, relocation, provision of flood free land, notification of flood hazards, flood warnings, and flood proofing. See *Floodplain.*

Floodplain: A relatively smooth portion of a river valley, adjacent to the river channel, built of sediments carried by the river and which is covered with water when the river overflows its banks. Floodplains are described on the basis of how often they are likely to flood. Generally, that land which, on the average, is likely to flood once in every 100 years is known as the "100 year floodplain." The chances are about 1 out of 4 that a home built on the 100 year floodplain will be flooded before a 30 year mortgage is paid off.

Floodplains are also described in terms of the severity of a flood event. Floodplains can be divided into two parts, the floodway and the flood fringe. The floodway is that part which includes the channel and the lands immediately adjacent which carry the fast moving floodwaters. The flood fringe consists of the outer areas of the floodplain that are subject to flooding but have lower velocities and are not important in carrying flood waters.

Floodplains perform natural functions which, if properly managed, are beneficial. They support an abundance of plant life because of their moisture and fertile soils. The roots of the many plants and trees anchor the soil to retard erosion and prevent sediment from entering streams. This helps to maintain water quality and fish and insect populations. Plants and trees also cool the water which benefits fish populations. Floodplains are also a valuable recreational resource.

However, floods threaten the lives and property of people who live on floodplains. In addition, floodplains tend to have high water tables, even during normal stream flow levels. There is a higher risk of basement flooding and failure of on-site sewage disposal systems. Building on floodplains contributes to more frequent and more severe flooding, while covering the floodplain with impermeable surfaces decreases the recharge of groundwater. Furthermore, the removal of vegetation increases erosion and accelerates stormwater runoff.

Planning agencies may adopt measures to ameliorate the effects of floodplain hazards: floodplains can be identified and classified on maps made available to the public; dangerous floodway areas can be reserved for open space, recreation, or agriculture; floodproofing may be required in respect to construction in the flood fringe area; impermeable coverage may be minimized through control of building densities and parking lots; and storm runoff from the entire watershed may be controlled.

Florida, US: See *Acid Rain; National Wildlife Refuges.*

Focal Point: A locale or center of activity or structure on which attention is concentrated by arrangements of other features, patterns, or sightlines.

Food and Agriculture Organization (FAO): An Agency of the United Nations whose essential function is to combat the poverty, malnutrition and hunger which afflict about half the people in the world. The creation of FAO was recommended by the United Nations Conference on Food and Agriculture held at Hot Springs, Virginia, USA in 1943, and an interim Commission was set up to plan the organization. In October 1945, delegates from 42 countries met in Quebec, Canada, and founded the Organization. By 1978, FAO had 131 member nations. Since 1951, the headquarters has been in Rome, Italy. The Secretariat consists of about 3,500 professional staff.

The improvement of the food situation is bound up, inevitably, with the process of economic development as a whole including the development of agriculture, forestry and fisheries. FAO has developed a program comprising many elements, including the provision of information and advice, aimed at many aspects of the food and agricultural problem and financed from a variety of sources. One initiative has been promoting the building up of national food stocks against future crop failures. Vigorous attention has also been given to the promotion of agricultural productivity, to the raising of the availability of protein foods, the reduction of waste, and rural unemployment.

About one-third of the recommendations of the Stockholm Conference on the Human Environment in 1972 were addressed, in whole or in part, to FAO. This illustrates the significance of the degree of interaction between agricultural and environmental factors. With financial support from the United Nations Environment Program, FAO has embarked on a broad range of study, monitoring and field action on environmental matters such as the problems of soil degradation, rangelands, tropical forest cover, wildlife management, integrated pest control, conservation of animal and plant genetic resources, protection of fisheries resources, and food safety and quality standards.

Food, Drug and Cosmetic Act (FDCA), 1938, US: US federal legislation, with subsequent amendments, under which the US Environmental Protection Agency (q.v.) (EPA) has authority to establish tolerance levels for pesticide residues in or on food or feed crops in the United States. A "tolerance level" is the pesticide residue legally permitted on food or feed and represents the maximum residue allowable in the harvested commodity. Such tolerances have been established in the United States since 1954 (from 1970, by the EPA).

A "tolerance" must be established before a pesticide can be applied to a food or feed crop that is intended for sale, distribution, or consumption in the United States. A tolerance is derived from an evaluation of laboratory toxicity tests and crop residue data. Tolerances are set at levels considered adequate to protect the public health and which are consistent with good agricultural practice. See *Insecticide, Fungicide, and Rodenticide Act, US; National Human Monitoring Program, US.*

Footloose Industry: Industry of all kinds which is not "tied" to existing locations by virtue of its relative independence in respect to specific transport links or markets. Also called mobile industry.

Foreground: The total landscape which is perceived by an observer from a particular viewpoint, and which ranges from 0 up to a distance of 400 to 600 meters. Within this range, the observer experiences maximum discernment of details such as shape, color, and contrast. See *Background.*

Forest and Rangeland Renewable Resources Planning Act, 1974, US: An Act, subsequently amended, directing the US Secretary of Agriculture to prepare a comprehensive long-range assessment of the nation's renewable forest and rangeland resources, and develop a program for forest service activities. The assessment must embrace timber, range, water, fish, wildlife, outdoor recreation, and wilderness resources. The future supply and demand for each of these resources must be projected together with the potential opportunities to meet the nation's future needs. The Act requires a full assessment every ten years, with a program review and update every five years. The 1985 review and assessment was transmitted to the Congress, accompanied by a statement of policy from the President.

Forest Products: A wide range of commodities and services valuable to affluent industrial and poor rural communities alike: timber for housing and construction; pulp for packaging and newspapers; timbers for mining and railways; fuelwood for villages; fodder, fruits, honey, pharmaceuticals, meat, fibers, resins, gums, dyes, skins, waxes and oils; and finally beauty, amenity and recreational qualities. Forests also play a vital part in the preservation of genetic diversity, essential to the future of all mankind.

In developing countries, the heaviest demand on forests and woodlands is for fuel and as a base for shifting cultivation. It has been estimated that more than 1,500 million people in developing countries depend on wood for cooking and keeping warm. The effect of this intense and enlarging demand has been to denude the land of wood over wide areas. In Gambia, fuelwood is now so scarce that gathering it takes 360 woman days a year per family. Some of the projects of the World Bank have been directed towards augmenting fuelwood supplies.

India's forest industries have expanded enormously; wood products make up a sizeable proportion of the incomes of individual states—12.5 per cent in Madhya Pradesh and 8.4 percent in Himachal Pradesh (*The Economist*, 5 March, 1983). Furthermore, in Uttar Pradesh a quarter of the tree cover has been lost because of illegal felling.

Deforestation appears to be worst in the Himalayan areas, involving some of the world's most beautiful forests. The problem has been intensified in Kashmir as a result of excessive resin tapping; this has been the main cause of the large-scale death of chir pines.

According to the Indian forestry department, 3.4 million hectares of forest have been lost in the past 20 years out of a total of 75 million hectares. The immediate effect of a deforestation is felt by the people, particularly women, in the countryside. Firewood is an essential; women, who do the collecting, regularly have to walk 15 kilometers a day to find wood. See *Table 7*.

Table 7: Rate of Consumption of the World's Forests

	Total ....Closed Forests..... Area Projected To Be Lost 1975-2000, 1,000 ha	% of 1975 Area	"Operable" ....Hardwood Forests.... Area Projected To Be Lost 1975-2000, 1,000 ha	% of 1975 Area
West Africa	6,600	47.1	6,600	54.7
Centrally-planned tropical Asia	6,300	29.1	6,000	35.3
South Asia	16,400	23.0	13,600	27.9
East Africa and islands	3,300	17.8	3,200	50.4
Insular Southeast Asia	21,600	16.5	20,000	26.3
Central America	10,900	13.4	4,600	23.9
Tropical South America	64,200	12.0	57,300	13.3
Continental Southeast Asia	4,100	10.6	4,000	13.3

Source: International Union for Conservation of Nature and Natural Resources (1980) *World Conservation Strategy: Living Resource Conservation for Sustainable Development.* Geneva: IUCN.

France: See *Acid Rain; International Commission for the Protection of the Rhine; Nature Conservation Act, 1976; New Towns, France; Oil Spills; Paris Convention; US-France Memorandum of Understanding, 1984.*

Franchise Agreements: Agreements reached directly between developers and governments, often endorsed by an Act of Parliament at state, federal or national level. The effect usually is to exempt a developer from meeting normal planning and environmental requirements and procedures, while making the basic decision immune from review by the courts. Such agreements are sometimes referred to as "indenture agreements."

The agreement may make some reference to pollution control or environment protection measures but exempts the developer from surveillance by the appropriate environment protection agencies. Indeed, in the making of such agreements, the state environment protection agen cies are often ignored.

Freeways: Selected arterial roads in urban areas which, by the restriction of access and the absence of all at-grade intersections, provide for the rapid movement of large quantities of goods and large numbers of people free of conflict and, therefore, more safely than other arterial roads.

Freeways are needed in urban areas where they can provide for the present and future movement of goods and people more safely and at lower overall cost to the community than on free access or even limited access arterial roads.

They are also needed where they can remove traffic from other arterial, sub-arterial and local residential roads or streets so that these roads can carry out their correct function of distributing goods and people, and thus not become hazardous nor be barriers to community activity.

Urban freeways must be located, and their design and construction such, that in the long term the attributes of the communities through which they pass and which they serve are improved. Furthermore, any likely adverse impacts of freeways on the communities should be reduced to a minimum.

Urban freeways and public transport should be complementary and they should contribute collectively to the total transport system. See *Roads, Hierarchy of.*

Frost Commission, Australia: A Royal Commission set up by the Australian Government in 1977 to investigate the Australian whaling industry. The report recommended that Australia should:

- Ban whaling in its 200-mile offshore fishing zone and pursue internationally a policy of opposition to whaling and ban the import of all whale products;
- Seek to classify both male and female sperm whales as protected stocks and amend the present catch limits to zero;

- Continue membership of the International Whaling Commission and support its efforts.

The Cheynes Beach Whaling Company, the only Australian whaling company, had announced at the beginning of proceedings at the Albany hearing in July 1978 that it intended to cease operations in the near future.

In April 1979, the Australian Government announced a ban on whaling within its proposed 200 mile fishing zone and a ban, from January 1981, on the importing of whale products and goods containing whale products. Australia would also work towards an international ban on whaling.

Functional Pattern: The overall physical organization of urban land uses in a city or urban area. Included are the neighborhoods; retail, commercial or industrial concentrations; major institutions; and centers of government activity. See *Activity Pattern*.

G

Garden-City Movement, UK: A movement in England developed from the writings of Ebenezer Howard in the 1890s which proposed pre-planned new cities, to be constructed on land held by the community and limited to a population of some 30,000, complete with business services and employment centers, and surrounded by permanent greenbelts of rural land. The initial experimental cities were constructed on private initiative in a spirit of reform; Letchworth was commenced in the early 1900s, and Welwyn Garden City in the 1920s. The movement was a great and continuing influence on efforts elsewhere to improve the urban environment.

Gdansk Convention: See *Convention on Fishing and the Conservation of Living Resources in the Baltic Sea and the Belts, 1973.*

General Agreement on Tariffs and Trade (GATT): An international commercial treaty which came into force in 1948. Some forty governments, accounting for well over 80 per cent of the world's trade, are now contracting parties to the Agreement. GATT provides a code of conduct for international trade and, in addition, seeks to help raise living standards, to develop the world's resources, to promote economic development, to expand production and exchange of goods and to bring about full employment. In becoming parties to the Agreement, governments pledge to work together towards these goals.

The code of conduct for international trade includes in its principles that trade should be conducted on the basis of non-discrimination, and existing preferential arrangements should be gradually reduced through negotiations and finally eliminated. As a result of a series of tariff bargaining conferences, customs duties have been lowered on products accounting for about half of the world's trade. Notable among these conferences were the Kennedy Round negotiations of the 1960s, and the Tokyo Round negotiations of the late 1970s.

Machinery has been established in GATT for the examination of the trade policy aspects of anti-pollution measures. It is known as the Group on Environmental Measures and International Trade.

General Improvement Areas (GIAs), UK: Areas designated by British local authorities under the Housing Act, 1969, where the external environment of a housing area could be improved. Later, the Housing Act, 1974, led to the designation of Housing Action areas where it was intended that the improvements should be carried out to both houses and environment within a period of 5 years; the local authorities were given special powers of acquisition. In 1977, a system of Housing Investment and Strategy Programs was set up whereby local authorities were required to present a comprehensive picture of the housing situation in their respective areas, and formulate strategies to meet the needs of the situation. Annual programs were to be submitted to the Secretary of State for the Environment.

General Plan, Los Angeles, US: A comprehensive plan (q.v.) adopted by the Los Angeles City Council, California, as the official guide to the intended future development of the City of Los Angeles. This comprehensive declaration of purposes, policies and programs for the development of the city includes a land use element, a circulation element, a service-system element, and an environmental element. The General Plan now serves as a basic and continuous reference in planning, coordinating, and regulating public and private development in the city. The objectives of the General Plan are:

- Preserve the low-density residential character of Los Angeles, except where higher density centers are encouraged; protect stable single-family residential neighborhoods from encroachment by other types of uses; rehabilitate and/or rebuild deteriorated single-family residential areas for the same use; help make single-family housing available to families of all social and ethnic categories.

- Provide maximum convenience for the occupants of high and medium density housing (apartments); locate the bulk of such housing within, or near to, concentrations of urban facilities and employment opportunities; help make high and medium density housing available to persons of all social and ethnic categories.

- Provide employment opportunities and commercial services at locations convenient to residents throughout the city; reserve suitable and adequate lands for industrial and commercial uses; help make Los Angeles a desirable location for industry and business.

- Provide adequate transportation facilities for the movement of people and goods; provide a choice of transportation modes; alleviate traffic congestion; optimize the speed and convenience of all transportation modes; achieve economy and efficiency in the movement of goods.

- Provide needed public services to all persons and businesses; achieve economy, flexibility and efficiency in the provision of services, both those furnished by the City of Los

Angeles and those furnished to Los Angeles citizens by other governmental jurisdictions; provide suitable sites for public facilities at locations convenient to their users.

- Provide facilities for leisure time activities at locations readily accessible to all persons; furnish local recreational services; develop specialized recreational facilities; preserve the ocean shoreline and other comparable recreational resources for public use.

- Conserve the city's natural resources and amenities; preserve open space; protect outstanding geographical features; minimize all forms of environmental pollution including air pollution, water pollution, noise, and visual pollution.

- Enhance the quality of the city's physical environment; integrate all aspects of the city's development through the application of urban design principles; establish the identity of the various communities of the city; preserve historical and cultural features; control the placement of commercial signs; provide landscaping where it serves or enhances the physical environment.

- Balance population growth with available facilities, services and amenities for a productive, healthy and desirable environment.

See *Central City Community Plan; Los Angeles City Planning Commission; Transport Plan*. See *Figure 6*.

Geneva Convention: See *Convention on Long-Range Trans-Boundary Air Pollution*.

Gentrification: The movement of people into run-down neighborhoods who invest money in rehabilitating what were once fine houses or apartments; good homes have been created in improving neighborhoods, often at a cost much lower than that of a home in the suburbs or exurbs; gentrification has reversed the earlier wholesale flight from city to suburb. However, gentrification has been criticized for destroying established communities, and making houses too expensive for lower-income groups.

Georgia, US: See *Litter Laws, US*.

German Democratic Republic (East Germany): See *Council for Mutual Economic Assistance (COMECON); Environment Policy Act, 1970*.

Ghana: See *Environmental Protection Council*.

Ghetto: An enclave or defined part of a city occupied by members of some identifiable group, e.g., people of a certain racial composition who feel compelled to live together in one district by outside pressures and discrimination against them, or because of mutual support and cultural

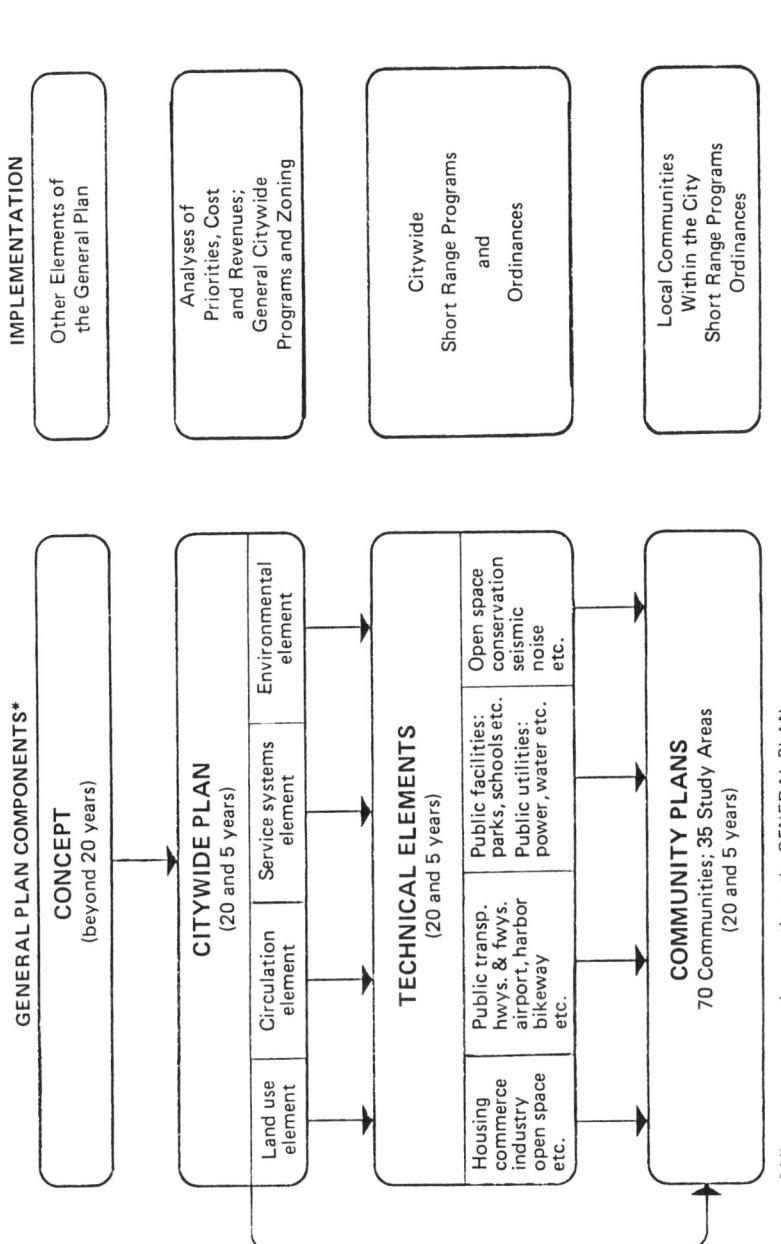

Figure 6: City of Los Angeles, U.S.—General Plan.

togetherness. The word is usually associated with congestion and poverty; it is not normally associated with districts and suburbs which simply have a high proportion of people of a particular derivation.

It appears that the word is of European origin being connected in particular with Jewish communities; it is in vogue in the United States usually in connection with "black ghettos." Despite instances that appear to meet the definition, the word "ghetto" does not appear to have come into common use in Britain, Canada, Australia or New Zealand.

Glasgow Eastern Area Renewal Project, 1976-87, Scotland: Considered the most determined public sector attempt to achieve inner city urban renewal (q.v.) in the United Kingdom, a multi-agency project coordinated by the Scottish Development Agency to achieve a comprehensive social, economic and environmental regeneration of the east end of Glasgow, Scotland's largest city. The aim is to create conditions for the development of a balanced and thriving community, providing substantial additional employment, new housing, and improved social services.

Global Atmospheric Research Program (GARP): See *International Council of Scientific Unions (ICSU)*.

Global Environment Monitoring System (GEMS): Part of the United Nations Action Plan for the Human Environment and one of the components of Earthwatch (q.v.), the Global Environment Monitoring Program comprises five closely linked major programs each containing various monitoring networks with provision for training, technical assistance, and detailed evaluation and review: health-related monitoring, climate-related monitoring, long-range transport of pollutants monitoring, renewable natural resources monitoring, and ocean monitoring.

Within the framework of GEMS, air pollution (SO_2 and total suspended particulates) is being monitored in some 200 cities in 50 countries, food contamination at 21 national centers, and water quality in 70 countries. Nitrogen oxides, photochemically-produced oxidants and air-borne metals are sampled in many places. Monitoring of ionizing radiation is carried out in 21 countries.

The ecological monitoring program focuses on the conditions of soil, water, vegetation, wild and domestic animals, and humans, as observed on the ground and from air and space by remote sensing. Comprehensive reviews of the state of knowledge of tropical forest ecosystems and grazing land ecosystems were published in 1978; the GEMS assessment of the global extent of tropical forest resources and their rate of depletion became available in 1981.

The WMO Background Air Pollution Monitoring Network (BARMON) and the WHO Program of Monitoring Air Quality in Urban and Industrial Areas have been developed as part of GEMS.

In 1980, GEMS completed a methodology for assessing the state and risk of world soil degradation, assessment of the capability of various areas to grow particular crops, while outlining the elements of a world soils policy.

Gordon-Below-Franklin Dam, Australia: A hydroelectric project proposed to be constructed in the South-West Tasmania Wilderness area; the campaign to stop the dam made conservation history. On a world scale, south-west Tasmania is a wilderness area of major importnce; it is one of only three major remaining southern temperate wildernesses and one of the most important archaeological sites. Tasmania generates most of its electricity from its hydro resources. Prior to the Gordon-Below-Franklin proposal, a major conservation battle had been waged and lost over the flooding of Lake Pedder. The next proposal by the Hydro-Electric Commission of Tasmania was a further utilization of the resources of the River Gordon which would have involved a flooding of part of the Franklin River, an area of outstanding natural beauty.

The issue wracked Tasmanian politics for over two years, bringing down a premier and later the State government; however, after the May, 1982, election the Tasmanian Parliament voted approval of the scheme. The battle then switched to the federal scene, for the federal government has sought the inclusion of south-west Tasmania in the UN World Heritage List. The federal government introduced the World Heritage Properties Conservation Act 1983 with appropriate regulations, and launched High Court proceedings for an injunction to stop construction of the dam. The High Court of Australia ruled on 1 July, 1983, that the federal Act was constitutionally valid and that work on the dam must stop immediately. The decision was important also in establishing the principle that the federal government could intervene in what would normally be a state matter if an international convention, such as the World Heritage Convention, was involved.

Compensation was paid to Tasmania to cover the possibly higher cost of obtaining electricity from other sources, and to sustain employment. The battle over the Gordon-Below-Franklin Dam galvanized opinion throughout Australia; it became a truly national issue involving a national alliance of conservation groups. It was characterized by meetings and street marches, while the "No Dams" triangle was to be seen from Perth to Cairns.

Grade Separation: The separation of two roadways, by a fly-over or under-pass, to allow traffic on the one road to proceed without interfering with traffic on the other road.

Great Barrier Reef, Australia: See *Royal Commission on the Great Barrier Reef*.

Great Lakes Water Quality Agreement: An Agreement signed in 1972 by the United States of America and Canada to clean up the pollution of the Great Lakes, particularly Lakes Erie and Ontario. The Agreement was initially binding for five years only. Some progress was made under the surveillance of the existing International Joint Commission for the Lakes.

Following the lapse of the 1972 Agreement, a new Agreement was negotiated and signed in 1978. The new Agreement was more comprehensive and more stringent in its requirements. Municipal control programs were to be completed by 1983; industrial pollution control programs by the end of 1984.

There was a marked emphasis on limiting toxic substances. Concentration standards are set for arsenic, cadmium, chromium, copper, iron, lead, mercury, nickel, selenium, zinc and fluoride. A stricter standard is imposed on phosphorus: 0.5 milligram per liter for Lakes Erie and Ontario, compared with the previous 1.0 milligram per liter limit. The Agreement set specific objectives for nine categories of organic pesticides. Also, it set a standard of 0.1 microgram per gram for total polychlorinated biphenyls (PCBs) in fish tissues, "for the protection of birds and animals which consume fish." The concentration of total dissolved solids in Lakes Erie and Ontario, and the international section of the St. Lawrence River, was set at 200 milligrams per liter.

The Agreement called for the construction of waste treatment plants to minimize eutrophication problems, and prevent degradation from the presence of phosphorus. It introduced an annual public inventory of discharges and pollution control requirements; and called for an improved monitoring and surveillance system. The new program involves also the monitoring of airborne pollutants.

It appears that the deterioration of the Great Lakes has been halted, and that improvements have been noted. Further improvements, particularly in respect of Lakes Erie and Ontario, are now confidently anticipated.

A 1983 addendum to the Agreement committed the United States and Canada to further improvements in controlling the flow of phosphorus into Saginaw Bay (Lake Huron), and into Lakes Erie and Ontario. Between 1972 and 1985, the two countries have spent a total of $7.6 billion in improving or replacing over one thousand sewage treatment plants along the shores and tributaries of the Great Lakes.

Great South Western Industrial Estates, US: Claimed to be the largest planned industrial development in the world, an industrial estate based at Arlington, a town midway between Dallas and Fort Worth, Texas, which some 20 years ago had a population of only 8,000 and had now grown to over 80,000. The site was selected by a group of New York private developers.

Greater Copenhagen Regional Plan: A regional plan adopted by the Greater Copenhagen Council, Denmark, in August, 1975, to promote the more efficient control of urban development in the metropolitan area and provide a better environment in both new and old urban areas; the development of this plan had a long and frustrating history. As early as 1947, a proposal for a regional plan for Greater Copenhagen (better known as the Finger Plan) was developed. Future urban growth was to be concentrated in linear towns ("fingers"), the existing city being regarded as the palm of the hand. By choosing this form it was possible to preserve open green wedges in between the fingers.

The Finger Plan never received any official confirmation, though it became a source of guidance and inspiration for the Ministry of Housing and the local municipalities; the existence of the Finger Plan probably saved Copenhagen from the chaos and the conflicts which have characterized the development of many other European cities since the Second World War.

However, the fingers of the plan tended to grow longer and become thicker, so that in 1958 the Ministry of Housing initiated a resumption of regional planning, setting up a special office for this purpose. The new plan which emerged suggested two new towns of a quarter million inhabitants and a new urban center in order to relieve pressure on the City of Copenhagen. This plan met with opposition from many sides. The plan was modified and by 1963 a revised Finger Plan had been agreed on. Even this modified, though constructive, plan did not fulfill its promise largely due, it appears, to a lack of coordination between housing development and rail and road facilities.

In 1967 a Regional Planning Council was established comprising representatives from the three metropolitan counties concerned; the following year a new attempt was made at regional planning, and much needed research was put in hand. By 1971, a draft regional plan had been prepared suggesting a new regional structure and four different ways of locating future urban growth in relation to the new structure. Much lively public debate followed. Some 52 municipalities put forward written opinions. The idea of transferring some of the traffic volume from individual traffic to the public transport system received considerable support.

The revised structure plan for the region comprises four elements:

- A super-structure indicating transportation corridors and A-zones. The one-kilometer-wide transportation corridors would be secured in their total width for traffic purposes and regional supply lines (electricity, oil, gas, water); the one to two kilometer wide A-zones would be kept free from housing and preserved generally for regional activities related to trade and commerce or for recreational purposes.

- Regional parks landscapes in the northern and western part of the region would be preserved for leisure activities. No new areas for summer cottages would be permitted; the coast and its hinterland would be gradually opened to the public. Linear parks between valuable landscapes would be secured. Agriculture in the affected areas would be retained.

- Existing urban areas east of the super-structure. A finger town would be incorporated in the new structure. The green wedges between the town fingers would be preserved and their recreational value improved. Enough land was available to accommodate urban growth and improve the quality of the environment.

- Urban growth areas between the super-structure and the regional parks. These areas would also accommodate limited urban development and some new townships would be possible. However, considerable areas would still be preserved for agricultural purposes in this part of the region.

Eventually, the Greater Copenhagen Council adopted the regional plan as its own. In 1976, the plan was approved by the Minister of the Environment. Half a century had elapsed since the first pioneers began to advocate a general planning of what was then called the Copenhagen area.

Greater London Development Plan (GLDP), UK: The structure plan (q.v.) for London, England, administered by the Greater London Council (GLC). The first plan was approved by the UK Secretary of State for the Environment in 1976. Proposed alterations to the GLDP were completed by the GLC and submitted to the central government at the close of 1984. The revised GLDP would provide further statutory guidance to the many London boroughs in preparing local plans for their respective areas; it would also help the GLC and the boroughs to set priorities for their programs to improve housing, transport, public facilities, and the environment. The GLDP and the borough local plans provide the main basis for deciding the 40,000 planning applications which are made each year in Greater London.

In London, all applications for planning permission are submitted to the London borough councils; however, under town and country planning regulations certain of the more significant applications must be referred by the borough councils to the Greater London Council for decision or direction. During 1985, some moves were made by the central government to abolish the GLC and have it replaced by a London Planning Commission. See *Abercrombie Plans for London, UK*.

Greater London Plan, 1944: See *Green Belt*.

Greater Stockholm Finger Plan: A plan developed, though never formally adopted, for the expanding city of Stockholm, Sweden, by an advisory Regional Plan Commission in the early 1950s; in 1957, a Greater Stockholm Planning Board was created to coordinate the location of new housing. In 1959, the Swedish Government passed the Lex Bollmora which allowed the city to develop beyond its borders by mutual agreement to meet both its own needs and those of other communes. Ten agreements followed and a series of substantial new satellite towns grew beyond the city borders, the largest—Jarvafaltet—being designed for 100,000 people.

Stockholm's famous finger plan for city growth was based on major extensions of the subway system and the creation around each station of a neighborhood for about 10,000 people. The housing was arranged in tiered densities around the station. Groups of neighborhoods were related to a district center serving some 60,000 people. High standards of social planning were adopted, including clubs for the aged, facilities for the handicapped, play centers and other facilities to help working wives. Traffic was minimized and a network of safe walkways was established. Trees and other natural features of the landscape were preserved. A whole series of suburban townships was constructed along these lines, including Vallingby, Farsta, Arsta, and Grindstorp.

The process was not free of conflict. The other communes resented the city's large and secret land purchases; the Lex Bollmora aggravated these disagreements until in 1966 the government insisted that a common housing exchange be set up as a condition for central housing loans. Transportation conflicts also increased until in 1964 a Greater Stockholm Traffic Association was formed jointly by city and county so as to unify the ownership and planning of all public transport systems.

Stockholm provides an outstanding example of the ability of a city government to integrate public transport, housing, shopping centers, social facilities, and environmental protection through a system of public land management and coordinated programs. The Swedish government's insistence on a common housing exchange and its financial support for the public transportation system were critical elements in success.

A final structural reform in 1971 created a directly elected Greater Stockholm Council responsible for regional planning, transporttion, health, water supplies, and main sewerage; the city and suburban communes kept local planning and housing for which cooperative arrangements were continued.

Greece: See *Athens Charter; Council of Europe; Ekistics; National Council on Physical Planning and the Environment.*

Green Alliance: A London, England, based umbrella group for environmental bodies in 19 countries funded by the World Wildlife Fund (q.v.). In respect to environmental issues, the Alliance accuses the industrial nations of a "persistent failure to act effectively on remedies which are well understood, widely communicated, and for which there is strong public support." The Alliance calls on industrial countries to join the "30 per cent Club" formed to achieve a one-third reduction in sulfur dioxide (q.v.) emissions, a major cause of acid rain (q.v.).

Green Ban Movement, Australia: Originating in Sydney in 1971, a movement based upon close cooperation between citizen groups and the trade unions. In central Sydney and the historic area of The Rocks, many major developments were blocked in the name of environmental and heritage considerations as well as the welfare of low-income groups. In another part of Sydney, a group of women citizens appealed to the Builders Laborers Federation for assistance in saving "Kelly's Bush," a beauty spot used by women and children for walking and playing; the appeal was successful and the area was saved. The struggles have been bitter.

Green bans were imposed in other parts of Australia. Bans were imposed on the construction of Newport power station in Victoria over a number of years, until the matter was substantially resolved through a public inquiry chaired by this author.

Green Belt: Belts of land of irregular shape, usually several kilometers wide, around urban or metropolitan areas. Their purpose is to prevent further expansion; prevent the coalescence of neighboring conurbations; perhaps to channel the growth of a conurbation in a certain direction; preserve the special character of some historic cities; and certainly to bring fresh air and unspoiled countryside within reach of as many urban dwellers as possible. The concept has been employed in London, Paris, and Moscow.

The Green Belt (London and Home Counties) Act, 1938, enabled local authorities in the London area to acquire land and establish a green belt or girdle of open space. The matter was again reviewed by Sir Patrick Abercrombie during his preparation of the Greater London Plan in 1944. Essentially, Sir Patrick recommended that congestion in the crowded "inner ring" should be relieved by building new towns and expanding existing towns in the fourth or "outer country ring," outside the existing green belt ring.

Within the green belt, normal rural and other activities would continue undisturbed and the belt itself would be widened by up to 16 kilometers. The London green belt, finally approved in all its parts in 1959, has proved a very important factor in keeping land for recreation within the reasonable reach of Londoners. The whole process was assisted by transferring the liability for compensation from the local planning authorities to the central government.

Although Birmingham, Leeds and Sheffield had made valiant attempts to establish green belts, it was not until the Town and Country Planning Act, 1947 (q.v.) that it became possible to establish green belts generally around other cities. The advantage of this legislation was that it was no longer necessary to acquire land for the purposes of a green belt. It was sufficient then to define the green belt on a map and refuse planning permission for anything within it incompatible with its purpose.

Under the 1947 Act, Bristol and Bath, Cambridge, Chester, Nottingham, Oxford, the Potteries of north Staffordshire, and York have established green belts, together with the above cities. See *Abercrombie Plans for London*.

Green Great Wall: A desertification control (q.v.) measure introduced by the People's Republic of China. In the late 1970s, the State Council decided to plant protective forests in the northern region and create a 'Green Great Wall' of benefit to present and future generations. According to the plan, some 5.3 million hectares were to be covered by various kinds of protective forests by 1985. When the plan is eventually fulfilled, shelterbelts in the agricultural and pastoral districts menaced by sandstorms will occupy about 10 per cent of the total areas involved, compared with only about 4 per cent in 1979. See *Figure 7*.

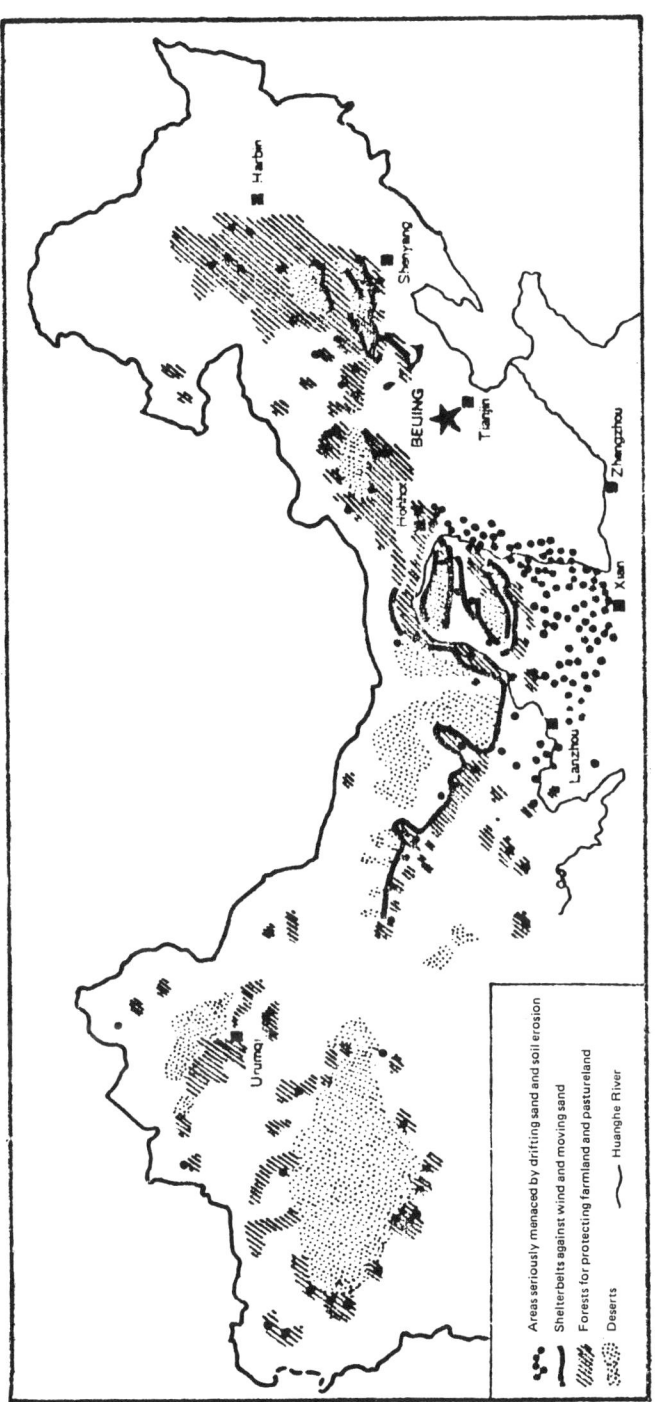

Figure 7: The proposed green great wall of Northern China. Source: *Beijing Review* No. 36, 7 September 1979.

In addition, the forest areas in the middle reaches of the Huanghe (Yellow) River where there has been serious water loss and soil erosion will be raised from 5 per cent in 1979 to 18 per cent of the total. This will protect local farmland and pastures, transforming the environment and improving economic conditions.

Green Summit, 1984: A summit conducted in December, 1984, by the leaders of the industrial nations following their earlier London economic summit. The six countries taking part were Britain, Canada, Italy, Japan, the United States of America, and West Germany. Items on the agenda included acid rain, pollution by pesticides, stricter controls on the export of chemicals, shipping of hazardous wastes, and threats to wildlife habitats.

Green Tracery Features: The network of street trees, private gardens, landscaped areas, and small open spaces that interweaves the built-up city.

Greenhouse Effect: The property of selective absorption used in the design of greenhouses which finds a parallel in the general atmosphere. Water vapor and carbon dioxide, although only a minute fraction of the mass of the atmosphere, exercise considerable influence over the heat balance of the atmosphere and ground. While relatively transparent to incoming shortwave solar radiation they are relatively opaque to long wave back-radiation from the earth, hence they exercise a warming or greenhouse effect. A secular increase in carbon dioxide in the atmosphere, arising from progressive industrialization and the combustion of fossil fuels, could raise the mean temperature of the atmosphere, effecting profound climatic changes.

Carbon dioxide enters the atmosphere as a waste product of animal and plant respiration, through the decomposition of living tissues, through naturally-occurring fires, and through the burning of fossil fuels such as coal and oil. In the 1890s the carbon dioxide content of the atmosphere was around 290 parts per million (ppm); by 1960 it was 317 ppm and by 1969, 324 ppm. By 1980, it had reached 338 ppm. In other words, the carbon dioxide content of the atmosphere had risen by about 10 per cent between 1890 and the 1972 UN Conference on the Human Environment, half of the rise taking place since 1945. See *Table 8*.

At the present rates of increase, the atmospheric concentration of carbon dioxide may produce a significant warming of the lower atmosphere before the middle of the next centrury, particularly in the polar region. This warming would probably change temperature patterns throughout most of the world, benefiting some regions and damaging others, possibly severely.

Gross National Product (GNP): A comprehensive measure of a nation's output. It may be measured in two ways: from the demand side as the sum of consumer expenditures, private investment, government pur-

Table 8: Carbon Dioxide Concentrations in Air, 1958-1981
(parts per million)

Year	Mauna Loa, Hawaii	South Pole, Antarctica
1958	NA	315
1959	316	316
1960	317	316
1961	317	317
1962	318	318
1963	319	318
1964	319	319
1965	320	319
1966	321	320
1967	321	321
1968	322	322
1969	324	322
1970	325	324
1971	326	325
1972	327	326
1973	329	327
1974	330	328
1975	331	329
1976	332	330
1977	334	332
1978	335	333
1979	337	335
1980	338	337
1981	339	338

Concentrations are average annual concentrations.
NA — Not available.

Source: 1958-1980: R.B. Bacastow and C.D. Keeling, "Atmospheric Carbon Dioxide Concentration, the Observed Airborne Fraction, the Fossil Fuel Airborne Fraction, and the Difference in Hemispheric Airborne Fractions," *in Scope 16; Global Carbon Modeling.* B. Bolen, ed. (London John Wiley and Sons, 1981).

chases, and net exports; and from the supply side as the sum of employee earnings, proprietors' earnings, corporate profit, net interest, rental income, capital consumption allowances and indirect business taxes. The results should be identical as each dollar measures both demand and supply; actual discrepancies are of a statistical and estimating nature. GNP is estimated in both current and constant dollars; a series in constant dollars is a measure of real output (the physical volume of economic activity) over time, and of changes in that output. The measurement of GNP in the United States is undertaken by the Bureau of Economic Analysis in the Department of Commerce, the results being published in the *Survey of Current Business* on a quarterly basis.

Growth Pole Theory: A concept of economic polarization which inevitably leads to geographical polarization; through a continuing process of internal and external economies of scale, resources flow to a limited number of centers within a region. Within this context, the rapid growth of leading industries (offering "propulsive growth") induces the polarization of other economic units into the pole of growth.

In some instances, there may be just one single dominant propulsive firm at the heart of an industrial complex, which it effectively influences. Within the concept may be seen the "trickling down" of "spread" effects so attractive to regional planners.

Guidelines of State Environment Policy, Indonesia: Guidelines introduced into Indonesia by Decree IV of the General Session of the Consultative People's Assembly in 1978. Some relevant extracts are:

- The activities of stock-taking and evaluation of natural resources need to be stimulated, with the aim of understanding better the forest, land, water and energy resources which are vital to development.

- Surveys, exploitation, and the use of natural resources must have regard to the environment, thus ensuring the maintenance of the quality of natural resources and the living environment.

- The realization of development must always be coupled with a conscientious evaluation of its impacts on the living environment, so as to ensure the best possible safeguards for the realization of development. The evaluation must be done sectorally and regionally, and to this end it is necessary to develop criteria for the qualities of the living environment.

- The rehabilitation of damaged natural resources including soil and water must be further intensified, through integrated approaches to the river basin and land-use generally.

- The utilization of coastal regions and water resources needs

intensification, without inflicting harm to the living environment's quality and its conservation.
- In founding new settlements, priority must be given to the improvement of the living environment of low-income people.

H

Habitat Evaluation Procedure (HEP): A system devised by the US Fish and Wildlife Service in 1976 for use in the evaluation of habitats in the vicinity of major federal water projects. It determines the quality of a habitat type using functional curves relating habitat quality to quantitative biotic and abiotic characteristics of the habitat. Habitat sizes and quality are combined to assess project impact. As shown in Figure 8 the HEP involves six steps for the evaluation of impacts of a development project.

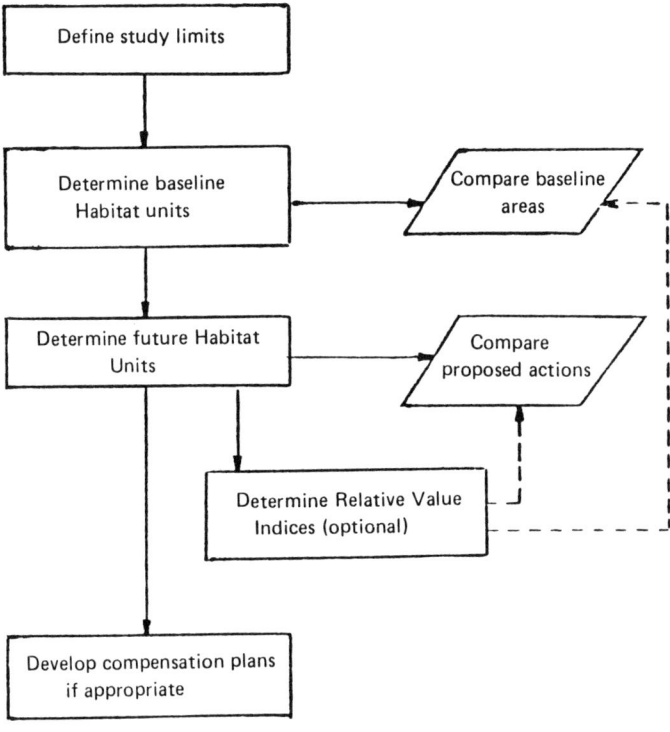

Figure 8: Generalized evaluation process using HEP. Source: U.S. Fish and Wildlife Service, 1980.

Habitat: The United Nations Conference on Human Settlements: An international conference held in Vancouver, Canada, in 1976; 131 governments were represented. A Habitat Forum for non-government organizations and interested individuals ran concurrently.

The Conference brought to fruition two years of preparatory work and studies with the adoption of a Declaration of Principles and of a series of recommendations to assist in meeting the urgent problems of housing shortages, crises of urban and rural communities, the proper use of land, access to essential services, and public involvement and participation in remedial action. The "blueprint" for national and international action to improve the living places of people throughout the world comprised:

- A "Vancouver Declaration on Human Settlements, 1976" which proclaimed that "The improvement of the quality of life of human beings is the first and most important objective of every human settlement policy." This declaration is also known as the Declaration of Principles.
- A "Vancouver Plan of Action," a set of 64 recommendations suggesting concrete ways in which people might be assured the basic requirements of human habitation—shelter, clean water, sanitation and a decent physical environment, plus the opportunity for cultural growth and the development of the individual.

The report on the Habitat Conference was considered by the United Nations General Assembly late in 1976, in New York. The Assembly responded by creating a Commission on Human Settlements and Habitat, a Centre for Human Settlements. The Assembly resolved that there should be close links between the new Centre and the United Nations Environment Program (UNEP) and that accordingly the Centre would also be located in Nairobi.

Han River Basin Environmental Master Plan, Korea: A plan prepared by the Seoul Metropolitan Government to drastically improve and develop the Han River in the Seoul area. The length of the Han River in Seoul is 36 km while the basin covers an area of 1,206 sq. km. Swimming in the Han River downstream of Seoul has been prohibited since the late 1960s. A new source of water became necessary, and since 1982 water has been supplied to Seoul from the Paldang intake station at the Paldang Dam located some 40 km upstream from Seoul. New sewage treatment plants are being constructed.

The Master Plan extends into the late 1980s. The riverbed is to be dredged, with many recreational facilities including sports grounds and resorts being established. Large grassed areas are to be established with nature study gardens, fishing grounds and ponds. Some 26 km of riverside roads are to be constructed with navigation extended from the Kimpo bridge to the Paldang Dam. In order to collect the domestic sewage and industrial wastewater from the areas not served by sewage treatment plants, interceptors some 145 km in length are to be established, cutting off the inflow of wastes into the river; these wastes will be discharged downstream until new treatment capacity is ready in the period 1987-2001. See *Environmental Conservation Law (ECL), 1977, Korea; Saemaul Undong (New Community Movement), Korea.*

Hawaii, US: See *Litter Laws, US; State Plan, Hawaii.*

Hazard and Problem Areas: Areas that due to their natural conditions prohibit new development, or that are difficult to develop, and where care must be taken to avoid environmental damage and economic loss.

Hazardous Air Pollutants: See *Clean Air Amendment Act, 1977, US.*

Hazardous Chemicals: See *Environmentally Hazardous Chemicals.*

Hazardous Substance Response Trust Fund, US: See *Comprehensive Environmental Response, Compensation, and Liability Act, 1980, US.*

Hazardous Waste: Any kind of waste which may constitute a danger to the life or health of living organisms when released, or to the safety of humans, or equipment, if incorrectly handled during disposal operations. It includes toxic, infectious, explosive, flammable, corrosive, oxidizing, and radioactive substances.

Special precautions have to be taken in the handling and disposal of hazardous wastes, such as tarry liquids, waste paint, oil impregnated rubbish, arsenic wastes, beryllium wastes, wastes containing cyanide, spent sheep dip, sulfides, acid tars and sludge from leaded gasoline. The UK Deposit of Poisonous Waste Act 1972 places upon the disposers of any poisonous, noxious or polluting waste a duty to notify responsible authorities—the relevant local and water authorities—of any intention to remove or deposit the waste. It also made it an offense to deposit on land such waste in any way which might give rise to an environmental hazard, that is causing danger to people or animals, or polluting water supplies. See *Environmentally Hazardous Chemicals; Heavy Metals.*

Health: Defined in the preamble to the constitution of the World Health Organization (q.v.) as " . . . a state of complete physical, mental and social well-being and not merely the absence of disease or infirmity." While this definition has been widely accepted in broad principle, techniques for measuring health as so defined have not evolved; hence most assessments of health still rely upon mortality and morbidity statistics.

The strong relationship between health and environmental improvement was stressed in the 1979 World Development Report published by the World Bank (q.v.):

> "The health of the population... benefits substantially from improvements in nutrition and sanitation. The secular improvement in health

standards in Western Europe and North America, for example, followed from the rise in living standards and better social conditions, rather than from improved medical care; the incidence of cholera and typhoid fell in the United Kingdom and the United States of America long before effective methods of treatment were available for these diseases. Similar results have been observed in the developing countries: in the Philippines, for example, improved water supply and sanitation have reduced the incidence of cholera by about 70 per cent. Significant improvements in the health status of the poor may not be forthcoming until water supply and sanitation are adequate." (*World Bank*, 1979, p. 94).

Health and Safety at Work Act, 1974, UK: An Act which created a Health and Safety Executive for Britain, and placed a general duty on those concerned to ensure that their work places were made and kept safe and healthy. Apart from supervising work places, the Executive was given an additional role in the assessment of public safety from potentially hazardous industrial development. For the first time, risk assessment was introduced to the main line of planning debate. The Executive and the planning authorities became responsible for ensuring that the public was properly protected from the external effects of works activities.

One of the key roles of the Executive is to provide information and guidance to planning authorities in cases where proposed development could prejudice community safety. Where plans and policies are being prepared to guide development, indicative standards are provided only. However, where specific proposals classed as hazardous developments are proposed, a compulsory notification system allows the Executive to consider each case and provide the planning authority with the detailed physical separation and safeguarding distances that should be observed. The standards adopted are kept under review by the Hazardous Substances Policy Group of the Executive. The Major Hazards Branch of the Executive carries out risk appraisal work on large-scale hazards and provides detailed advice to the authorities. See *Department of the Environment, UK; Planning System, UK*.

Hearing Examiners, Los Angeles, US: Professional staff members of the Los Angeles Planning Commission who assist the Commission in conducting hearings concerning zoning changes, conditional uses, building lines, or height districts. An examiner is empowered to hold hearings as a representative of the Commission; his or her findings are included in a formal report. The Chief Hearing Examiner or the Planning Director may disagree with those findings; the report may not be changed, however, any dissenting opinions being conveyed with the report to the Planning Commission. See *Board of Zoning Appeals; Public Hearings and Inquiries; Zoning Administrator*.

Hearings: See *Public Hearings and Inquiries*.

Heavy Metals: A term used very loosely for a wide range of elements not all of them "heavy" or even metals; however, toxicity is always clearly implied when the term is used in a pollution context. Since the 1960s, attention has focussed especially on the heavy metals lead, cadmium and mercury; they have been shown to present in certain situations an environmental hazard, in each case mainly through the ingestion of excessive quantities of the metal.

Lead is the most widespread potential hazard since quite small increases in lead consumption can raise blood lead levels to a point where biochemical changes are detectable. The risks from cadmium appear limited mainly to groups of people consuming food produced in areas when the soil or irrigation water has been contaminated. Mercury has become a problem where populations eat large amounts of fish taken from contaminated waters or waters with a high natural mercury content. In all cases, children appear to be more sensitive to exposure to heavy metals than adults. At moderate levels of exposure to lead, impairment of the central nervous system may occur while severe lead poisoning can cause major damage to the brain.

Probably the most notorious case of cadmium toxicity was the disorder known as itai-itai disease which appeared in Japan in the late 1940s. The disease arose from an increased uptake of cadmium in locally consumed rice grown in paddy fields irrigated with cadmium-contaminated river water. The disease was essentially an osteomalacia, associated with kidney damage and proteinuria. The problem was intensified by the nearly regular daily rice diet at the affected site.

The two major epidemics of methyl mercury poisoning also occurred in Japan, in Minamata Bay and in Niigata; they were caused by the industrial release of methyl and other mercury compounds into Minamata Bay and the Angano River, followed by the accumulation of the mercury by edible fish.

Epidemics resulting in a large number of cases of poisoning and fatalities have been caused by the ingestion of contaminated bread prepared from wheat and other cereals treated with alkyl mercury fungicides. The largest recorded epidemic occurred in Iraq in 1971-72 resulting in the admission of over 6,000 patients to hospitals and over 500 deaths.

Not all mercury at significant concentrations is the result of industrial operations. A five-year study of mercury pollution in the Mediterranean revealed that tuna and sword fish have mercury levels ten times higher than those in the Atlantic. However, this is the result not of man-made pollution but almost entirely of natural pollution from mercury rich rocks and the outgassing of volcanoes.

Helsinki Convention: See *Convention on the Protection of the Marine Environment of the Baltic Sea, 1974*.

Herbert Report, 1960, UK: The report of the Royal Commission on Local Government in Greater London which recommended that an elective Greater London Council should be created and charged (as one of its most important tasks) with the planning of Greater London. Town planning in London since the Second World War had been guided by the Aberecrombie plans (q.v.); a Greater London Council would be able to address more adequately the issues of the larger metropolitan area and to review the 1944 plan. The twin problems of physical planning and transportation were considered to need urgent attention.

Heritage Act, 1977, New South Wales: An Act introduced by the New South Wales Government, Australia, for the purpose of creating an 11-member Heritage Council and the introduction of interim and permanent conservation orders in respect to heritage items whether buildings, objects, or sites. The Heritage Council maintains a register of buildings and other property subject to orders; advised the Minister for Planning and Environment in respect to new orders and other measures; and publishes an annual report on progress in conserving the state's heritage. An interim or permanent conservation order prevents the demolition of buildings and brings the approval of alterations under the control of the Council. Provision is made for appeals and public inquiries.

Heritage Buildings List, UK: Lists of buildings of special architectural or historic interest compiled, as required by the planning Acts, by the Secretary of State for the Environment and the Secretaries of State for Scotland and Wales; over 166,000 buildings are already listed. It is an offense to demolish or alter the character of any listed building without special consent from the local planning authority or the appropriate Secretary of State; where consent is given to demolish a building, the Royal Commissions on Historical Monuments have an opportunity to make a photographic record of the building. Emergency "building preservation notices" can be served by the local planning authority to protect buildings not yet listed. The respective Secretaries of State (on the recommendation of the appropriate Historic Buildings Council) can make grants and loans for the repair or maintenance of buildings (or groups of buildings) of outstanding interest, and local authorities can make grants and loans for any building of architectural or historic interest even if it is not listed. Over 2,000 buildings in Britain have received grants and loans under the Historic Buildings and Ancient Monuments Act 1953. Local planning authorities have designated for special protection some 2,000 "conservation areas" of special architectural or historic interest. Special studies have been made in Bath, Chester, Chichester and York of the difficulties of preserving features of historic interest in ancient towns in the face of commercial development and increasing motor traffic. The Secretaries of State for the Environment, Scotland and Wales, are responsible for the maintenance of royal parks and palaces and for the protction of ancient monuments of which over 700 are in their care. The Royal Commission on Historical Monuments (for England), and similar bodies for Scotland and Wales, survey monuments and specify those worthy of preservation. In Northern Ireland some 80 historic monuments are maintained and protected by the state.

Heritage Coast, UK: In Britain, areas of coast identified as of outstanding beauty and merit. The Countryside Commission has identified over 40 stretches of the coastline of England and Wales, totalling over 1,260 kilometers, as "heritage coast." These stretches, representing some 30 per cent of the total length of the coastline comprise undeveloped and outstanding national coastal scenery. The Commission has, in cooperation with the concerned maritime local authorities, defined many of these heritage coasts in detail as a first step towards the preparation of comprehensive long-term planning and management programs for these areas. The programs are prepared by the local authorities while the Commission provides a coordinating role coupled with technical and financial assistance. A full range of coastal planning and management methods is being explored.

Heritage Conservation: Measures adopted to restore, preserve and maintain buildings and townscapes of heritage value. In simple terms, conservation of buildings and townscape involves:

- Retaining the existing fabric and setting of the building including detailing, finishes, verandas, joinery, roofs, chimneys etc.
- Reinstating missing components of the building's fabric including detailing chimneys, color schemes, verandas, finishes etc.
- Blending in new alterations and extensions to relate sympathetically with the existing building and townscape.
- Reuse of existing heritage buildings by strata titling or recycling in preference to demolition and redevelopment. Additional floor space may be achieved with new extensions consistent with the design of the building.
- Sympathetically designed new "infill" houses or development in terms of siting, massing, scale, character and materials to relate harmoniously with surrounding buildings.

Demolition is not the only way of destroying historic or architecturally valuable buildings. This loss frequently occurs in a more gradual and less dramatic process through lack of maintenance. This results in deterioration of detailing and the building fabric generally; mediocre additions, substitution of materials, veranda enclosures, inappropriate color schemes and subdivision; unsympathetically designed carports, garages, fences, walls and swimming pools; plus poorly sited and designed new development. See *Conservation Zones; Convention for the Protection of the World Cultural and Natural Heritage; Environmental Heritage; Facadism; Heritage Act, 1977, New South Wales; Heritage Buildings List, UK; Heritage Coast, UK; Heritage Conservation; Heritage Criteria; National Historic Preservation Act,*

1966; United Nations Educational, Scientific, and Cultural Organization; Williamsburg Restoration.

Heritage Criteria: Criteria which are used to identify significant buildings in heritage terms. These include:

- Association with an historic event, person, phenomenon, or institution.
- Being a good example of an architect's work.
- Being a good example of a particular style, which includes regional variations and vernacular buildings.
- Being a landmark or focal point in a townscape or stretch of countryside.
- Forming part of a group of buildings of townscape or rural value.
- Displaying technical or planning evolution such as the early use of concrete.

See *Heritage Buildings and Conservation Areas; Heritage Conservation; Williamsburg Restoration.*

Highrise Folly: Descriptive of the highrise developments of the 1950s and 1960s in both the inner areas and new suburbs of many cities. It proved unpopular among families with children, sometimes structurally unsafe, and led to extensive vandalism. Highrise development took place in Britain, Sweden, in state-supported projects around Paris, and in public projects in the largest American cities.

British cities began to pull down tall housing blocks after less than 20 years' life (or allocated them to students or the homeless). The Greater London Council declared its wish to eliminate all tall apartment buildings as quickly as possible. Much of the highrise development had arisen from a shortage of land, opposition from agricultural interests, the prospect of using new mass production techniques, and the large extra subsidies provided as a consequence for highrise building.

Holford Rules: Basic rules proposed by the late Lord Holford, Professor of Town Planning, University College London, as a guide to minimizing the environmental effects of transmission lines; these rules are set out in *Table 9.*

Housing Act, 1949, US: An Act which created the Urban Renewal Administration with the principal objective of creating better urban housing conditions in the United States. Programs have been directed primarily at slum clearance, although the scope of activities has been progressively widened to embrace commercial area renewal, support for urban mass transportation, and general community planning. Under these

Table 9: Holford Amenity Rules for Transmission Lines

(1) Avoid altogether, if possible, the major areas of highest amenity value, by so planning the general route of the line in the first place, even if the total mileage is somewhat increased in consequence.

(2) Avoid smaller areas of high amenity value or scientific interest by deviation, provided that this can be done without using too many angle towers.

(3) Other things being equal, choose the most direct line, with no sharp changes of direction and thus with fewer angle towers.

(4) Choose tree and hill backgrounds in preference to sky backgrounds whenever possible, and when the line has to cross a ridge, secure this opaque background as long as possible and cross obliquely when a dip in the ridge provides an opportunity.

(5) Prefer moderately open valleys with woods, where the apparent height of the towers will be reduced and views of the line will be broken by trees.

(6) In country which is flat and sparsely planted, keep the high-voltage lines as far as possible independent of smaller lines, converging routes, distribution poles, and other masts, wires and cables, so as to avoid a concatenation or "wirescape."

(7) Approach urban areas through industrial zones, where they exist, and where pleasant residential and recreational land intervenes between the approach line and the substation, go carefully into the comparative cost of undergrounding, for lines other than those of the highest voltage.

Source: Central Electricity Generating Board, London, England.

programs vast projects of urban redevelopment and renewal have been completed in many American cities. See *Community Development Block grant; Urban Development Action Grant Program.*

Housing Densities: Numerical definitions of densities of housing which range from the "very low" to the "very high;" the definitions adopted by the City of Los Angeles, California, are shown in *Table 10.*

Housing Improvement Grants, UK: Grants provided by the British Exchequer under the Housing Act, 1949, and the Housing Repairs and Rent Act, 1954, by way of assistance to improve existing housing. Improvements eligible for assistance included the provision of a bath in a bathroom, hot-water supply, water closet, wash hand basin, and facilities for storing food; up to half the cost might be met. The program, after a slow start, gathered considerable momentum and did much to rehabilitate and extend the useful life of dwellings.

Table 10: City of Los Angeles—Housing Density Definitions

Density	Dwelling Units per Gross Acre**	Persons* per Gross Acre**	Preferred Locations
Very Low	3 or less	12 or less	Remote Suburbs and mountains
Low	3+ to 7	10–30	Suburbs
Low-Medium	7+ to 24	20–75	Fringes of Centers; Regional Core; Suburbs near commercial areas
Medium	24+ to 40	50–100	Centers; Suburbs near commercial areas and on some highways
High	40+ to 80	80–160	Centers
Very High	80+	150 & above	Core of Centers

*This column does not indicate policy but is for statistical purposes only. It illustrates the approximate range of population densities which could be expected.
**Gross Acreage includes streets.

Hungary: See *Council for Mutual Economic Assistance (COMECOM)*.

Hydrological Cycle: The continual exchange of water between the earth and the atmosphere. Since the height of the ocean surfaces remains essentially unchanged from year to year, evaporation from the oceans must equal the rainfall over the oceans plus the run-off from rivers and streams and effluent discharges. See *Figure 9*.

142   Environmental Planning

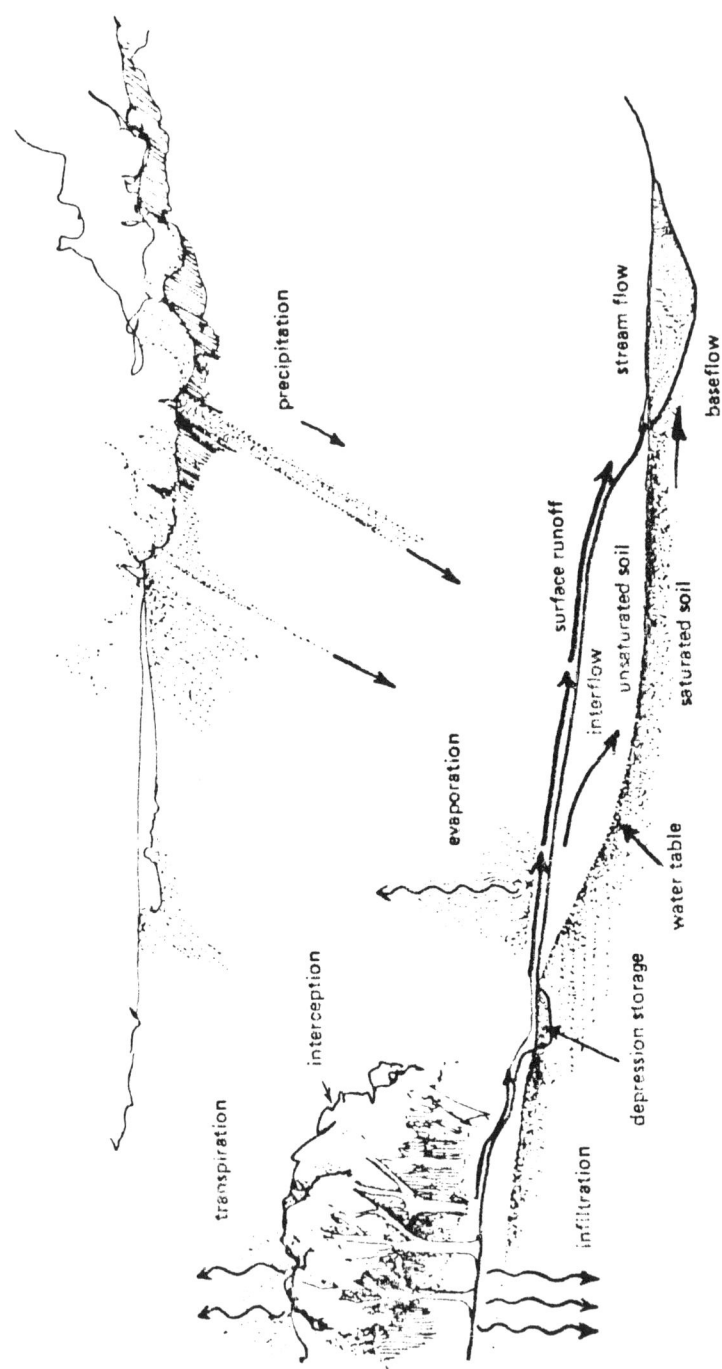

Figure 9: The Hydrological Cycle. Source: *Cooks River Environmental Survey and Landscape Design*, 1976, New South Wales Department of Environment and Planning.

I

Incompatible Uses: Land uses, situated in relation to each other in such a way as to be regarded by many as in conflict, e.g., reducing the environmental and social quality of residential areas. Examples are inappropriate commercial uses side by side with residential ones, mining and parks, heavy traffic in residential streets, junkyards on scenic highways.

Prior to 1920, Washington D.C., for example, had no comprehensive zoning controls; as a result many areas of the District developed with commercial, industrial, and other non-residential uses scattered in residential neighborhoods. The adoption of Euclidian zoning controls, with their segregation of land uses, rendered many existing establishments non-conforming, that is, not in accord with land use law.

India: See *Bhopal Disaster, 1984; Chipko Andolan Movement; Forest Products; Jhumming; National Committee on Environmental Planning and Coordination; Overgrazing; Social Forestry.*

Indonesia: See *Guidelines of State Environment Policy.*

Industrial Air Pollution Inspectorate: A British central government inspectorate which for over 100 years has enforced air pollution control legislation relating to the heavy and chemical industries. The legislation (until recent years, the Alkali Act) requires that:

- Works scheduled under the Act may be operated only after a certificate of registration, renewable annually, has been obtained;
- As a condition of first registration of such works, the owners must satisfy the chief inspector, that the plant is furnished with the "best practicable means" for preventing the discharge of noxious or offensive gases to the atmosphere, and for rendering such gases, when discharged, harmless and inoffensive;
- After registration, the apparatus which constitutes the "best practicable means" must be adequately maintained, and the process must be properly supervised by the owner;
- For certain processes the concentrations of acids in the waste gases must not exceed the maximum values laid down.

- The owner of a registered works must provide all necessary facilities for entry, inspection, examination, and testing by an inspector, and must also disclose in confidence such details of the process as are required by the inspector;
- The chief inspector must report annually to the minister on the work done by the district inspectors.

In 1975, the Alkali Act was repealed and embodied in the Health and Safety at Work Act 1974. This has meant virtually no change in the actual legislation controlling emissions from the scheduled processes. The alkali inspectorate has retained its identity, to a large extent, within the Health and Safety at Work Commission. The principle of the "best practicable means" has been preserved.

The Report of HM Alkali and Clean Air Inspectorate for 1982 estimated that some £400 million was being spent every year on combating air pollution in England and Wales, despite the effects of economic depression. About half of this was spent by firms registered with the inspectorate. Improvements included marked reductions in lead and vinyl chloride emissions. The report also noted some impressive improvements in mineral works operations and encouraging progress in the control of pollution from coke works.

However, the report also commented that too many planning applications resulted in polluting industrial developments being sited near houses, shops, schools or hospitals. The report supported the recommendation of the Fifth Report of the Royal Commission on Environmental Pollution that there should be a mandatory obligation on planning authorities to consult the inspectorate on applications to build or alter registrable works. See *Best Practicable Means*.

Infant Mortality Rate: The annual deaths of infants under one year of age per thousand live births. An early crude indicator of social welfare, it continues to be a valuable indicator of those aspects of the quality of life impinging on health, both personal and environmental, and the development of health services generally. The infant mortality rate in western industrialized countries is about 15 per thousand live births; in communist countries generally about 25; and in poorer developing countries about 25-130. Countries with a rate exceeding 100 during the late 1970s included Brazil, India, New Guinea, Nigeria, the Sudan, and Turkey. See *Environmental Health; Health*.

Infill Development: The development of vacant land or the carrying out of additions or extensions to existing development, within localities that are substantially developed.

Information and Documentation System for Environmental Planning (UMPLIS), West Germany: A highly coordinated, systematic procedure developed to achieve purposeful planning, and monitoring of government-sponsored environmental research.

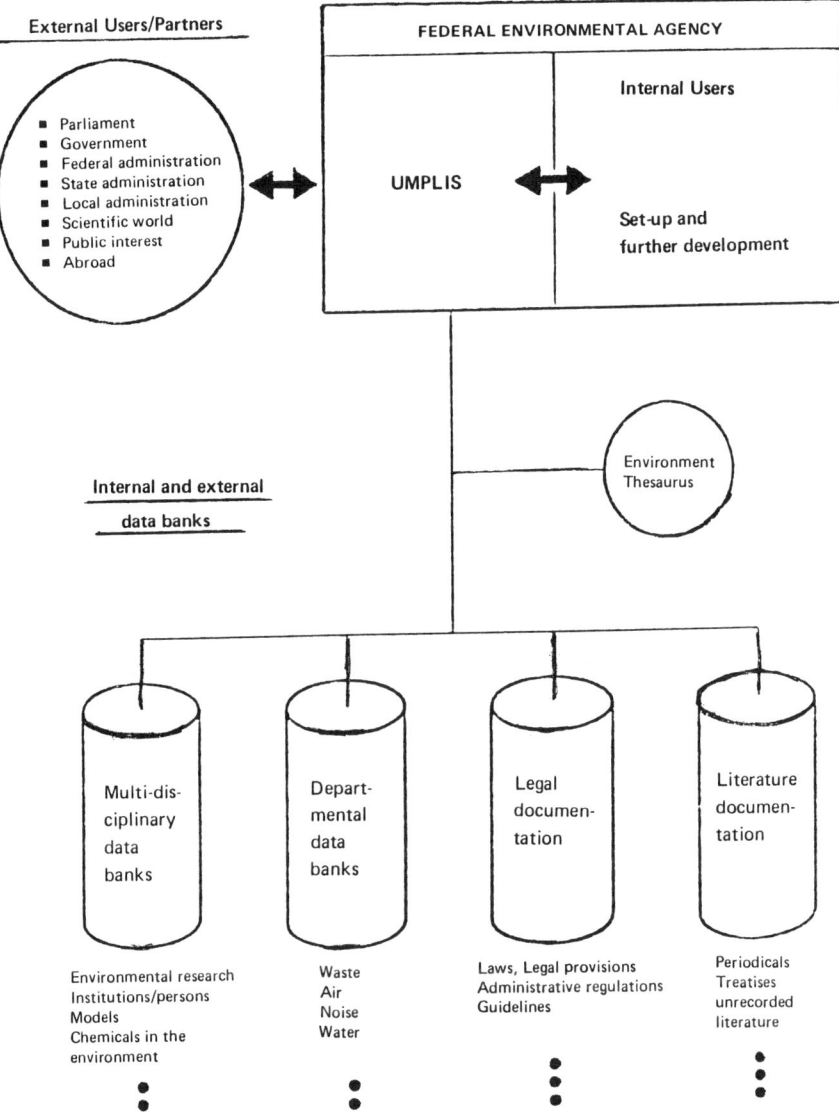

Figure 10: West Germany–UMPLIS. Source: Federal Environmental Agency, West Germany.

The data used to record and describe a project is prepared in accordance with standard instructions using standard data sheets and is stored in the Information and Documentation System for Environmental Planning (UMPLIS) data bank for environmental research. The technical content of the research and development projects is indexed separately according to activity and science in a standard classification system upon which the different parties involved agreed at the beginning of 1976 so that objective information can be obtained for coordination purposes. This data bank contains all the information necessary for official financial and administrative clearance and coordination of government sponsored projects.

A Federal Government directive of 1976 ensures that a contract for a project is only to be awarded subject to all relevant information being entered in the UMPLIS system.

UMPLIS should be able to provide: Information aids for coordination and cooperation in environmental research and development; information services in the environmental field; and planning aids and instrumental aids for administration and determining policy for the benefit of the users. See *Figure 10*. See *Federal Environmental Agency, West Germany*.

Infrastructure: The system of essential services, utilities, public and community facilities, e.g., water, sewerage, power, roads, schools, health facilities and so on, which are necessary to enable urban development to function.

Injurious Affection: A term referring to the depreciation in the value of land caused by the adverse effects of public works through noise, vibration, smell, smoke, fumes, artificial lighting, overshadowing, loss of support, and restriction or loss of access. Injurious affection is usually associated with the carrying out of substantial public undertakings such as the construction of a freeway or an airport. Land surrounding an airport may well diminish in value because of noise resulting from aircraft. In some cases, injurious affection may result in compensation though usually only when the land is acquired for the project. However, under British legislation compensation is not so limited, being payable outside the actual acquisition area.

No compensation is payable to the community for injurious affection of community facilities. Nor do the law reformers suggest there should be. Indeed, it is hard to imagine how there could be compensation. The community must simply suffer the adverse affects of a road upon its facilities, including open space. The road, after all, is built because it serves the public interest. It will be important, therefore, to consider whether the community which suffers a degradation of its facilities (and especially open space) by the proximity of the road, nonetheless derives an equitable share of the benefits arising from the road. See *Land Compensation Act, 1973, UK*.

Inner Urban Areas Act, 1978, UK: An Act which provides for the designation of areas where local authorities, sometimes in partnership with central government, may participate in the economic development of their area by making loans and grants for industrial development, including the improvement of the industrial environment through the declaration of Industrial Improvement Areas.

Insecticide, Fungicide, and Rodenticide Act, US: A federal measure, amended by the US Congress in 1978, which provides for the regulation by the Environmental Protection Agency (q.v.) of pesticide substances; such substances include products to control insects, weeds, and disease in agricultural production, disease vector control (e.g., mosquitoes or rabid animals), and products used as hospital or home disinfectants and sterilants. The Act directs the EPA to regulate such pesticide substances to ensure that they do not cause "unreasonable adverse effects" on humans or the environment, requiring a balancing of risks against benefits in agriculture, public health, and the economy. Every pesticide marketed in the United States must obtain premarket clearance and registration from the EPA; and pesticide residue tolerances (legally acceptable levels) or exemptions must be established for pesticides used on food or feed. In 1985, a major review of this legislation was initiated. Among the major issues were whether states should have the right to enforce standards for pesticide use that are tougher than the federal government's standards; whether the public should have access to confidential health and safety data filed by pesticide manufacturers; and whether the EPA should share pesticide data with foreign governments.

Institutional Planning: Planning by important public bodies not central to the planning process yet having significant implications for environmental planning as a whole. One of the functions of the central planning body is to accommodate and coordinate proposals to enable agency objectives to be reconciled with overall environmental planning objectives. The planning of institutional agencies often embraces new schools and colleges, hospitals, communication facilities, highways and public transport facilities, public utilities, and national parks.

Interaction Matrix: One of the earlier methodologies used in the United States of America and elsewhere to examine the environmental impacts and effects of development projects. The matrix consists of a display of project actions or activities along one axis, with appropriate environmental factors listed along the other axis of the matrix.

When a given action or activity is anticipated to cause a change in an environmental factor, this is noted at the interaction point in the matrix and further described in terms of magnitude and importance. Many variations of the simple interactive matrix have been utilized in environmental impact studies.

Interim Zoning: An emergency measure that may be adopted by a county or municipality by way of a temporary interim zoning map or temporary interim zoning ordinance for the purposes of controlling a proposed use or development; the purpose of interim zoning is to protect the public health, safety, and general welfare of the community. Generally such measures can only be adopted if a county or municipality is conducting, or in good faith intends to conduct, studies within a reasonable time for the purpose of considering a comprehensive plan (q.v.) or official controls (q.v.) in respect to any territory for which no zoning has been previously adopted.

Internal Rate of Return (IRR): A method of assessing the economic viability of a project; it is defined as the rate of return or discount rate which makes the investment in question marginal. A commonly used discount rate of 10 per cent may produce a benefit/cost ratio for a project of say 1.39. The question then is: "What discount rate would have to be used to reduce the benefit/cost ration to 1.0 (at which point the project would be marginal)." In this instance, the figure would be 17 per cent; that is the "internal rate of return." The higher the discount rate, the more attractive the proposition in economic (if not environmental) terms. The method is valuable when priorities need to be established in relation to a limited budget.

International Atomic Energy Agency (IAEA): An autonomous intergovernmental organization with headquarters in Vienna, Austria. It is related to the United Nations by the terms of an Agreement which recognizes it as "the agency under the aegis of the United Nations responsible for international activities concerned with the peaceful uses of atomic energy." The agency was established in 1957.

International procedures exist to verify that nuclear material is not diverted from a country's peaceful nuclear industry to nuclear weapons. The main systems of safeguards are administered by the International Atomic Energy Agency. IAEA safegaureds are applied under an agreement between the IAEA and the country being inspected. They operate by requiring the country to make periodic reports of quantities, disposition, and usage of nuclear material it possesses. IAEA nuclear inspectors correlate what they find, by inspection and sampling, with what the country declares. To the extent necessary to obtain proper confidence in its findings, the IAEA also applies the complementary techniques of containment (e.g., seals on locks, safes) and independent surveillance (e.g., automatic cameras under the control of IAEA) to the material it safeguards.

The term "safeguards" is also commonly used in a wider sense to encompass additional conditions which a nuclear supplier country might require to be included in a bilateral agreement with an importing country. For example, an exporting country such as Australia may impose conditions in relation to high enrichment of uranium or reprocessing of spent fuel, since these activities can result in a country's acquisition of types of nuclear material which have potential weapon use. Australia's policy, announced by the Prime Minister in 1977, combines safeguards of this type with international safeguards in order to provide the highest degree of confidence that uranium supplied to other countries for peaceful purposes is not used for nuclear weapons.

International Biological Program (IBP): See *International Council of Scientific Unions (ICSU)*.

International Commission for the Protection of the Rhine: An international commission created by the riparian countries in 1950 for the protection of the River Rhine against pollution. The commission includes representatives from France, Luxembourg (in connection with the Moselle), the Netherlands, Switzerland, and West Germany. The secretariat is based in Koblenz, West germany. Specifically, the responsibilities of the commission are to: determine the amount and nature of pollution in the river at various stages of its course and at various times of the year; identify the sources of pollution; compare the legislation of the member countries relating to water pollution control; collate national inventories of discharges and receive reports on monitoring results; prepare and submit recommendations to the governments concerned.

The Commission set up a monitoring system for the waters of the Rhine; sampling is conducted at selected points in accordance with an established schedule and a laboratory testing program. The member countries also conduct monitoring and report to the commission.

The Commission was given legal status by a Convention signed by five countries in 1965. For some years, however, there was no international convention committing the riparian countries to take concerted action against water pollution. National legislation alone was relied upon. Nevertheless, the commission obtained much closer cooperation between the countries involved. Furthermore, the water supply authorities revealed a direct interest in the detection and prevention of dangerous contamination.

International Commission on Radiological Protection (ICRP): A commission founded in 1928 to provide technical guidance and promote international cooperation in the field of radiation protection. About 50 countries are represented on the Commission. The Commission has standing committees on radiation effects, internal and external exposures, and on the application of recommendations.

The Commission makes the cautious assumption that any exposure to radiation, however small the dose (or low the dose-rate), carries some risk for the development of leukemia, other forms of cancer, and of the induction of genetic effects. That is, there is no wholly safe dose of radiation. The commission believes that the policy of assuming some risk at low doses is the most reasonable basis for radiation protection.

Nevertheless, the Commission has published recommendations of allowable exposure both for radiation workers and for the population at large. For the public, the Commission recommends an annual dose limit for individuals of 500 millirems to the most sensitive organs, or one-tenth of the maximum allowable for radiation workers. The Commission has also made a recommendation for the limitation of genetic dose to the whole population amounting to an average of 5,000 millirems per individual over a thirty year period, additional to natural background radiation.

It has suggested that, for planning purposes, the average concentration of radioisotopes in air or water, applicable to the population at large, should not exceed one-thirtieth of the maximum concentration allowed for occupational exposure. This reduced concentration is equivalent to an annual exposure of 170 millirems for members of the public in continuous contact with such concentrations. The United States Federal Radiation Council had adopted this criterion, though some investigators consider this level too high.

International Commons: See *Commons, The*.

International Council for the Exploration of the Sea (ICES): Created under the auspices of the United Nations, an agency concerned primarily with all aspects of marine exploration; in addition it has coordinated marine pollution studies in the North Atlantic, the North Sea and the Baltic Sea. The baseline studies conducted involved all the countries bordering these seas; in 1976, it published a *Comprehensive Plan for the Global Investigation of Pollution in the Marine Environment* (GIPME). See *Regional Seas Program, UNEP; United Nations Conference on the Law of the Sea*.

International Council of Scientific Unions (ICSU): A non-government body with a membership drawn from universities and research institutes throughout the world; many members are environmental scientists. Since the 1950s, the International Council, often in concert with one or more intergovernmental organizations, has planned, promoted and successfully developed a series of international field research programs of diverse character and far reaching scientific and environmental importance. These have included the very notable achievements of the International Geophysical Year (IGY), the International Biological Program (IBP), the Global Atmospheric Research Program (GARP), as well as other long-term investigations.

In 1966, the International Council, in recognition of the rapidly developing environmental crisis, decided to bring into being a broadly based body of scientific expertise especially designed to make a scientific review and assessment of the more urgent environmental problems and to examine them by means of studies in depth supported by field investigation, encompassing not only naturally occurring man-made environmental changes, but also the effects of these changes on man and living ecosystems. In 1969, ICSU set up a permanent body within its organization, the Scientific Committee on Problems of the Environment (SCOPE), charging this body with the development of these studies through the full deployment of all relevant sources of expertise and in close working relationship with other non-governmental bodies and intergovernmental organizations both within and external to the United Nations System.

In 1984, the Council unanimously endorsed a resolution calling for a worldwide project to study the interactions of the earth's physical, chemical, and biological processes. The ultimate goal of this federation of 20 scientific organizations and 71 national academies of science is to address such global problems as acid rain, desertification, erosion of the ozone layer of the atmosphere, and the possible climatic effects of a build-up of carbon dioxide in the atmosphere. The resolution had the support of the United States and the Soviet Union. See *US-USSR Joint Committee for Environmental Protection*.

International Drinking Water Supply and Sanitation Decade (IDWSSD): The decade 1980-1990 declared by the United Nations to be the International Drinking Water Supply and Sanitation Decade; the aim is to bring clean water and adequate sanitation to everyone by 1990. This involves bringing a water supply to 1,800 million people, and providing sanitation for 2,400 million people. The program will cost more than $600 billion at 1978 prices. However, the World Health Organization estimates that poor sanitation is responsible for some 25 million deaths a year.

International Energy Agency (IEA): An autonomous body established within the framework of the Organization for Economic Cooperation and Development (q.v.) to implement the international energy program, adopted by the participating countries in November 1974. The Agency is supported by over 20 countries.

Major decisions are made by a governing board composed of ministers from each of the participating countries. Four standing groups report to the board, each group dealing with an important segment of the program. In addition, a high-level committee on energy research and development promotes cooperation in these areas among the countries involved.

The Agency has issued a list of recommended energy conservation measures. These include: the pricing of energy at world market levels; changes in utility marketing practices and price structures to reward conservation; comprehensive public education measures; programs to increase the use of waste heat from electricity generation and from industrial processing; priority for government funding of energy efficient public transport systems; changes in building codes and standards to promote thermal and lighting efficiency in new commercial and public buildings, and in new residences; energy efficiency labeling for all major consumer appliances, e.g., water heaters, air conditioners, refrigerators, freezers, and automobiles; speed limits (e.g., 90-110 km/h) on all highways; programs to stimulate energy efficiency in industrial production; policies and programs to improve the efficiency of electricity generation such as peak load pricing and thermal storage.

International Labor Organization (ILO): Established in 1919, an intergovernmental agency in which representatives of governments, employers and workers participate; it seeks through international action to improve labor conditions, raise living standards, and promote productive employment. The ILO was recognized by the United Nations in 1946 as a specialized agency. It has 123 members.

The ILO is concerned, therefore, with the working environment. Within this context, it has obtained international support and implementation of labor conventions. Among these may be mentioned the 1974 convention on occupational cancer; and the 1977 convention on air pollution, noise and vibration in the working environment. Many model codes have been established.

Reports issued prior to meetings of the International Labor Conference review subjects such as vocational training, hours of work and rest periods, prevention and control of risks caused by exposure to carcinogens, the effects of noise, and the impact of recent technical and scientific advances on the health and safety of employees.

In 1978, ILO hosted a UNEP convened inter-agency consultative meeting on the working environment. Its purpose was to identify gaps and overlaps in the activities of the agencies and to explore the scope for inter-agency coordination.

In order to promote the effective involvement of the trade unions and workers in the development and implementation of environment protection policies, UNEP supported the ILO in the preparations for a seminar, held early in 1980, on trade union attitudes, policies and programs in relation to the environment. The ILO is currently participating, with the World Health Organization (q.v.), in an International Program on Chemical Safety, involving actual experimental and epidemiological studies.

International Maritime Organization (IMO): An arm of the United Nations system, the International Maritime Organization (previously the Intergovernmental Maritime Consultative Organization) has far-reaching responsibilities for maritime safety and regulation. In the environmental arena, the principal objectives of IMO have been to: control the operational discharge of pollutants from ships, working towards ultimate prohibition; prevent accidents at sea from which massive pollution can arise; establish conditions for the safe carriage of polluting cargoes; reduce the release of pollutants if accidents should occur; recommend practical measures for dealing with pollution when it occurs; facilitate action by sovereign countries outside their territorial seas, to mitigate or eliminate pollution damage resulting from casualties; provide means of compensation for pollution damage.

In the pursuit of these objectives, IMO has initiated six major international agreements. A conference on Maritime Pollution was held in 1973. The International Convention for the Prevention of Pollution from ships was a convention concluded in London in November 1973, at the end of the Conference. The conference—the largest ever held on the subject— was attended by 500 representatives of the leading maritime nations.

International Referral System (IRS) (INFOTERRA): An information system created by the United Nations Conference on the Human

Environment, 1972 (q.v.). The system, which became fully operational in 1977, aims to put in touch those who seek information on environmental issues with those best able to provide such information.

Although the central unit of the System is located in Nairobi, it functions through a worldwide network of national, regional and sectoral information systems. National offices appointed by governments of participating countries form the hub of the system. They have among their principal tasks to identify and to register sources of environmental information within their respective countries as well as to act as a switchboard between users and sources of information. Some 113 countries participate.

Once the sources are identified, details about them are fed into an International Directory, somewhat similar to "Yellow pages" of a telephone directory. Available on magnetic tape, as well as in printed form or microfiche, the Directory can be quickly scanned to identify those sources likely to be of help on a particular environmental problem. Each national office also maintains a file of their local sources. The International Directory has about 9,000 reference sources, classified under 1,000 different subject headings.

An interesting example of how the system works is provided by Egypt. Work is underway in Egypt to transform Suez City into a major industrial area. To prevent environmental problems arising from industrialization, the Egyptian Government's Suez Canal Regional Plan called for a detailed environmental impact analysis. An expert attached to this venture contracted the IRS to obtain sources of information on air and water pollution, resulting from industrialization. The IRS promptly replied with some 20 sources dealing with the topics. Project experts and the Egyptian Government contracted seven of these sources and found them extremely useful. As a result, the Egyptians have embarked on a major effort to protect air quality in Suez City.

International Register of Potentially Toxic Chemicals (IRPTC): A part of Earthwatch (q.v.), a world information center for toxic chemicals. Effective since 1975, the center was established under the Action Plan adopted by the United Nations Conference on the Human Environment, 1972 (q.v.).

International Security and Development Act, 1985, US: An Act authorizing foreign aid for the United Nations Environment Program (q.v.) and making provision for international population assistance.

International Union for the Conservation of Nature and Natural Resources (IUCN): An independent, international body whose main objective is promoting or supporting action to ensure the perpetuation of wild nature and natural resources in as many parts of the world as possible. Formed in 1948, it has its headquarters in Switzerland. Membership comprises states, government agencies, private institutions, and international organizations.

A Review Panel appointed by the Council of IUCN, in which UNEP and the World Wildlife Fund (WWF) participated, met from 12 to 16 June 1978 at Morges, Switzerland, to discuss the preparation of a World Conservation Strategy (q.v.), which is intended to present a comprehensive and coherent picture of the planet's conservation needs.

International Whaling Commission (IWC): An international body formed in 1946 with the purpose of framing regulations for "the conservation, development, and optimum utilization of whale resources." It operates under the terms of the International Whaling Convention. It meets each year to review whale catches and consider the conditions of whale stocks. Its original membership of 14 nations has gradually increased to include a number of non-whaling nations.

Prior to 1972, quotas were expressed in terms of an overall "blue whale unit," rather than by separate species. One blue whale unit equalled one blue, or two fin, or two-and-a-half humpback, or six sei whales, the equivalence being based on relative oil yields. This approach resulted in the successive depletion of one species after another. In 1972 this unit was abolished and quotas in the Antarctic and North Pacific were then set by individual species. In the same year an Iternational Observer Scheme was brought into operation; under the Scheme member-nation observers were stationed at each others' factories. Thus a measure of enforcement of regulations was brought into being.

Several species of whale had undoubtedly been hunted to the point of commercial extinction—notably the blue, humpback, and finback whales.

At the meeting of the International Whaling Commission in Brighton, England, in July 1981, Japan (the world's leading whaling nation) was outvoted by 25 to one (with three abstentions) when delegates imposed an indefinite ban on the killing of sperm whales. The small, minute whale remains unprotected because it is in no danger of extinction. At that time whaling nations accepted commercial bans only on the hunting of genuinely endangered species such as the blue, humpback and bowhead. In July, 1982, however, the Commission voted to ban commercial whaling worldwide beginning in 1986.

Intractable Waste: Any type of waste which is potentially hazardous or persistent in the environment and which requires special treatment and/or disposal to destroy it. High temperature incineration is the most effective means of destroying intractable wastes.

Inversion: Temperature inversion in the atmosphere in which the temperature, instead of falling, increases with height above the ground. With the colder and heavier air below, there is no tendency to form upward currents and turbulence is suppressed. Inversions are often formed in the late afternoon when the radiation emitted from the ground exceeds that received from the sinking sun. Inversions are also caused by katabatic winds, i.e. cold winds flowing down the hillside into a valley, and by anti-

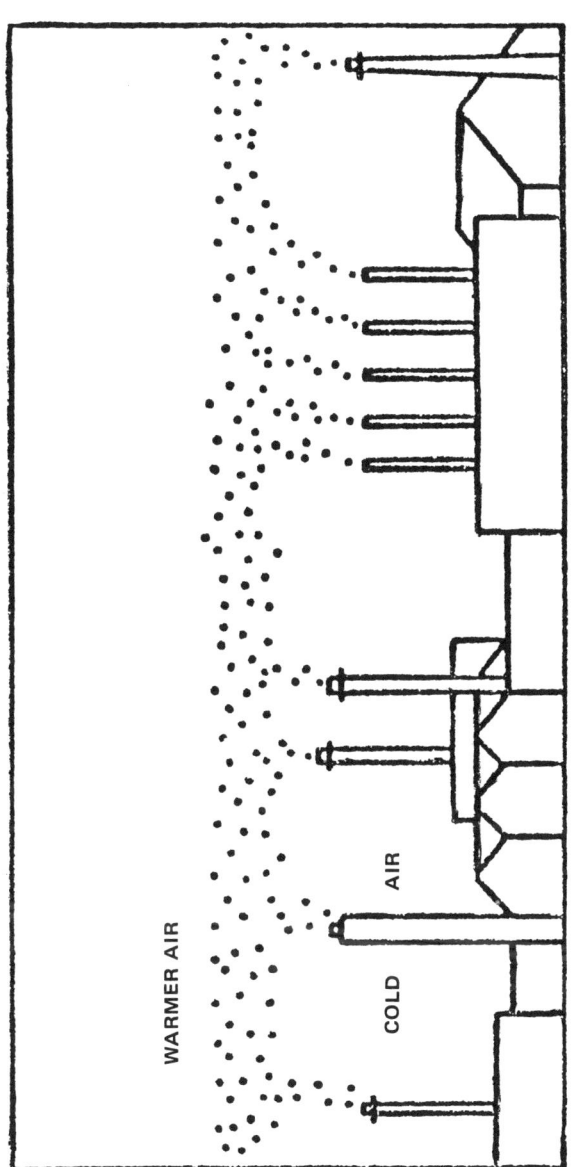

Figure 11: Temperature inversion. A pall of pollution trapped within a stable layer.
Source: Warren Spring Laboratory, Stevenage, England.

cyclones. In inversion layers, both vertical and horizontal diffusion is inhibited and pollutants become trapped, sometimes for long periods. Low-level discharges of pollutants are more readily trapped by inversions than high-level discharges; furthermore, high-level discharges into an inversion tend to remain at a high level because of the absence of vertical mixing. See Lapse Rates. See *Figure 11*.

Iowa, US: See *Litter Laws, US*.

Iraq: See *Heavy Metals*.

Ireland (Eire): See *Council for the Environment*.

Israel: See *Desertification Control; Environmental Protection Service*.

Italy: See *Air Pollution; Facadism; Green Summit, 1984*.

J

Japan: See *Australia-Japan Migratory Birds Agreement; Clean Lakes Law, 1984; Environment Agency, Japan; Heavy Metals; National Ambient Air Quality Standards; Natural Parks.*

Jhumming: A form of shifting cultivation practiced in the northeast of India. A tribe cuts down a patch of trees, cultivates the soil for two or three years, then plants trees and moves on. Once the same patch would be used only once every 20 or 30 years; now the cycle is down to five or six years, which does not give the forest time to regenerate. The reasons are that much of the forest has been progressively cleared for other purposes and the population of the tribal people has increased.

Joint Committee for Environmental Protection, US-USSR: A committee established in 1972, just before the United Nations Conference on the Human Environment (q.v.) to ensure continuing cooperation between the Soviet Union and the United States of America in the area of environmental protection. The joint committee was created formally under an Environmental Protection Agreement.

The Agreement aimed at working out measures to prevent pollution and generally at developing the basis for controlling the impact of human activities on nature. Cooperation has taken the form of an exchange of scientists and research scholars, organization of symposia, exchange of information, and the joint development of programs and projects.

The Agreement was renewed in 1982. By then some 42 specific projects were under way in the areas of air, water, and marine pollution; urban and agricultural pollution; nature conservation, biological/genetic effects, climatic effects, earthquake prediction, arctic-subarctic ecosystems, and legal-administrative measures. By 1985, more than 2,000 American and Soviet specialists had participated in exchange visits.

Joint Group of Experts on the Scientific Aspects of Marine Pollution (GESAMP): A body sponsored by the United Nations Environment Program (UNEP) and other UN agencies. The United Nations Conference on the Human Environment, 1972 (q.v.) recommended that GESAMP should assemble scientific data and provide advice on the scientific aspects of marine pollution. Later this brief was broadened to embrace the health of the oceans, a subject of report to UNEP in 1982.

The effects of wastes, it was reported, could be seen in semi-enclosed seas, shelf seas and coastal zones, e.g., the Gulf of Mexico, the Mediterranean Sea, the North Sea, and the Baltic Sea. Commercial fishing has been stopped in certain limited areas, and in some areas pollution has been implicated in reducing the populations of some marine animals. The use of coastal zones for sewage disposal is worldwide and the input is increasing; a reassessment of sewage disposal practices has become necessary. Also the management of coastal zones and semi-enclosed seas needs to be improved.

Joint Planning Unit, US: A combination of planning operations by the legislative bodies of cities, municipalities and counties, by agreement, for the purpose of carrying out a joint city-county planning program. The arrangement may well be temporary and falls short of a regional planning commission (q.v.).

Junkyard Control Act, 1967, Tennessee, US: A measure introduced by the State of Tennessee to regulate and restrict the establishment, operation and maintenance of junkyards in areas adjacent to state highways; the Act was intended to promote public safety, health, welfare, convenience and the enjoyment of public travel; to promote the public investment in public highways; and to preserve and enhance the natural scenic beauty of lands bordering public highways.

The Act makes it unlawful to operate any vehicle junkyard or automobile graveyard within 1,000 feet of the right-of-way of any interstate, federal or state roadway without a state permit. It is unlawful to operate any other junkyard within 1,000 feet of the right-of-way unless it is screened in at the owner's expense.

K

Kansas, US: See *Business Improvement District Act, 1981; Redevelopment of Central Business District Areas Act, 1976.*

Karst Topography: A limestone region marked by sinkholes (q.v.) and interspersed with abrupt ridges, rock outcrops, caverns, and underground streams.

Keep America Beautiful: See *Litter Laws, US.*

Keep Singapore Clean Campaign: An ambitious plan of action launched in the late 1960s to transform Singapore into one of the cleanest cities in the world whereby litter, dirt, and other forms of waste pollution would be banished. The attainment of this objective was declared a matter of national priority second only to defense and economic development.

A National Campaign Committee, headed initially by the Minister for Health, was formed to run the campaign. The Committee was composed of representatives from a whole host of organizations which were selected for the special services they could render. They included organizations like the Chambers of Commerce; Employers' and Trade Union organizations; governmental organizations such as the Ministries of Education, Interior and Defense, and Culture; the Police, Vigilante Corps and the Public Works Department; and quasi-governmental organizations such as the Housing and Development Board, the Public Utilities Board, Tourist Promotion Board, and the Hurong Town Corporation (which is a statutory corporation in charge of industrial estates). Thus the whole campaign was given a national character.

After elaborate preparations had been completed, the national campaign was launched in 1968 with a fanfare of publicity at an opening ceremony to which all state dignitaries, civic leaders and representatives of mass organizations were invited. It was then followed by a month-long intensive program of activities. The "Keep Singapore Clean" campaign sought to educate every single individual—man, woman or child; employer, worker or housewife—on the importance of not littering the streets, drains and public places.

The follow-up action was to be the strict enforcement of the anti-litter laws. It was realized that although the majority of the public would become quite aware of the need not to litter there would be some who would still persist in their bad old habits. The full force of the law had to be brought to bear on this recalcitrant minority.

As the majority of the population had been won over to accept the objectives of the campaign and the supporting legislation, recalcitrants were strictly prosecuted and their names published in the press. School children offenders were, however, reported to their school principals who would discipline them by making them sweep their classrooms or school compounds. Thus indiscriminate littering and dumping became successfully imprinted in the minds of the public as anti-social acts. A seven day working week was introduced whereby streets were swept and refuse removed every day of the week, including Sundays and public holidays. An efficient cleansing service was vital to the success of the whole plan of action.

The streets and surroundings in Singapore today are generally tidy, clean and green, and the environment healthy. They were not so some fifteen years ago when the indiscriminate throwing and dumping of garbage and refuse in public places was almost a way of life of the average Singaporean.

Kentucky, US: See *Litter Laws, US*.

Korea: See *Environmental Conservation Law, 1977; Han River Basin Environmental Master Plan; Saemaul Undong (New Community Movement)*.

L

Lake Baikal, Soviet Union: Located in eastern Siberia, a lake which contains one-fifth of the world's surface fresh water. It functions as a huge biofilter, for the water in the rivers which empty into it are much more polluted than that which flows from it. This purification is brought about by biological processes.

This unique lake may be utilized in various ways. The problem for biologists has been to understand the ecological process which takes place when industrial waste enters it, and to assess the extent to which wastes might be acceptable. Chemists, on the other hand, have had the parallel task of developing technological processes to ensure wastes will comply with the requirements of the biologists and planners.

Public concern at the possible adverse effects of development on Lake Baikal led to the slogan *Hands Off Baikal*. It appears that in the Soviet Union also, the public may fear the irresponsibility of the big developer.

The task of developing general principles and organizing the management of the Lake rests with the State Planning Commission and the USSR Academy of Sciences. Not everyone shares the confidence of Dr. Pyotr Kapitsa that "in a socialist system, the state can fully guarantee the coordination between science and industry needed to utilize such natural features as Lake Baikal in the right way."

A report issued in March, 1981, indicated that the condition of Lake Baikal was deteriorating, despite strict pollution control legislation. Factories discharging effluents into rivers and water-courses serving the Lake were blamed, together with paper mills, ships carrying oil, and the city of Slyudyanka which discharged raw sewage into the Lake.

Land and Environment Court, New South Wales: A Court created by the Land and Environment Court Act, 1979, of New South Wales, Australia, for the purpose of hearing appeals against local authority decisions in respect to development applications and other matters relating to pollution control and heritage conservation. The Court is headed by a Chief Judge.

Land Capability: The uses to which an area of land may be put, taking account of all constraints. The concept of "land capability" was developed by an American landscape architect and planner, Ian McHarg, in a seminal work *Design With Nature*, first published in 1969.

The assumption underlying this work is that a particular land formation may be compatible with one or more of a number of uses, and incompatible with others. By examining systematically a number of variables, planners and landscape architects can determine the appropriate use to which the land should be put. In this way they can harmonize development with nature.

With sloping land, for instance, the use to which it can be put will obviously depend upon the degree of slope. If it is flat, or practically flat, the land would be compatible both with urban development and the development of open space (sporting fields and so on). The soil composition, the geology, the hydrology, or the natural features may indicate that one use is rather better than another. If, besides being flat, it is, for instance, also swampy with a high water table, it may be unsuitable for residential development but ideal for a bird sanctuary, etc.

Land Commission, UK: A commission which came into being in Britain in 1967. It had two main functions:

- To collect for the Treasury a special tax on property development values known as a "betterment levy." See *Betterment*.
- To acquire, manage, and dispose of land which, in its opinion, was "suitable for material development." The commission could acquire land by agreement or by compulsory purchase (q.v.). It could sell its land to any developer, public or private, and sell land for housing at less than the market price.

The betterment levy was abolished in 1970, followed by the gradual winding up of the Land Commission.

Land Compensation Act, 1973, UK: An Act of the British Parliament which provides both for the granting of compensation for loss of value caused by the physical factors arising from the use of certain public works or land, and grants for insulation from noise. The principal works concerned are roads and aerodromes; while the physical factors are noise, vibration, smell, fumes, smoke and artificial lighting. The noise must arise from the use of the works and the source of the physical factors must be on or in the new or altered public works.

Thus, where a road is widened, the noise must arise from the traffic along the widened stretch of road; there is no compensation for the effects of increased traffic further down the road where no improvements have taken place. There is also no claim for the loss of view, for personal inconvenience, or the effect of increased traffic on unaltered roads. The loss of value is measured by the difference in value resulting from any new or increased physical factors arising from the use of the works, i.e., the depressed effect on the market value of the property or the tenant's interest. The date of valuation is one year after the start of use. Noise or other physical factors during construction are separate matters for compensation.

The eight principles behind this legal measure were set out in the paper *Development and Compensation—Putting People First* (HMSO Cmnd. 5124) as follows:

(i) Harmful impact on the immediate surroundings must be alleviated by comprehensive planning and remedial measures.

(ii) Noisy and unattractive public developments must, by better planning, be separated from people and their homes.

(iii) Damage to visual amenity by large-scale public works must be minimized by good pleasing design.

(iv) Noise, smell and other forms of pollution must be reduced to a minimum at source—if it is practicable, eliminated.

(v) Where, in spite of these efforts, damage still is done to individual amenities, reasonable compensation must be provided for those who suffer noise and other harmful effects.

(vi) The processes of inquiry and decision on project, compulsory purchase and payment of compensation must be thorough but concentrated in time and must be conducted so as to minimize blight and the hardship this entails.

(vii) No time must be lost in carrying through the new approach to design and planning, to remedial works and sound insulation, to acquisition and compensation.

(viii) People threatened by, or suffering from, the effects of public works must be told, in an understandable way, their rights and the help which is available to them.

In addition, the Act is a measure which, taken in conjunction with the Noise Insulation Regulation, 1975, requires compensation in the form of insulation (and perhaps accompanying ventilation) once noise levels exceed 68 dB(A) measured one meter in front of the building facade; a standard which more or less approximates the results of surveys carried out to determine the level at which people are annoyed by noise.

In London, a survey was carried out between 1961 and 1963 resulting in a report from the Urban Motorways Project Team. It concluded:

"From this survey it has been possible to show that, taking all sites together and median responses, people were as likely to be satisfied as dissatisfied at the L_{10} noise level of 70 dB(A) at external facades when averaged over 18 hours of the day from 6 a.m. to midnight."

A further study produced a somewhat lower figure. This survey suggested that the point at which people begin to become annoyed by noise (where dissatisfaction begins to outweigh satisfaction) is around 66 dB(A) for the L_{10} (i.e., intrusive noise) over the 18 hour day between 6 a.m. to midnight.

Landfill: The most common form of disposal of refuse and garbage; the types of sites used for such disposal include: mineral excavations, low-lying land; valleys; areas involving the reclamation of land from water; and flat land where a feature is planned. Generally, in a sanitary landfill scheme, refuse is tipped in trenches or cells prepared to such a width that the daily input of refuse can be effectively covered, presenting a clean face each day. The refuse can be tipped either at the bottom of the face and dozed into the face, or tipped on top of the previous fill and dozed over the face. It is essential that the refuse be adequately covered and compacted to allow traffic over the fill. The landfill method is known in the United Kingdom as "controlled tipping" and in the United States by the title of "sanitary landfill;" while the content of both expressions is identical, the former emphasizes the system by which the waste is deposited while the latter emphasizes the hygiene aspects. The landfill technique is often used constructively to provide facilities for sport; its use for urban development involves many years of settlement. Under environmental legislation in many industrial countries landfills need to be authorized and carefully managed to avoid, for example, leachates reaching streams or the breeding of vectors and rodents. The method is sometimes used for the disposal of hazardous wastes, using particularly rigorous control methods. See *Table 11*.

Land Use Plan: A compilation of policy statements, goals, standards, and maps, and action programs for guiding the future development of private and public property. The term includes a plan designating types of uses for an entire municipality, as well as a specialized plan showing specific areas or specific types of land uses, such as residential, commercial, industrial, public or semi-public uses or any combination of such uses.

Land Use Planning: Traditionally, a technical or physical approach to the segregation of incompatible activities, such as housing and industry, through systems of land use and development controls; it has certainly been effective in preventing generally the worst aspects of mixed development in new areas and has served individual as well as community interests. The individual benefits from a reduction in the degree of uncertainty surrounding his/her investment; while the community may gain in the protection of open space and recreational opportunities and in a reduction of infrastructure costs against those which might arise in respect to haphazard developments.

Urban planning considered in terms of physical space and control of land use has often been perceived as the central objective of planning, as if through this instrument alone the objectives of planning can be wholly achieved. Indeed, the concept of land use as the end result created the "land use planner."

Table 11: Topographical Location and Environmental Impact of Landfill Disposal Sites

Topography	Characteristics	Likely Environmental Impact
Floodplain sites	Flat, poorly drained, often close to or at water table. Tendency to be in fine-grained material.	Danger of rapid production of leachate, and often liable to flood and pollute surface water. Should never be used for hazardous waste disposal.
Valley-side slopes	Sloping, usually well-drained, above level of water table. May be in solid rock or soils.	Free from flood danger, quite well drained usually, infiltration rates low because water runs over surface. May present a visual impact, but usually easily screened.
Ridges and Summits	Sloping, likely to be well above water table. Usually in solid rocks.	No flood danger, usually well drained. Likely to present a strong visual impact and to suffer from wind dispersal of dust and paper. Not easily accessible sites in many environments, possibly necessitating new road construction.

Source: Gemmell, A. "Solid and Liquid Wastes and the Impacts of Their Disposal." Proceedings of International Seminar on Environmental Impact Assessment, University of Aberdeen, Scotland, UK, 8-21 July, 1984.

However, in democratic societies planning must be consumer-oriented and this involves taking account of social and economic factors as well as the physical land use aspects. Land use planning becomes not a revered end but simply a principal instrument to serve higher planning objectives. The aim is to improve the quality of life (q.v.) for the consumer of planning services and this involves such concepts as social space as well as physical space. It is for these reasons that planning has matured from the phase of "land use planning" to "environmental planning" (q.v.). Land use planning may then be viewed as an important means, not to be confused as in the past with the end.

Today, planning must draw upon geography, geomorphology, economics, sociology, political science, psychology, public and social administration, and the management sciences. Increasingly, these ingredients are tested through the processes of public participation and the public inquiry. Through these processes, the land use issue becomes at times quite subordinate to larger public interest considerations. Sometimes, when the final decision is made it is the land use plan which becomes amended to the larger benefit of the community. See *Environmental Planning, Objectives of; Zoning.*

Land Use Prediction Models: Theoretical models which seek to describe and predict the locational decisions of households, firms and other types of urban activities; such models may be based upon the principle that such decisions are to a large extent dictated by the costs of overcoming distance which separate interrelated and interactive activities. See *Transport Model.*

Landscape: The total visual environment. See *Landscape Analysis, Factors to be Considered; Landscape Evaluation, UK; Visual Impact Measurement.*

Landscape Analysis—Factors to be Considered: Factors relevant to landscape analysis which may be applied systematically to environmental blocks or sub-areas of a total study area, with each factor rated on a scale from zero to ten and linked with a weighting system. Factors which may be considered include:

- Slope;
- Vegetation quality;
- Vegetation extent;
- Visual significance;
- Flooding;
- Mine Subsidence;
- Potentially erodible area;
- Watershed;
- Geology;
- Natural features;
- Fauna;
- Weedbeds and mangroves;
- Severity of inversion;
- Mineral resources;
- Salt spray;
- Aspect, and
- Wind exposure.

One of the major benefits of this kind of analysis is that it sets out in quantitative terms the degree to which factors influence a final land use recommendation.

Landscape Character: A description of the appearance of broad areas of the total visual environment where certain visual elements predominate.

Landscape Evaluation, UK: An assessment of the quality of landscape for development planning and preliminary site selection purposes, as distinct from visual impact assessment of specific proposed developments. In the United Kingdom, landscape evaluation has been a long-established practice. The Town and Country Planning Act, 1947, the Countryside Act, 1967 (Scotland) and 1968 (England and Wales) made the conservation of natural beauty and amenity of the countryside a basic objective. The main approach used by the planning authorities to protect and manage the visual resources of the countryside has been the preparation of a county or regional landscape evaluation. Landscape evaluation has had the greatest influence and success in rural environments where development pressure is low, several alternative development areas are available, and the type of development is small-scale. With increasing pressure in respect to large site-specific industrial developments, the traditional landscape evaluation can lose much of its effectiveness as a controlling element in planning. In such circumstances, comprehensive visual impact assessment during detailed project appraisal can be a significant factor in minimizing the adverse visual impacts of a project that may inevitably proceed.

Landscape or Amenity Conservation: The safeguarding for public enjoyment of scenery or landscape, and of opportunities for outdoor recreation, tourism, field sports, and similar activities; the concept includes the preservation and enhancement not only of what has been inherited but the provision of new amenities and facilities as well.

Landscape Unit: An individual part of the landscape within the total visual environment containing certain uniform characteristics or homogeneity which distinguish it from other areas; used as a classification system.

Lapse Rates: The rate of decrease of temperature with increasing height in the atmosphere. From the surface to a height of about 11 kilometers, known as the troposphere, the lapse rate is, on average about 6-8°C per kilometer. There are considerable departures from this average at all times and at all levels, and it is these departures which characterize the type of weather experienced. If the temperature shows no change with height, the condition is described as isothermal. In the stratosphere, the layer of the atmosphere above the troposphere, the lapse rate is isothermal, or very nearly so.

Law of the Sea: See *United Nations Conference on the Law of the Sea (UNCLOS)*.

Lead in Gasoline: See *European Economic Community; Heavy Metals; National Ambient Air Quality Standards, US; Royal Commission on Environmental Pollution, UK; Sulfur Dioxide.*

L'Enfant Plan for Washington, D.C., 1974: The design and plans for the original City of Washington, District of Columbia, which were promulgated by President George Washington and recognized by the US congress as the work of Pierre Charles L'Enfant. These plans were subsequently laid out by the Office of the Surveyor of the District of Columbia government according to the "King Plats of the City of Washington in the District of Columbia, 1803."

The city design focused on locating grand edifices on prominent sites; these sites were connected with broad avenues and a formal street and open space network which imposed a majestic sense of scale and order within the city. L'Enfant chose to locate the Capitol on Jenkins Hill, the most prominent site in the new city. The plan also chose a separate axial street with similar prominence for the President's Palace, known as the White House.

The L'Enfant Plan responded to the topographic and environmental conditions in the city by calling for "a city within a garden." This concept, subsequently termed "city in a park," combines formal state squares, grand avenues, and natural parks to give a wide range of landscape opportunities within the city.

The original plan has for almost two centuries been embraced by individuals and organizations who have served as guardians of the plan, including presidents and senators. Senator James McMillan of Michigan is especially noted for the 1901 plan of the city which extended the original concept. The American Institute of Architects also contributed with their "city beautiful movement" during the same era. See *Comprehensive Plan, Washington D.C.; National Capital Planning Commission.*

Limnetic Zone: The region of open water beyond the littoral zone (q.v.) of a lake, down to the maximal depth at which there is sufficient sunlight for photosynthesis. This is the depth at which photosynthesis balances respiration, known as the compensation depth. Rooted plants are absent in this zone, but there is a great abundance of phytoplankton.

Linkages: A frequently used term in planning offices, a reference to the interdependence of many aspects of urban life in physical, social, political, economic, or environmental contexts. Linkages connect units that are distributed in time as well as space and may relate to people, goods and services, movement or communication. A single business decision in one area of life may have far-reaching consequences for others not directly involved. The closing of a railroad may affect freight movement as well as commuters in numerous ways, while the closing of a mine or steel plant may destroy a community. At the other end of the spectrum, a theater may be a viable and popular venture conveniently adjacent to a transit stop; but what if the transit ceases to run before the theater closes?

Litter Control: See *"Do the Right Thing" Campaign; Keep Singapore Clean Campaign; Litter Control, UK; Litter Laws, US; Safe Growth Plan, Tennessee.*

Litter Control, UK: A series of measures to control litter in Britain. The Litter Act 1958 made it an offense to leave litter on land to which the public has free access; under the Civic Amenities Act 1967 local authorities were empowered to remove abandoned cars and other bulky rubbish; the Countryside Act 1968 allows the Secretaries of State for the Environment and Wales to make grants to the authorities for the removal of objects disfiguring the countryside; the Dangerous Litter Act 1971 improved provision for the prevention of dangerous litter; and the Deposit of Poisonous Waste Act 1972 introduced more stringent controls to prevent illegal dumping of poisonous waste materials.

Litter Laws, US: A variety of legal measures to discourage littering and promote recycling introduced at state level. Seven states have laws which require a minimum refundable deposit on certain types of beverage containers: Oregon, Vermont, Maine, Michigan, Iowa, Connecticut, and Delaware. The normal refundable deposit is five cents. Five of these states require the distributor to reimburse the dealer or operator of a redemption center a minimum of 20 per cent of the value of the returned containers.

Thirteen states have banned the pull-tab or flip-top beverage container. Hawaii has banned all plastic beverage containers. Four states have banned the plastic loop retainer which holds beverage containers, usually 6-packs, together. Two states, Vermont and Delaware, have banned non-refillable glass containers altogether.

Eight states have laws which tax those businesses whose products may end up as litter: Washington, Vermont, Virginia, Colorado, California, Kentucky, Connecticut, and Ohio. These taxes vary in their base and their incidence; all apply to all manufacturers, but some also apply to wholesalers, distributors, retailers and others.

The State of Georgia, for one, has formally endorsed the clean community system for picking up litter. This system was developed by Keep America Beautiful and is a standardized and systematic approach to picking up litter in urban areas and in smaller communities.

Littoral: Relating to or taking place on or near the shore.

Littoral Drift: Sand moved under the effect of longshore current; it carries sand which has been stirred into suspension by the turbulence of the breaking waves. The direction and strength of the waves determine the direction and magnitude of the littoral transport at a given time. Determining the direction and average net annual amount of the littoral drift is important in developing shore protection plans. Onshore and offshore sand movements caused by low swells and steep waves respectively, coupled with littoral drift, help to explain the major shoreline changes on the open coasts of the world.

Littoral Zone: The shallow, marginal region of a lake, characterized by rooted vegetation; the area where sunlight is able to penetrate to the lake bottom, the zone sustaining a high level of photosynthetic activity. A small shallow pond may consist entirely of littoral zone, while a deep lake with an abruptly sloping basin may possess an extremely restricted littoral zone. Per unit volume of water, the littoral zone yields more biomass than either the limnetic or profundal zones.

Living Resource Conservation: As defined in the World Conservation Strategy (q.v.), a proposed conservation policy with three specific objectives: the maintenance of essential ecological processes and life support systems (such as soil regeneration and protection, the recycling of nutrients, and the purification of waters) on which human survival and development depend; the preservation of genetic diversity (the range of genetic material found in the world's organisms) on which depend the breeding programs necessary for the protection and improvement of cultivated plants and domesticated animals, as well as innovation and scientific advance in many industries; and to ensure the sustainable utilization of species and ecosystems (notably fish and other wildlife, forests, and grazing lands) which support rural communities as well as major industries.

Local Environmental Plan, New South Wales: A plan formulated by a local government within the terms of the Environmental Planning and Assessment Act, 1979 (q.v.) relating to the whole or to any part of the local government area. Such a plan must be preceded by an environmental study, the results of which must be exhibited and open to public submissions; the draft plan must also be publicly exhibited. A local environmental plan must be consistent with state environmental planning policies (q.v.), any regional environmental plan (q.v.), and any other ministerial directions.

Each local environment plan must state the intentions and purposes of the plan and set out the prinicipal elements of land use by zones. The purpose of each zone must be stated, such as residential, business, commercial, industrial, etc., and the range of uses compatible with the intention of each zone. Basic development standards are also stated, e.g., minimum allotment sizes, floor space ratios, building heights in environmentally sensitive areas, densities and site coverage.

A primary aim is to define the least number of zones possible, each permitting a reasonably wide range of uses, thereby reducing the number of amending local environmental plans which are likely to be needed in the future. Detailed development standards are included in subsequent development control plans.

A local environmental plan developed by the Council of the City of Sydney for the Ultimo/Pyrmont/Haymarket area contains the following major aims, objectives and policies:

- To upgrade the Ultimo/Pyrmont/Haymarket area as a mixed residential, light industrial, commercial and service area;

- To arrest and reverse the population decline of the area by regenerating areas on the spine of the Ultimo/Pyrmont peninsula for predominantly residential purposes;
- To zone land for industry so as to provide for the continuation of existing, consolidated light industrial operations;
- To consolidate and allow for the limited expansion of established neighborhood centers so as to ensure that they can remain commercially attractive and viable, and service the local population;
- To encourage a limited component of local retail and commercial uses as part of major residential redevelopments to serve the local population;
- To zone land containing woolstores and similar buildings for a range of diversified mixed uses so as to encourage their re-utilization;
- To encourage the conservation of identified items of the environmental heritage, including both residential and non-residential buildings;
- To conserve the quality, scale and character of conservation areas, such as those containing terrace housing;
- To control the demolition and alteration of buildings so as to enable proper assessment of their significance and ensure that new development and alterations are compatible with the character of the area;
- To provide local open space through utilization of street closures and acquisition of vacant lands, such as in foreshore areas;
- To maintain and allow for mixed commercial, entertainment and service functions in the areas on the southern perimeter of the plan.

See *Department of Environment and Planning, New South Wales.*

Local Government Act, 1972, UK: Effective in 1974, an Act introducing a new two-tier system of local government into England and Wales, with a similar system for Scotland. England, outside Greater London, was divided into six metropolitan counties and 39 non-metropolitan counties. All these counties were further divided into 369 districts. Wales was divided into nine counties and 36 districts; Scotland into nine regions and 53 districts.

The elected councils of the counties and the councils of the districts became local planning authorities for their respective areas. In Greater London, the Greater London Council had already become the local planning authority for the whole area, while within certain limits the individual London boroughs became planning authorities for their respective areas.

In general, county councils became responsible for the preparation of broad planning policies, i.e., the preparation of structure plans; for related

172   Environmental Planning

development control; for the acquisition and disposal of land for development or redevelopment; and for highways and traffic. District councils became responsible for the preparation of local plans and for most development control; and assumed responsibility for housing including construction of dwellings; slum clearance schemes, improvement schemes and housing management; also for the maintenance of unclassified roads.

The "new style" developments plans were to be prepared in two stages: (a) Structure Plan, and (b) Local Plans . The Secretary of State for the Environment would approve Structure Plans, but local plans could be adopted by the local planning authorities. See *Local Plans; Structure Plan, UK*.

Local Government, Planning and Land Act, 1980, UK: See *Enterprise Zones, UK; Examination in Public, UK*.

Local Plans, UK: A responsibility of district councils under the Local Government Act, 1972 (q.v.), plans which may cover the whole of a council's area or may deal only with particular planning problems within a district. Local plans may be finalized as soon as possible after a structure plan (q.v.) has been approved by the Secretary of State. Local plans consist of a map and written statement together with other illustrations and supporting material.

The main purpose of a local plan is to guide authorities and private developers in the possible development of the land concerned. Draft plans must be exhibited for public comment and objection; this may be followed by hearings and inquiries.

Local plans must be in accord with the overall policies of the structure plan for the area. Hence their main functions are:

- To develop the policy and general proposals of the structure plan and to relate them to precise areas of land;
- To provide a detailed basis for development control;
- To provide a basis for coordinating the development and other use of land; and
- To bring local and detailed planning issues before the public.

Statutory local plans are of three types:

- District plans—these may be prepared for the whole or part of a district council's area. The district plan should amplify the broad strategy of the structure plan.
- Action area plans—these are quite detailed plans prepared when there is likely to be a major physical change in an area. they may give great detail in respect to the layout of the area and the actual form of building development.

- Subject plans—these treat topics in isolation from other development policies such as mineral extraction or green belts.

At the end of the process, the local planning authority may formally adopt a plan by resolution of the authority.

Location of Industry, Optimal: The most efficient location of industry in accountancy terms; factors influencing the location of industry include:

- Availability and cost of raw materials of satisfactory quality and security of future supplies;
- Delivery costs of sending the final product to markets of sufficient size and certainty, and competitiveness of the product in those markets;
- Availability of labor having regard both to number and degree of competence required;
- Transport and communications;
- Level of local costs generally (rents, rates, wages, etc.);
- Adequacy of local resources such as water supply and facilities for waste disposal;
- Availability, reliability and cost of fuel;
- Availability of local specialized industries and services;
- Load-bearing requirements for site;
- Cost of site development (land cost, levelling, filling, drainage, roads, etc.);
- Possibility of development of site for additional plant in due course;
- Availability of housing, local transportation, schools, churches, hospitals and recreation facilities;
- Planning and development restrictions and inducements.

Clearly the "weighting" given to each of these items will vary according to the type of business activities involved. The primary reason for a plant location study by a company is to find a site at which a plant of suitable size can produce the highest return on invested capital; the aim is to establish the optimum location. The problem of location involves highly complex studies concerned not only with present but also future conditions and changing patterns, and often involves difficult decisions.

Planning and industrial development cannot be dissociated from the concept of the socially-responsible utilization (or conservation) of resources, and with larger objectives which may be regarded as desirable by some, such as "zero population growth" or "zero industrial growth" and a stable standard of living. These concepts, accepted as social objectives, would restrict industrial development to industrial change in a qualitative sense associated with only small or zero quantitative expansion. Even well short of these objectives, if a Barrier Reef or Santa Barbara shore is to be preserved unimpaired, then it may be desirable to forego the potential benefits of oil-drilling in such areas. There is certainly an increasing tendency in public policy towards the exclusion of certain industries from certain areas on the simple proposition that the social penalties would outweight any social gains. The nature of a "social penalty" and a "social gain" is bound to shift with any shift in "social values," but the general thrust today is undoubtedly in the direction of some restraint on unbridled industrial expansion. Industry today is careful to pay lip-service to such ideals, while not always matching promise with performance.

Location, Theory of: A branch of the theory of the firm embracing an analysis of locational decisions; a body of theory which attempts to explain and predict the locational decisions of firms, and the spatial pattern of industry, commerce, and agriculture which results from the sums of individual decisions. Early work in the area of agriculture was undertaken by Johann Heinrich von Thünen (1783-1850), and in the area of industry by Max Weber (1854-1920).

London, England: See *Abercrombie Plans for London; Barlow Report, 1940; Distribution of Industry Act, 1945; Green Belt; Herbert Report, 1960; Highrise folly; London Smog Incidents; Roskill Commission; Strategy Plan for Southeast England, Urban Development Corporations.*

London Convention: See *Convention for the Prevention of Pollution from Ships, 1973.*

London Docklands Development Corporation: See *Urban Development Corporations, UK.*

London Smog Incidents: Acute episodes of heavy pollution associated with natural fog covering the Greater London area. The fog which covered the Greater London area during the four days, 5th to 8th December 1952, was on a much different plane compared with those previously experienced. An anticyclone reached London from the northwest in the early hours of 5th December and then became stationary. On 6th and 7th December, London Airport had a minimum temperature of -5° and -6° C and a maximum just under 0° C. Soundings showed two inversions, one close to the surface and the other caused by descending (anticyclonic) air higher up. By 7th December, the two inversions were very close together. The final result was that London was at the bottom of a pool of cold stagnant air, with a very effective "lid" overhead. The atmosphere contained a great deal of water in the form of very small droplets. Nearly 4000 people died.

Long Shore Drift: The movement of sedimentary material in the littoral zone (q.v.) under the influence of waves and currents; generally in a predominant direction determined by waves breaking at an angle to the shore.

Los Angeles, California, US: See *Los Angeles City Planning Commission; Los Angeles Smog; Transport Plan; Transportation Study.*

Los Angeles City Planning Commission, US: A five-member, part-time planning commission (q.v.) appointed by the Mayor with the concurrence of the Los Angeles City Council. The commission reviews the recommendations of the planning director and advises the Mayor and the City Council; it has been responsible for recommendations in respect to the adoption of the General Plan (q.v.). In respect to some matters relating to maps, zone changes, conditional use, and zone boundary adjustments, the Commission is empowered by the City Charter to make final decisions; however, these decisions may be appealed to the City Council. All meetings of the Commission must be open to the public and there are restrictions on commissioners in relation to the discussion of planning matters in private.

The City Planning Department serves the Planning Commission, the Mayor and the City Council, and has been responsible for the actual preparation of the General Plan.

Los Angeles Smog, US: A smog of photochemical nature, largely attributable to the effect of sunlight on motor vehicle exhaust gases. The smog problem of Los Angeles persisted after a drastic reduction of major stationary sources of air pollution; this left the motor vehicle as the major probable source. However, the "cause and effect" mechanism was not at all easy to explain. During smogs considerable increases of ozone and oxidant material occur. The oxidant concentration that can cause eye irritation is regarded as being in the range 0.21 to 0.32 mg/m^3 (0.10 to 0.15 ppm), levels reached very frequently in Los Angeles. The highest value ever recorded in the city was 1.75 mg/m^3 (0.82 ppm) on 16 December 1966. These are not found in appreciable concentrations at night but only during daylight hours, beginning to form simultaneously throughout the Los Angeles basin in hazy air shortly after dawn.

These facts suggested a photochemical formation of ozone or oxidants from air impurities by the action of sunlight. Sulfur dioxide, nitrogen dioxide and aldehydes absorb ultraviolet radiation in the wave lengths present at ground level and react in their excited states with molecular oxygen to produce atomic oxygen. Reactions with SO_2 and aldehydes are irreversible, but not so in the case of NO_2. In the latter case the absorption of ultraviolet light leads to the bond rupture to form atomic oxygen and nitric oxide. Reaction of the products with molecular oxygen leads to the formation of ozone and the regeneration of nitrogen dioxide. Thus NO_2 is available for the repetition of the process, unless converted to nitric acid or used up in organic substitution reactions. Hydrocarbons also play an essential part in this reaction, olefins being the most reactive.

Ozone formed during the smog accounts for accelerated rubber cracking; while the oxidation of SO_2 to SO_3 with the formation of H_2SO_4 as an aerosol, as well as existing smoke, dusts and fumes, still further reduce visibility. Scientists in Los Angeles appear certain that the ozone and PAN (peroxyacetyl nitrate) in Los Angeles smogs have caused the serious decline in the citrus and salad crops in the area. Ozone and PAN produce eye irritation, coughing and chest soreness experienced by many Los Angeles residents on smoggy days.

Love Canal, US: An example of the consequences of environmentally hazardous chemicals (q.v.). It was revealed in 1980 that many residents in the Love Canal area of Niagara Falls township had suffered severe chromosome damage from toxic chemical waste buried there. Indeed, an abnormally high proportion of the 3,000 members of the community had been found to be suffering from various forms of cancer, high rates of birth defects, liver and kidney damage, and respiratory ailments.

Between 1947 and 1952, two corporations began to use the Love Canal site as a chemical waste dumping ground. For 50 years the site had remained derelict after William T. Love had begun construction of a canal in 1894 and then abandoned the project. After the dumping of toxic wastes into the large excavation, the site was dedicated to the Niagara School Board in 1953. The board built a school on part of the landfill site and some residences on the rest.

In 1977, a toxic substance began bubbling to the surface of the school playground. The substance was found to be a mixture of 82 identifiable industrial chemicals, 11 of them carcinogenic. In 1978, Love Canal was declared a disaster area and President Carter approved emergency aid for the area; the state began buying abandoned homes, eventually purchasing 239 of them. In May 1980, the area of emergency was extended to embrace an additional 710 homes.

Love Canal has become a symbol for the problem of finding safe disposal sites for toxic chemical wastes.

Luxembourg: See *International Commission for the Protection of the Rhine*.

L_x **Noise Levels:** Noise levels in dB(A) which are exceeded for a specific percentage of the measurement period. For example, L_{10} and L_{90} noise levels mean that the noise level in dB(A) was exceeded for 10 percent and 90 percent of the measurement period, respectively. The Noise Advisory Council of the United Kingdom has recommended the adoption of the L_{10} index for measuring disturbance by traffic noise. It has recommended also that existing residential development should in no circumstances be subjected, as an act of conscious public policy, to more than 70 dB(A) on the L_{10} index unless some form of remedial or compensatory action is taken by the responsible authority.

M

Macroeconomics: A branch of economics concerned with the analysis of the economy in the large, i.e., with such large aggregates as the volume of employment, savings and investment, the national income, and so on. More specifically, it is concerned with the economy's total output of goods and services, the growth of that output, the price level and rates of inflation, unemployment, the balance of payments, exchange rates, booms and recessions. It embraces the study of economic policies and policy variables that affect national economic performance, monetary and fiscal policies, money stock and interest rates, the public debt, and federal and central government budgets and deficits. In sum, macroeconomics is vitally concerned with the interactions among the goods, labor, and assets markets of the national economy. See *Microeconomics*.

Macro Environmental Problems: Regional, national and international problems relating to the environment including:

- Dereliction, slums and blight in most of the cities and rural districts of the world;
- Inadequate sewerage systems and the pollution of major waterways;
- Floods and the associated threat to life and property;
- Traffic accidents and congestion;
- Population growth in relation to resources in many areas;
- Mortality and morbidity arising from controllable environmental sources;
- Increasing dereliction and blight due to secular decline in industrial activity in particular regions;
- Transboundary and global air pollution problems;
- Threats to the regional seas and marine resources;
- Use and misuse of uranium and the disposal of nuclear wastes;
- Increasing noise levels;
- Disposal of toxic wastes;
- Incidence of drought and famine;

- Location of hazardous industries;
- Visual pollution;
- Threats to natural resources including forests, woodland and mangroves;
- Endangered flora and fauna.

See *Micro Environmental Problems*.

Maine, US: See *Litter Laws, US*.

Man and Biosphere Program: See *United Nations Educational, Scientific and Cultural Organization (UNESCO)*.

Mandatory Dedication: A measure that local governments can incorporate in subdivision regulations to acquire open land in new residential developments; as a condition for subdivision plan approval, the municipality can require the developer to dedicate a portion of the development site for use as a recreational or school site. This requirement is somewhat similar to the retention of open space areas in planned residential developments. It encourages developers to cluster buildings in the most suitable areas and to retain hazardous or environmentally sensitive areas for open space. See *Easements; Subdivision*.

Mangrove Parks, Bangladesh: A national program for the conservation of mangrove areas through the creation of parks or protected areas. Bangladesh has larger areas under mangroves than any other country in the world. Mangroves are important not only for fisheries and timber purposes including fuel wood, but as an important resource for wood-based industries. The main supply of pulp for paper is met from a species of mangrove which was regarded as useless and a menace only 15 years ago. Mangroves are even more important to Bangladesh than perhaps to other parts of the world as they form the main bulwark against natural disasters such as cyclones and tidal waves and floods with which Bangladesh is so frequently plagued. It is significant that the cyclonic and tidal wave damage occurs only in the eastern part of Bangladesh where the mangroves have been removed and not in the western part covering the districts of Jessore and Barisal which are still clothed with mangroves. There is an urgent need to develop local expertise in mangrove ecology and management. The Forest Department is currently undertaking a program of mangrove propagation. However, a much larger effort is needed in this regard, while at the same time a study in the various aspects of mangrove ecology and mangrove management also needs to be undertaken both in the Forest Research Institute at Chittagong, and in appropriate universities. See *Mangroves*.

Mangroves: Plant communities and trees that inhabit tidal swamps, muddy silt and sand banks at the mouths of rivers and other low-lying areas which are regularly inundated by the sea, but which are protected from strong waves and currents. Mangroves are the only woody species which will grow where the land is periodically flooded with sea water; individual species have adapted themselves to different tidal levels, to various degrees of salinity and to the nature of the mud or soil. Mangroves vary in size from substantial trees up to thirty meters in height down to miniature forms less than waist high. Mangrove swamps and thickets support hundreds of terrestrial, marine and amphibian species.

Mangroves are important for a number of reasons:

- They are major producers of organic material and may have a special role in supporting estuarine fisheries (finfish, crustaceans and shellfish);
- They are involved in nutrient recycling;
- They help to reduce water pollution;
- They provide shelter, refuge and food for many forms of wildlife;
- They help prevent bank erosion and provide protection from storm surge;
- They act as visual screens along industrial foreshores, improving the amenity of the waterway;
- Mangrove habitats act as important nursery areas for many economically important (commercial and angling) fish species.

Manitoba, Canada: See *Federal Environmental Assessment Review Office.*

Marine Protection, Research, and Sanctuaries Act, 1972, (MPRSA), US: An Act of the US Congress which established a permit system for the dumping of materials into the ocean, and for the transportation of materials to be dumped.

The Environmental Protection Agency evaluates applications for permits to determine the need for the proposed dumping; the effect or likely effect of the dumping on human health and welfare (including economic, aesthetic, and recreational values); the effect on marine ecosystems; the concentrations of materials dumped and the likely persistence of environmental effects; and alternative means of disposal.

The Act authorizes the Secretary of Commerce to designate as marine sanctuaries those areas which he determines should be preserved or restored for their conservation, recreational, ecological, or aesthetic values.

In 1977, the MPRSA was amended to prohibit the dumping of sewage sludge into ocean water after December, 1981; industrial waste was also banned. Through the Surface Transportation Assistance Act, 1982, Congress further amended the MPRSA placing a moratorium on the disposal of low-level radioactive waste into ocean water, save for research purposes. The MPRSA was reviewed and reauthorized in 1985.

Marpol Convention: See *Convention for the Prevention of pollution from Ships, 1973.*

Maryland, US: See *Communities Program; State Development Plan.*

Master Plan: An early concept in city planning of a single and authoritative plan for the physical environment. However, experience has demonstrated that as the future is not precisely and sometimes not approximately predictable, that any plan must be flexible. The ideals and objectives themselves change; there needs to be stress, therefore, on the process and direction of change involving the coordination of a set of interrelated programs influencing physical, social, and economic conditions.

McMillan Plan, Washington, D.C., US: The body of plans for the systematic improvement and extension of parks and public buildings sponsored by Senator James McMillan on behalf of the US Senate in 1901 and set forth in the *Report of the Park Commission* in 1902; the plans were subsequently realized under the guidance of the Architect of the Capitol, the Commission of Fine Arts, and the National Capital Park and Planning Commission. See *L'Enfant Plan for Washington.*

Megalopolis: A designation for a number of coalescing metropolitan areas to form a huge massing of people and economic activities. The term was introduced by the French geographer Jean Gottmann who referred to the northeastern area of the United States as perhaps "the cradle of a new order in the organization of inhabited space." Gottmann referred to the area extending from Boston to Washington, DC. The term megalopolis may also be applied to such areas as metropolitan Los Angeles and San Francisco, metropolitan Dallas and Houston, metropolitan Chicago and Milwaukee.

In Europe, the term megalopolis may be applied to the region around London and the Black Country, to the area of Greater Paris, to the Ruhr industrial basin in West Germany. In Asia, the term may be applied to the Tokyo-Osaka complex.

Merseyside Development Corporation, UK: See *Urban Development Corporations, UK.*

Metropolis: A major city, often the result of the coalescence of a number of smaller towns and villages which become suburbs of the new entity. A major school of thought, following the pioneer British city planner Sir Patrick Geddes (1854-1932) and such architects as the American Frank Lloyd Wright (1869-1959), has maintained that the huge metropolis was an inhuman institution. The evils of noise, congestion, traffic, tension, and impersonality, even more than considerations of cost, suggested policies designed to check metropolitan growth. Others, however, have asserted that positive human values are to be found in the rich complexity of the metropolis. Certainly, attempts by a number of countries to check the drift to metropolitan areas have enjoyed only limited success; where metropolitan growth has been effectively curbed this has been, as often as not, due to industrial and commercial factors and cost levels. See *New Towns*.

Mexico: See *Air Pollution; Mexico City Industrial Disaster, 1984*.

Mexico City Industrial Disaster, 1984: A fire and explosion at a natural-gas storage and distribution facility at San Juan Ixhuatepec, Mexico City, in November, 1984, which resulted in several hundred deaths, many more injured, and 10,000 homeless; it was the worst industrial disaster in Mexican history. The disaster involved a private gas-bottling plant adjacent to the storage tanks of the government-owned oil corporation, PEMEX. The plant was situated in the very heart of a poor congested district, ignoring all environmental protection principles.

Such disasters are not inevitable, nor the necessary price of progress. There are two important means for protecting the community from these major hazards. The first is through engineering controls such as better control systems, reduced inventories of hazardous materials, and better protection of vessels such as sand-mounting. The second is through planning controls such as the introduction of exclusion zones or buffer zones with restricted developments around such facilities. Such a dual policy will not preclude incident, but adequate engineering controls will reduce the likelihood of incident and the magnitude; while planning controls reduce the consequence to the public. These restrictions do ensure that large numbers of people are not affected. Progressive nations have been introducing these principles over some years. See *Bhopal Disaster, 1984, India; Buffer Zone; Risk and Hazard Assessment*.

Michigan Episode, US: An example of the consequences of error involving environmentally hazardous chemicals (q.v.) and serving to emphasize the importance of strict controls. The Michigan Chemical Corporation had been manufacturing a substance known simply as PBB (trade name Firemaster) used in plastics as a fire-retarding chemical compound. Later the company introduced another product, magnesium oxide (trade name Nutrimaster) to add to stock feed to make it more palatable and less acidic. The two products were similar in appearance.

In June, 1973, a mix-up occurred when a number of bags of Firemaster were dispatched with Nutrimaster stencils. Nutrimaster was sold in quantity to the Farm Services Bureau for mixing with stock feed; the feed was then sold to dairy farms all over Michigan. It appears that for six months stock were fed on contaminated feed. Feed consumption and milk production fell. People suffered a wide range of disabilities.

The error was traced back to the Michigan Chemical Corporation, the sole manufacture of PBB. Apparently, a partially filled bag of Firemaster had been found at the Farm Services Bureau. Two hundred farms were quarantined. In 1976, tests on over a thousand people showed that one-third had symptoms of PBB poisoning—muscular weakness, nausea, acute indigestion, swollen joints, skin lesions, memory failure, and debility. By then over 30,000 cattle and large numbers of pigs, sheeps, and chickens had been destroyed; together with large quantities of dry milk, products, eggs, butter and cheese.

In May 1974, the Food and Drug Administration set allowable limits for PBB in milk and meat, and then progressively tightened these standards. The effects of this error are still being felt.

Microeconomics: Or economics "in the small;" a branch of economics concerned with the analysis of the behavior of individual consumers and producers, particularly with the optimizing behavior of individual units such as households and firms. It examines the determination of prices in particular markets, and the effects of monopoly in such markets. This branch of theory is used in many areas of applied economics. See *Macroeconomics*.

Micro Environmental Problems: Immediate environmental problems that daily affect, in most cities, the lives of citizens. These include:

- The continuing existence of unfit and overcrowded dwellings and urban blight;
- Noise and hazards from traffic in local streets;
- Fumes and vibration from traffic and industrial processes;
- Poorly located industrial plants;
- Loss of light and over-shadowing from overhead roads;
- Severance of communities and neighborhoods and established patterns of local movement by limited access highways, traffic management schemes, or large scale developments;
- Danger and inconvenience caused by the juxtaposition of pedestrian and vehicular traffic flows on a street system evolved for entirely different conditions and relative speeds;
- Lack of space for play or recreation and the growing demands of leisure on limited amenities;
- Increasing visual squalor ranging from litter to clutter, including the overhead wirescape;
- Dereliction arising from abandoned and closed dwellings, abandoned or unmaintained business premises and factories;
- Progressive erosion of heritage buildings and the special character of areas.

See *Macro environmental Problems*.

Middleground: The total landscape which is perceived by an observer up to a distance of 4 to 6 kilometers. This forms a linkage between the foreground (q.v.) and background (q.v.) parts of the landscape. Within this range, the observer experiences overall shapes and patterns, and discerns the relationship between landscape units.

Mine Bonding: The posting of a bond by a miner before the commencement of earth moving at a surface mine, which will guarantee the reclamation of the area to be mined. If a miner goes bankrupt or refuses to reclaim the site, he forfeits the bond to the State which then becomes responsible for the reclamation of the mine site. There are several difficulties with this approach: inflation diminishes the real purchasing power of the bond, the real cost of reclamation may increase if concurrent reclamation is not maintained, and several years may elapse between the time an operator walks off the mine site and the time reclamation actually begins with increasing environmental degradation in the meantime.

Mining Wastes: Unwanted material arising from all classes of mining operations in two principal forms:

1. Rock waste, which may occupy valuable land and disfigure the landscape;
2. Tailings from mills, which as silt can impede the natural flow of streams.

Tailings may contain chemicals hazardous to vegetation and animal life. Modern mining methods utilize abandoned workings such as shafts, tunnels, and adits to store unwanted waste as backfill, thus conserving space. Great harm to stream life may occur where mill tailings have been piled in proximity to the drainage area of streams, e.g., copper and zinc compounds, lethal to fish, may be carried from the pile to the stream by rainwater. Filtrates may be carried for considerable distances underground to reach wells or other sources of water for livestock and human consumption. Sometimes old tailing dumps are reworked to recover the metal content. Piles of tailings and other wastes may become mechanically unstable, when saturated with moisture, forming a mud flow that can damage property and life, as at Abafan in Wales. See *Figure 12*.

Ministry of Environment, Norway: A ministry established by the Norwegian Government in 1972 to be responsible for:

- Coordinated planning of the use of water and land resources (municipal, regional, county and national planning);
- Pollution control and noise abatement;
- Waste disposal;
- Conservation of nature areas, of flora and fauna, of hunting and fishing resources;
- Open-air recreation areas;
- Preservation of cultural environments in town and country;

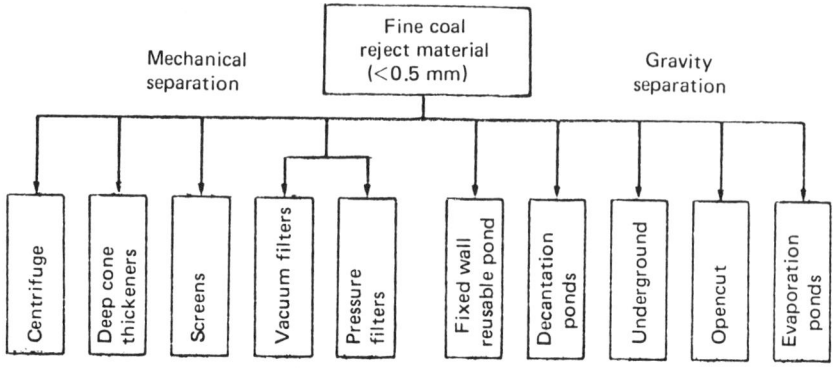

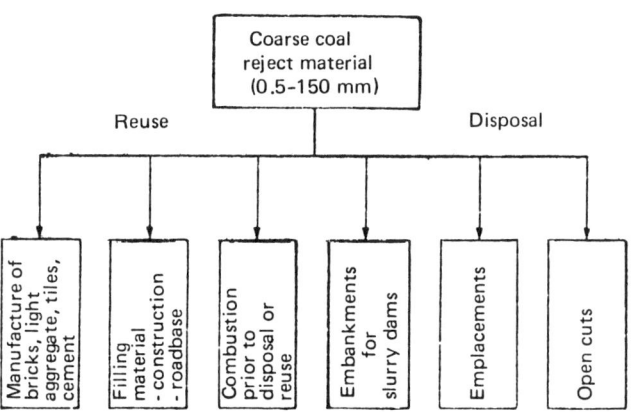

Figure 12: Some methods of treatment and disposal for coal washery reject material. Source: State Pollution Control Commission, New South Wales, Australia.

- Coordination of environmental research;
- International cooperation on environmental matters.

The Ministry comprises several sub-departments dealing with pollution control, nature conservation, planning, natural resources, and coordination. Institutions within the Ministry include the State Pollution Control Authority, the Smoke Control Council, the Council on Oil Pollution, the Directorate for Wildlife and Freshwater Fish, the State Council for Open-Air Activities, the State Council for the Conservation of Nature, and the Central Office of Historic Monuments. The Norwegian research institutions closely affiliated with the Ministry include the Institute for Air Research, Institute for Water Research, and the Institute of Urban and Regional Research. See *Acid Rain*.

Ministry of the Environment (Environment Ontario): A ministry created by the Ontario Government, Canada, in 1972 to bring the responsibility for all aspects of environmental protection, enhancement, and restoration under one agency of the Ontario Government. The operating legislation of the Ministry now comprises:

The Ontario Water Resources Act, 1970

The Environmental Protection Act, 1971

The Pesticides Act, 1973

The Environmental Assessment Act, 1975

To provide the eight million citizens of the province with effective environmental management, Environment Ontario has set four major long-term objectives: to control contaminant emission; to establish environmental safeguards to protect human health and the natural environment; to manage Ontario's water resources and to manage waste; and to develop and maintain measures to preserve, restore, and enhance the natural environment.

The Ministry is divided into three divisions, dealing with environmental assessment and planning, regional operations and laboratories, and finance and administration. The environmental assessment and planning division comprises six branches dealing with air resources, water resources, resource recovery, pollution control, environmental approvals and project coordination.

The Ministry is organized to provide a wide range of services through a regional structure, established in 1974. Direct services are provided from six major regional bases and 23 district offices which serve as key delivery points for inspection and pollution abatement activities and local approvals in order to bring service and control functions closer to the people they actually affect. See *Figure 13*.

In respect to acid rain (q.v.), it has been established by the Ministry that at least 140 lakes are adversely affected. Lakes which shift from an alkaline

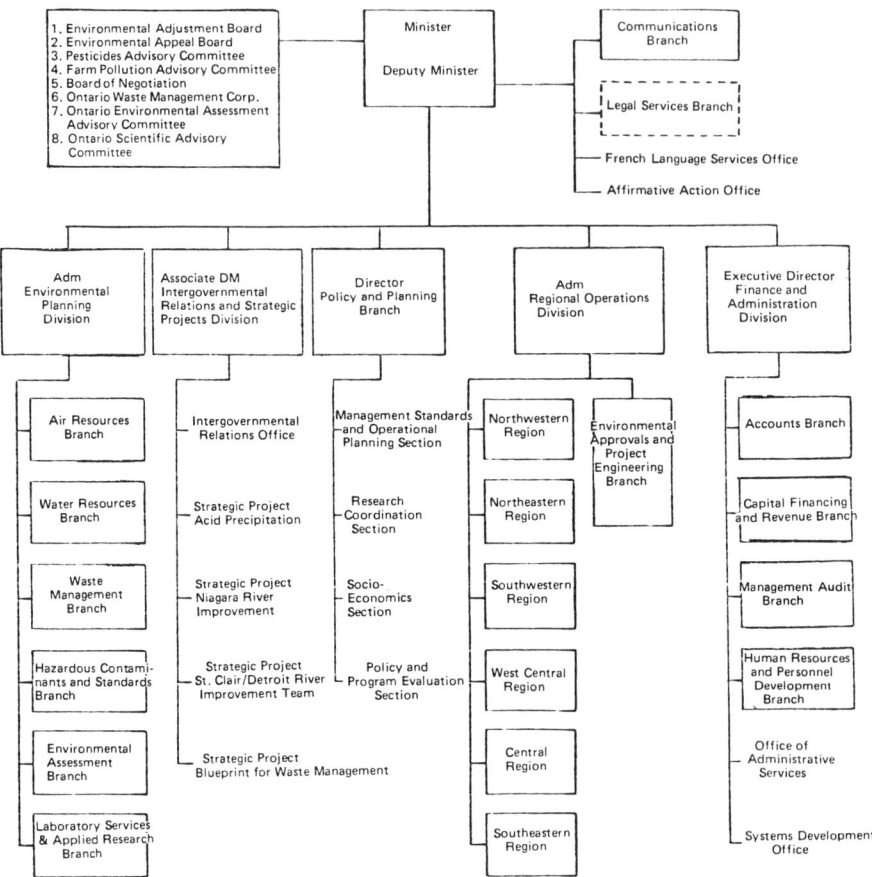

Figure 13: Ontario, Canada—Ministry of the Environment, 1985. Source: Environment Ontario, 1985.

to an acidic balance are increasingly unable to support a fish population. There is also strong reason to suspect that this shift in balance in these waters increases the load of mercury and other heavy metals which can be taken up by fish, metals which can either be toxic to fish or can render them unfit for human consumption.

Precipitation in Ontario is generally the result of the northerly transport of warm moist air aloft emanating from the Gulf of Mexico. The heavily moisture laden warm air mass rises as it moves north into cooler air. The air from which the cloud develops, has passed across the heavily industrialized Eastern United States prior to reaching Ontario, and still contains much of the pollutants emitted into it.

Acidic lakes were first identified in Ontario in the early 1950s, as a local problem resulting from their proximity to major smelting operations in the Sudbury area. Since 1975, data collected by Environment Ontario has established that acidic precipitation is having more widespread effects over a broader area that includes some of Ontario's most popular recreation areas.

Minneapolis-St. Paul, Minnesota, US: See *Skyways, Skywalks, Pedways*.

Minnesota, US: See *Acid Rain; Regional Development Commission, US*.

Missouri, US: See *Times Beach*.

Mitigating Measures: Physical measures and actions taken to prevent, avoid, or minimize the actual or potential adverse effects of a project or action. Measures may include:

- Avoiding an impact altogether by abandoning or modifying a project or by not taking a certain action or part of an action.
- Minimizing an impact by limiting the degree or magnitude of an action in its implementation.
- Rectifying an impact by restoring, repairing, or rehabilitating the affected environment.
- Reducing or eliminating an impact over time through adequate maintenance and efficient operation.
- Compensation for the impact through relocation, new facilities, sound proofing, air conditioning, and landscaping.

See *Pollution Control Strategies*.

Model Cities Program, US: During the 1960s, an experiment conducted in several dozen United States cities in attacking the problem of major blighted areas with massive federal financial aid. It included programs of physical improvement coordinated with social and economic upgrading through job training, school improvements, encouragement of

188   Environmental Planning

economic enterprise, and a whole range of self-help and outside-help measures aimed at reducing poverty and all its adverse effects. The program was introduced by the Department of Housing and Urban Development (HUD).

Under the Housing and Community Development Act, 1974, the Model Cities Program along with federal assistance for urban renewal (q.v.), neighborhood facilities, and rehabilitation was incorporated within the Community Development Block Grant (q.v.) which was intended to give greater control over expenditure to local government. See *New Communities Program; Urban Development Action Grant Program.*

Mt. Laurel Decision, 1983, New Jersey, US: A decision by the New Jersey Supreme Court in 1983 in which disappointment was expressed in the current municipal planning and legal process which had failed to address the issue of affordable housing for the lower income groups. The Court required forthwith positive efforts to produce affordable housing.

The decision obliged every municipality to provide for its own lower income population, except where it was disproportionately large; and municipalities in growth areas to provide opportunities for prospective housing needs. The decision resulted in the abolition of the Division of State and Regional Planning, and the creation in 1983 of a State Planning Commission (q.v.).

Multiple Use: A sharing of land, the range of uses including reserves, national parks, heritage areas, recreational areas, agriculture, urban and suburban developments, forestry, mining, and special use corridors; the management of various surface and sub-surface resources so that they are utilized in the combination that will best meet the present and future needs of the community.

Multiple-Use Sustained-Yield Act, 1960, US: In effect an overriding policy statement of the US Congress for the management of national forests; national forests "are established and shall be administered for outdoor recreation, range, timber, watershed, wildlife, and fish purposes." The US Secretary of Agriculture was authorized to develop and administer the resources of the national forests for multiple-use and sustained-yield in respect to a wide range of products and services. Due consideration was to be given to the relative values of the various resources in particular areas. The concept of sustained yield involved the achievement and maintenance of a high level annual or regular periodic output of the various renewable resources of land, without impairment of the productivity of that land.

Forest service lands have become important recreational playgrounds for hikers, campers, skiers, hunters, and fishermen. Hiking and horse trails, fishing lakes and streams, scenic rivers, and roadside campgrounds attracted over 233 million recreational visitor days in 1982. On the other hand, the Forest Service had 25 million acres (10 million hectares) of special wilderness land where access was limited and no public travel by motorized vehicles was permitted. See *Federal Land Policy Management Act, 1976; Forest and Rangeland Renewable Resources Planning Act, 1974; National Forest Management Act, 1976.*

Multiplier: A ratio indicating the effect on total employment or on total income of a specified amount of real capital investment. This important economic concept shows how fluctuations in the amount of investment, although small in relation to the national income, are capable of generating fluctuations of much greater magnitude in total employment and income; in other words putting one person into employment leads to more than one person's employment and earnings. This is simply due to the fact that while some of the additional earnings of the enlarged labor force will be saved, most will be spent on goods and services. In turn, others will be employed to provide those additional goods and services, both in manufacturing and in distribution. Theoretically, the employment and earnings multiplier may be very high, although in reality may often prove to be 2 or 3.

The concept of the multiplier was first introduced into economic theory by Richard F. Kahn in 1931. He laid down the general principles by which to estimate the actual quantitative relationship between an increment of net investment and the increment of aggregate employment which will be associated with it. The concept was incorporated into John Maynard Keynes' *General Theory of Employment, Interest and Money* published in 1936.

The concept of the multiplier is often used to refer only to the multiplier effects of an increase in investment; in fact, the concept is of general application and applies just as well to government expenditure and taxation. The multiplier effect of a proposed investment is often included in environmental impact statements.

Municipalities Planning Code (MPC), Pennsylvania, US: A statutory instrument empowering local governments in Pennsylvania to undertake comprehensive planning. The MPC requires that a comprehensive plan shall include:

- A statement of the objectives for future growth and development in the municipality;
- A plan for land use describing the amount, location and intensity of residential, agricultural, commercial and industrial uses in the municipality;
- A plan for future community facilities and services such as recreation, water supply and sewage disposal;
- A plan for transportation improvements;
- A map showing how the growth of the municipality relates to adjoining municipalities.

Comprehensive planning is undertaken by municipal officials with the assistance of professional planners from the county or regional planning agencies or a private consulting firm. The involvement of citizens through public participation and public information activities is encouraged to achieve acceptance and implementation of the plan. Environmental Advisory Councils comprising interested and dedicated residents may be elected.

The MPC grants the power to assure that future plans are carried out. The power to regulate the private use of land enables local governments to control the type, location, and intensity of environmental impacts resulting from land development. Ordinances may be enacted, to implement plans, and enforced.

A majority of the State's local governments had by 1984 produced comprehensive plans, although many of the earlier plans produced by the municipalities did not adequately address environmental issues. Only recently has the importance of municipal action to protect environmental quality been fully recognized. Today there is a better understanding of these matters by state, county, and local officials, and by elected representatives.

N

National Ambient Air Quality Standards (NAAQS), US: United States standards which define the maximum concentrations of certain air pollutants allowable in ambient air in order to protect public health and welfare. The Environmental Protection Agency (EPA) (q.v.) is required to set and periodically review these standards. Once NAAQS have been set by the EPA, individual state governments have the responsibility to determine how they can be met and maintained most efficiently at the local level. The main administrative mechanism used by state governments to characterize local air quality and define strategies to achieve national standards is the State Implementation Plan (SIP). Under the Clean Air Act, the EPA is required to approve, review and exercise surveillance over all SIPs.

Primary standards are aimed at the protection of the public health; while secondary standards are concerned with the protection of welfare, amenity, materials, and crops. Initially standards were set for sulfur dioxide, particulate matter, carbon monoxide, nitrogen dioxide, and hydrocarbons. A national ambient air quality standard to protect the public health from exposure to lead particulates became effective in 1978. The new standard set an upper limit of 1.5 micrograms of lead per cubic meter of air based on a three-month average. The new limit was based on preventing children from experiencing blood levels exceeding 30 micrograms of lead per deciliter of blood. It has been argued that levels higher than this have been associated with an impairment of cell function. The attainment of this standard has been associated with a progressive reduction in lead in gasoline. From January, 1986, the upper limit permitted became 0.10 gram lead per US gallon of gasoline.

In 1979, the Environmental Protection Agency (q.v.) announced an increase in the permitted level of ozone in the atmosphere from 0.08 part per million to 0.12 part per million. The states were required to meet this relaxed, though still exacting standard, initially by 1982. The petroleum and automotive industries considered this relaxation too small; environmentalists regarded it as too generous.

In 1985, the EPA undertook a reassessment of the standard for particulate matter (other than lead), to determine whether to switch from TSP (total suspended particulates) to PM_{10} (particulate matter 10 micrometers or smaller). Particles smaller than 10 micrometers are inhalable and widely acknowledged to be more potentially damaging to health. See *Table 12*.

Table 12: United States of America—National Ambient Air Quality Standards

Averaging Period	...Primary Standard...		..Secondard Standard..	
	($\mu g/m^3$)	(ppm)	($\mu g/m^3$)	(ppm)
Sulfur dioxide				
Annual arithmetic	80	0.03	None	
24-Hour[a]	365	0.14	None	
3-Hour[a]		None	1,300	0.5
Particulate matter				
Annual geometric	75		60[b]	
24-Hour	260		150	
Carbon monoxide				
8-Hour[a]	10,000	9	Same	
1-Hour[a]	40,000	35[c]	Same	
Ozone				
1-Hour[d]	235	0.12	Same	
Nitrogen dioxide				
Annual arithmetic	100	0.05	Same	
Lead				
Calendar quarter	1.5		Same	
Hydrocarbons				
3-Hour (6 to 9 a.m.)	160	0.24[e]	Same	

[a] Not to be exceeded more than once per year.
[b] The secondary standard of 60 $\mu g/m^3$ is a guide to be used in assessing implementation plans to achieve the 24-hour standard.
[c] Revision to 28,630 $\mu g/m^3$ and 25 ppm proposed 8/18/80.
[d] Standard attained when the expected number of days per calendar year with maximum hourly average concentrations above 235 $\mu g/m^3$ and 0.12 ppm is equal to one or less.
[e] Hydrocarbon 3-hour standard used only as a guide to develop plans for achieving ozone standard.

National Capital Development Commission (NCDC), Australia: A commission established by the Australian Parliament in 1958 for the purposes of planning, developing and constructing the City of Canberra as the national capital of Australia. The commission undertakes the strategic and general land-use planning of the city, determines planning policies and programs, prepares detailed project plans and carries out a great deal of physical construction. It provides the servicing infrastructure for the city and supplies serviced land to the Department of Territories and Local government with proposed conditions and land-use controls for releases for residential and commercial purposes. The result is a highly coordinated partnership between the planning and development authority (the NCDC) and the city's administration.

The establishment of the NCDC coincided with the development of a national policy to transfer federal government departments from other centers to Canberra so far as practicable. This policy gave great impetus to the growth of Canberra which has developed from a small town of some 20,000 to substantially more than 250,000 over about twenty years. The capital is in effect a new town, established on Crown land with a leasehold system operating throughout the Territory. It comprises a number of centers or suburbs each with its own shopping centers. Canberra is the home of the Australian National University, with its research schools. A new parliament house is due for completion in the bicentennial year 1988.

The original plan for Canberra was laid down by the American, Walter Burley Griffin. The great central man-made lake is named for him. The essence of the plan has remained intact. The site of the city is one of great natural beauty; the infrastructure of highways, roads, and bridges, together with public and government buildings and major community centers blend with this natural beauty which has been augmented by a massive tree planting program for individual streets and avenues; while finally the quality of housing is generally of a high standard. The average per capita income is the highest in Australia. Canberra is also the home of the National Library, the High Court of Australia, and the Australian National Art Gallery.

National Capital Planning Commission, Washington, DC, US: A planning agency created in 1926 to continue the task of enhancing the L'Enfant Plan (q.v.) for the capital of the United States of America, and its region. The initial title of National Capital Park and Planning Commission was changed to National Capital Planning Commission in 1930. Under the Home Rule Act, the Mayor of Washington, D.C. has been responsible for preparing the local elements of the comprehensive plan (q.v.) while the Commission has been responsible primarily for the federal elements. The Commission collaborates with the District of Columbia Office of Planning and the Department of Housing and Community Development. See *Comprehensive Plan, Washington, DC.*

National Committee on Environmental Planning and Coordination (NCEPC), India: An agency established by the Indian Government in 1972 to provide a forum in the structure of government where environmental considerations could receive considered and close attention. The Committee reviews, formulates and promotes policies and programs covering development projects, physical planning, legislation, administrative procedures, education and research. It operates in conjunction with the Department of Environment.

The work of the Committee has resulted in several important policy decisions, laying down standards and guidelines to safeguard environmental quality. It has also been involved in the environmental appraisal of projects of varied nature and magnitude. One of the earliest exercises of the Committee was an assessment of the Nhava-Sheva fertilizer project.

An early assessment in nature conservation was the Chilka Lake development study which threw up a number of suggestions on the location of a Naval Boys Training School on the shores of the lake, and the economic development of the lake area in general. Perhaps the most successful of efforts was NCEPC's intervention with the Tamil Nadu State Government on the question of constructing a hydroelectric power generation project right in the heart of the Madhumalai Wildlife sanctuary in Nilgiris. A team from the NCEPC visited the area and had detailed discussions with the state government representatives, finally convincing the state government that the economic benefits of the project were far outweighed by the ecological damage that would be caused. The project was given up by the state government.

The NCEPC has been in close contact with the Indian Planning Commission in evolving a mechanism for an assessment of the environmental implications arising out of development projects, and in providing necessary safeguards. Questionnaires for environmental impact assessment have been prepared for (i) industrial projects, (ii) construction of roads and railways, (iii) mining projects, (iv) hydroelectric power generation and irrigation projects. An Inter-ministerial Environmental Appraisal Committee has been set up for in-depth examination of environmental aspects in respect of hydroelectric and irrigation projects that are referred to the NCEPC by the concerned authorities, e.g., the Planning Commission, Central Electricity Authority, or Central Water Commission. All the proposals for thermal power projects are also referred to NCEPC for environmental appraisal.

The NCEPC has advocated the setting up of agencies at the level of State and Union Territory administration to ensure environmental concerns in the development planning of different regions. Most of the States and Union Territories have since constituted Environment Committees. These Committees are headed by the Chief Ministers. The membership of these Committees includes representatives of various development departments, namely, Agriculture, Industries, Forestry, Health and Town Planning, etc. Representatives from education and research institutions have also been nominated as members of the Environment Committees. The Environment Committees work in conjunction with State Planning Boards and concerned departments. Guidelines have been drawn up by the NCEPC regarding the areas of responsibilities for the State Environment Committees.

National Contingency Plan, US: Guidelines that set down the procedures private companies and federal and state agencies must follow in any clean-up operations under the Superfund law. See *Comprehensive Environmental Response, Compensation, and Liability Act, 1980.*

National Council on Physical Planning and the Environment, Greece: A central administrative authority created by the Greek Government in 1975 primarily responsible for the protection of the nation's environment; it is a ministerial body supported in its task by a permanent committee and specialized secretariat. The legislation creating the Council set out the main goals of a national environmental protection plan. These are to:

- Secure and control the evolution of the country's regional structure in order to assure the efficient use of natural resources and the proper coordination of the various population activities;
- Conserve and protect the nation's natural and cultural environment by initiating special environmental programs in accordance with developmental policies;
- Determine and provide the institutional and legal framework for the development of the necessary infrastructure for effective physical planning;
- Coordinate procedures for environmental planning in Greece with those adopted by other nations and those proposed by international organizations which are concerned with these issues; and
- Ensure the conditions for proper planning aiming at an uninterrupted improvement of the quality of life.

Greece is one of the few countries which have included in their constitution an article on environmental protection. The Greek Constitution of 1975 explicitly states in article 24 that the protection of the natural and cultural environment constitutes an obligation of the State and that the State is responsible to take special preventive or repressive measures towards its conservation. It is also emphasized that monuments as well as historical sites are under the State's protection, while necessary restrictive measures and compensation procedures for the realization of this protection are defined by law.

National Emissions Standards for Hazardous Air Pollutants (NESHAP): See *Clean Air Amendment Act, 1977, US.*

National Environment Board, Thailand: An agency created by the Thailand Government in 1975 at a time of rapid transition from an essentially rice village economy to a contemporary system involving all the environmental ills characteristic of urbanization and industrialization.

The responsibilities of the Board are defined in the Improvement and Conservation of National Environmental Quality Act (B.E.2518) of 1975, and its subsequent amendment act (B.E.2521) of 1978. The Board has the power and duty to:

- Submit policy and opinion concerning the improvement and

conservation of environmental quality to the Council of Ministers;
- Consider the implementation of policy in respect to the schemes or projects concerning environmental quality;
- Consider and submit opinion on the projects of government agencies, state enterprises, and private organizations, which may have an adverse effect on environmental quality to the Council of Ministers or agencies concerned;
- Submit plans for the development, improvement and conservation of environmental quality to the Council of Ministers;
- Recommend standards of environmental quality and measures for the safeguarding and conservation of environmental quality;
- Coordinate works between agencies and others concerning environmental quality;
- Submit reports on the national situation of environmental quality.

National Environment Protection Board, Sweden: In Sweden, the central administrative body for environment protection which was set up in 1967 initially under the Ministry of Agriculture. The Board has responsibilities in respect to nature conservancy, water and air pollution control, noise abatement, protection of wildlife, mobile open air recreation and environmental protection research. In 1968, the Environmental Advisory Committee, chaired by the Minister of Agriculture, was formed to provide the government with information and advice on environmental protection.

The Board among other things endeavours to: protect the interests of nature conservancy in the planning of buildings and industrial projects; preserve suitable areas for scientific purposes, open-air life and recreation, and promote open-air activities; protect the shores; preserve the landscape against destruction from gravel-pits, etc.; combat water and air pollution; combat noise; supervise public cleansing; and conduct, finance and apply research in the field of environment protection.

The Board comprises a Director-General and other persons, appointed by the government. The Director-General is also chairman of the Board. In its work, the Board is supported by three advisory bodies: the Nature Conservation Council, the Water Protection Council and the Air Protection Council. A Land Acquisition Delegation helps the Board and the County Administrations with the valuation of land and in negotiations on buying or selling land. A Delegation for support to sports establishments decides on grants to various recreation facilities. Grants for research in the field of environment protection are allocated by a special research committee responsible to the Board.

The organization of the National Environment Protection Board includes an administrative division; three divisions—for nature conservation, water protection and air protection; a research secretariat; a secretariat for planning; coordination and information; and a research laboratory. There is also a solid waste management unit dealing with questions of public cleansing and refuse. The Board has some 330 employees. A further 290 persons work in the environment protection sections of the County Administrations. To this may be added the staff of other state authorities dealing with questions of environment protection. See *Environment Protection Act, 1969*.

National Environmental Policy Act, 1969, (NEPA) US: An Act of the US Congress the aims of which were to: declare a national policy which would encourage productive and enjoyable harmony between man and his environment; promote efforts which would prevent or eliminate damage to the environment and biosphere and stimulate the health and welfare of man; enrich the understanding of the ecological systems and natural resources important to the nation; require the preparation of environmental impact statements for major federal projects; and establish a Council on Environmental Quality.

The Environmental Protection Agency (EPA) has a unique responsibility in respect to the process. Under Section 309 of the Clean Air Act, the Agency is required to review and comment in writing on the environmental impact of any matter proposed by another federal agency which would have a significant impact on the environment. Thus the Agency has a mandate not only to police its own activities for compliance with the National Environmental Policy Act, but must also carry out a quality control screening of environmental reviews prepared elsewhere in the federal government system.

In 1977, in an Environmental Message, the US President emphasized the importance of the National Environmental Policy Act to the development of sound federal decisions and to public involvement in government. To improve the implementation of the Act and, in particular, the use of environmental impact statements, he issued an Executive Order:

- Authorizing the Council on Environmental Quality (CEQ) to issue regulations which would ensure the effectiveness of environmental impact statements while reducing unnecessary paperwork;
- Directing the Council to prescribe procedures to help resolve interagency conflicts regarding projects being reviewed under the Act.

The Council's NEPA regulations were promulgated in November, 1978, and became effective in July, 1979.

During the first twelve years of NEPA's operation, 70 federal agencies prepared approximately 15,000 environmental impact statements, and many times that number of brief environmental assessments that help guide decisions on whether to prepare a full EIS. During that twelve year period 1,445 NEPA law suits were filed; this represented about 10 per cent of all federal proposals for which a full EIS was prepared. The most NEPA lawsuits were filed in 1974 (189 cases) and the least were filed in 1977 (108 cases).

National Environmental Policy Regulations, 1979, US: Regulations introduced under the US National Environmental Policy Act (q.v.) to establish uniform procedures for the implementation of the Act. The Regulations had three principal aims: to reduce paperwork, reduce delays, and produce better decisions in respect to the protection and enhancement of the quality of the human environment.

The Regulations limited the length of an environmental impact statement (q.v.) to normally less than 150 pages, and for proposals of unusual scope or complexity to less than 300 pages. Plain language was advocated. A new scoping (q.v.) procedure was established to assist agencies in deciding what the central issues were in any instance. In 1983, the regulations were reviewed and strengthened.

National Forest Management Act, 1976, US: An Act which directed the US Secretary of Agriculture to develop a land and resource management plan for each administrative unit of the National Forest System by 1985; regulations were developed in 1979 to guide this effort, these being revised in 1982. Planning was to embrace such resources as recreation, fish and wildlife habitat, water, timber, range, wilderness, and minerals. The plans for all 121 administrative units were published in 1985.

National Franchise Board for Environment Protection, Sweden: An agency responsible for hearing cases under the Swedish Environment Protection Act, 1969. Breaches of the Act may lead to fines or a maximum of two years in prison. The Board also examines license applications for discharges to the environment under the Act. The Board operates in a similar manner to a court of law. See *Environment Protection Act, 1969; National Environment Protection Board*.

National Heritage Memorial Fund, UK: A fund established by the British Government under the National Heritage Act, 1980, for the purpose of assisting in the preservation of the national heritage through the provision of financial asssistance for the acquisition, maintenance and preservation of land, buildings, and objects of historical or other importance to the national heritage.

National Historic Preservation Act, 1966, US: An Act authorizing the Secretary of the Interior to expand and maintain a national register of districts, sites, buildings, structures, and objects of local, state, and national significance and to grant funds for the purpose of undertaking comprehensive statewide historic surveys and preparing statewide plans for historic preservation. It established a program of matching grants-in-aid to the states for the preservation, acquisition, and development of national register properties and provided funding to the National Trust for Historical Preservation to implement its programs.

This Act was amended by the National Historic Preservation Act Amendments, 1980. The existing national register programs were continued; the amendments now required public and local government participation in the nomination process while prohibiting the listing of properties if the owner objects. The amendments authorized US participation in the World Heritage Convention and established procedures for the nomination of World Heritage properties.

National Human Monitoring Program, US: Administered by the US Environmental Protection Agency (q.v.), a program which monitors on a national scale the prevalence of selected pesticides and polychlorinated biphenyls (PCBs) in human adipose tissue, blood, and urine. The program, established since 1972, consists of two major monitoring networks: the National Adipose Tissue Survey, and the Health and Nutrition Examination Survey.

National Park: A relatively large area of land set aside for its features of predominantly unspoiled natural landscape, flora and fauna, permanently dedicated for public enjoyment, education and inspiration and protected from all interference, other than essential management measures, so that its natural attributes are preserved.

In the United States, the Yellowstone national park was created in 1872; while the Yosemite and Sequoia national parks were established in 1890. The idea of protecting outstanding scenic and scientific resources, wildlife and vegetation took root and developed eventually into a national policy.

The national park system was expanded gradually, with the addition of vast natural areas and the inclusion of places of significance in the exploration of the continent and the development of American Indian culture. By the 1960s, the system included 192 areas encompassing over 10 million hectares most of which were in federal ownership. Of these areas, about 30 were national parks, most of them large.

National parks are established by Congress; another act of Congress is required to change boundaries or to modify the basic protection arrangements, unless boundary adjustment provisions were specifically stated in the original act. National parks have been defined as spacious land and water areas of nationwide interest established as inviolate sanctuaries for the permanent preservation of scenery, wilderness, and natural vegetation and animal life in their natural condition. The national parks service was established by Congress in 1916.

By 1982, there were 48 national parks, comprising 47 million acres (19 million hectares). However, many of these lacked adequate facilities and safety measures, and suffered from general neglect. The Reagan Administration adopted a policy of not creating or acquiring new parks; the view was that the park system may have grown too fast, or at least that the resources including staff had proved to be too thinly stretched to cope with extensions of this magnitude. The position had certainly been aggravated by the bringing of more and more urban parks into the national system, as individual cities became short of money. Instead, a five-year park restoration and improvement program was initiated.

National parks and nature reserves in the Soviet Union are under the control of the central government, and form part of the national budget. The parks are directed by special committees set up by the Council of Ministers; in some instances, responsibilities may be imposed on the USSR Academy of Sciences.

Areas within parks may be set aside for special purposes and declared prohibited areas. Otherwise parks are open to visitors for recreational purposes within the management plan. The destruction of wildlife and plants, hunting and fishing, as well as mining and livestock grazing are prohibited. Popular parks include the Crimea, Caucasus, Ritza and Teberda. Nature reserves have been set aside also for the protection of particular features and resources.

Measures are taken to protect wildlife. The hunting of polar bears is forbidden. The population of the sable, through the use of forest reserves, has been restored having been at the point of extinction at the beginning of the 20th century. Herds of bison are steadily growing and the population of beavers has considerably increased. Of the three herds of fur seals still left in the world, two are to be found on Soviet islands in the Pacific.

The parks and avenues of Moscow cover 40,000 hectares. Every year, an average of 500 hectares of boulevards, flower gardens, and courtyards planted with trees and shrubs are added to existing ones. The greening of Moscow goes on.

National Parks in England and Wales are set up under the National Parks and Access to the Countryside Act which received the royal assent in 1949. Under the Act, a National Park commission selected areas which it considered suitable to become national parks. The passing of the 1949 Act was the culmination of a campaign dating from the second half of the 19th century, fought by various voluntary organizations which strived to preserve the countryside and its wildlife and to secure public rights of access to mountain and moorland. These included the Commons, Open Spaces and Footpaths Preservation society, the National Trust, the Society for the Promotion of Nature Reserves, the Ramblers Association, the Royal Society for the Protection of Birds and the Councils for the Preservation of Rural England and Wales. The total membership of the voluntary bodies today has been estimated at over 3 million.

National parks in Canada are established by the federal parliament. National parks in the Canadian Rockies, in the forests and lakelands of the plains and eastward to the Atlantic coast, provide a variety of environments, recreational facilities, and wildlife. Except for recognition of native Indian rights, hunting is prohibited. Most provinces have set aside provincial parks, many as magnificent as the national parks. Canada has also designated many game reserves and wildlife sanctuaries, some of great size. Regulations under the National Parks Act 1970 prohibit the polluting of streams and other bodies of water in national parks and require the deposit of rubbish in designated places.

Australia is fortunate that there has been time to dedicate land for the purposes of national parks and reserves and that the need has been recognized before population pressures made it impracticable within metropolitan areas.

The first National Park service to be set up as a largely independent government authority was established in New South Wales in 1967, and it was followed by similar organizations in every other state. The Federal Parliament enacted legislation to establish the Australian National Parks and Wildlife Service and also the Great Barrier Reef Marine Park Authority in 1975. Australia now has an admirable system of parks and reserves. There are over 500 national parks and other conservation reserves of 1000 ha or more in area. In total, the reserves covered more than 28 million ha, or more than 4 per cent, of Australia in June 1979 and in excess of 33 million ha, or almost 5 per cent, in June 1981. See *Natural Parks, Japan; Nature Conservance Act, 1975, Sweden; Nature Conservation Act, 1976, France.*

National Physical Planning: National planning concerned with physical location. Its aims are: to chart the claims which various activities may make on land and water resources, location and environment, and the assets which are available to satisfy the expected demand; and to the extent that is reasonable in the national interest to draw up guidelines for the management of such natural resources.

The main principle underlying national physical planning is that it considers, in relation to one another, the different claims made on important natural resources. It grapples, for example, with the problem of coastal areas which offer both favorable locations for certain industrial installations and yet include assets of particular value for different leisure-time activities. The implication is that some of the future expansion of industry inimical to the environment will have to take place in locations other than those most favorable from the point of view of the industry itself. Physical planning needs to work on the basis of a perspective of up to 20 to 30 years, in other words considerably longer than what is regarded as practical in, for instance, long term economic planning as carried out in Sweden at present.

National physical planning as conceived in Sweden contains only general guidelines regarding the disposition in certain cases of land and water resources, guidelines which must be taken into consideration in the course of future decisions regarding land use. The main preparatory work in support of this approach has been carried out within the Ministry of Physical Planning and Local Government which has been responsible for physical planning matters since 1969.

The aim of physical planning in Denmark is to ensure coherent regulation of all land use, in both urban and rural areas. The planning reforms implemented over the past few years apply at three levels: nationally, by regions (counties) and by local government areas (municipalities).

The doubling of the Danish population since the turn of the century has exerted a steady and continuing pressure on land earmarked for urban development, housing, industrial purposes, communications and public institutions. On the other hand, population growth and concentration have sharply increased the need for recreational areas, beaches, woodland and parks, and areas for holiday housing.

Land-use legislation is also a vital element in pollution control, since physical planning helps prevent some of the nuisances otherwise created. Physical planning may thus be considered generally an essential part of successful environmental policy.

The National and Regional Planning Act was the catalyst needed to intensify physical planning at a national and regional level. Planning on a nationwide basis had begun in 1961 when the government appointed an advisory committee, the National Planning Committee, charged with defining guidelines for land use and for future urbanization. The following year the committee submitted its recommendations in the form of a draft zoning plan governing land use throughout the country and the objectives to be achieved. The committee itself was disbanded but the work of its secretariat, which was maintained under the supervision of an interministerial supervisory group, resulted in 1974 in a report on land use with forecasts and assessments of future social development.

Although physical planning is technically the domain of the Minister of the Environment, all major issues must be submitted to a committee of ministers. This applies not only to approval of regional plans but also to the issuing of national planning directives of both general or special character. See *Nature and Monuments Conservation Act, 1974, Greenland.*

National Planning Guidelines, UK: Issued by the Department of the Environment (q.v.), national guidance to the planning authorities in Britain and other agencies to assist in the assessment, approval and control of potentially hazardous industrial and other developments. Coastal Planning guidelines were issued in 1974, and National Planning guidelines for Sites for Large-Scale Industry, Petrochemical Developments, Agriculture, Forestry, Nature Conservation, and Landscape and Recreation were issued in 1977. Their main purpose was to: (a) identify and define the kinds of development and locations which could raise national issues relevant to land use planning; (b) set out the national aspects of land use which should be taken into account by local planning authorities in their regional reports, structure and local plans; (c) suggest where there might be a need for interim development control policies in relation to national issues; and (d) explain the criteria which form the basis for directions requiring certain planning applications to be notified to the secretary of State. The Guidelines also instruct planning authorities to cooperate with the Health and Safety Executive to produce policies for potentially hazardous sites and their environs which take account of the likely nature of development and necessary safeguards. See *Health and Safety at Work Act, 1974; Planning System, UK*.

National Pollutant Discharge Elimination System (NPDES), US: See *Clean Water Act, 1977, US*.

National Priorities List, US: A list maintained by the Environmental Protection Agency (q.v.) of those hazardous waste sites considered to pose the greatest potential long-term threat to human health and the environment. In 1985, the list comprised 786 sites, of which 16 were in Silicon Valley. See *Comprehensive Environmental Response, Compensation, and Liability Act, 1980*.

National Survey of Air Pollution, UK: An air pollution sampling network established in Britain in the early 1960s. Measurements of air pollution had been made since 1914 by local authorities and other interested bodies, in cooperation with the Department of Scientific and Industrial Research. However, the original system of measurements, while providing a rough indication of local conditions, was not sufficiently comprehensive to provide an accurate picture of the distribution of air pollution in different types of area throughout the country. Since information could only be obtained from a survey planned on proper statistical lines, and as the Clean Air Act became fully operative and more local authorities showed interest, the need for a more systematic national approach sharpened. As a result, a working-party consisting of representatives of local authorities, the Medical Research Council, the Meteorological Office, the Ministry of Housing and Local Government and DSIR was set up to devise a new survey. The recommended scheme was accepted by the Standing Conference of Cooperating Bodies for the Investigation of Atmospheric Pollution at their meeting on 14th November 1960. It was recognized that the pollution caused by dust was very localized and that it would be inappropriate to include dust deposition measurements in the new survey. The National Survey itself is confined to measurements of SO_2 and smoke. It was recommended that SO_2 in the atmosphere should be measured by the daily volumetric hydrogen peroxide method and smoke by means of a filter. Both were conveniently combined in the same piece of apparatus.

About 1,200 monitoring sites for air pollution are maintained by local authorities and other bodies. The information is collated and published by the Warren Spring Laboratory at Stevenage, as part of the National Survey

204 Environmental Planning

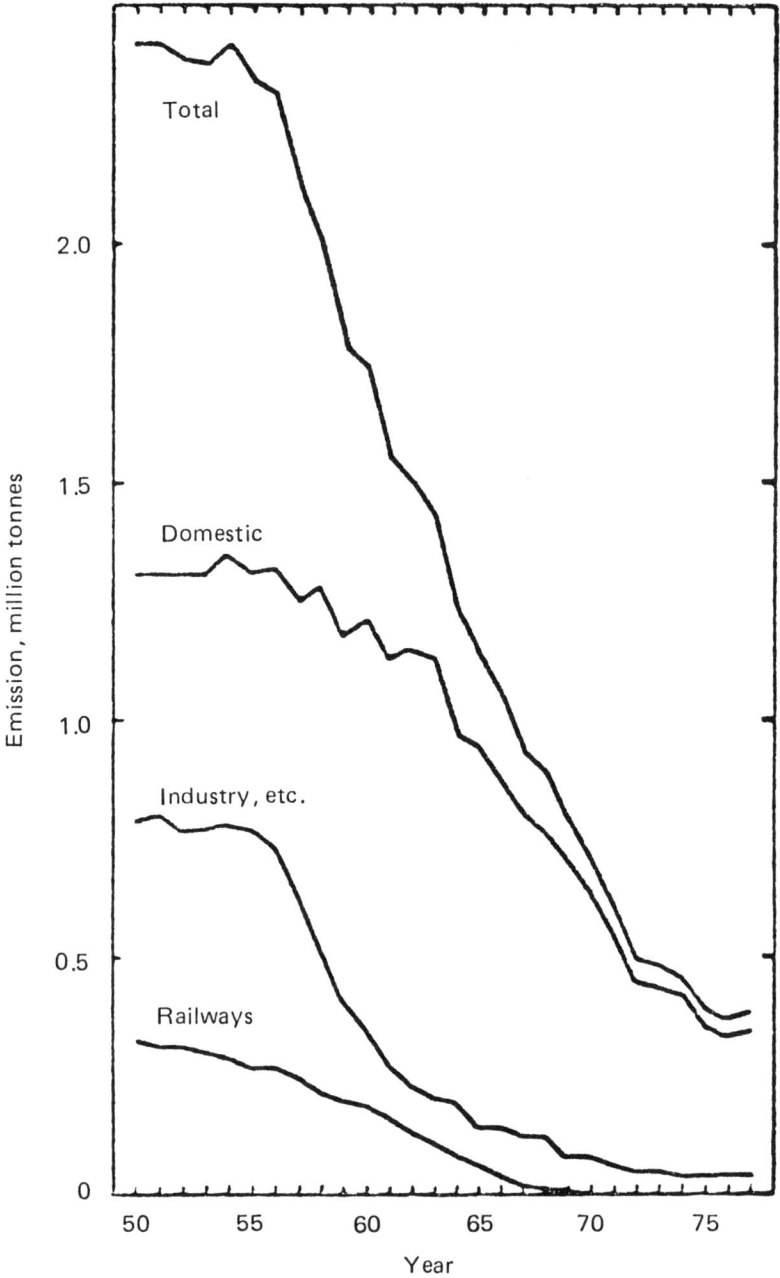

Figure 14: Emission of smoke from coal combustion in the United Kingdom 1950-1977.

of Smoke and Sulfur Dioxide. Other programs measure grit and dust deposition, acid particles, aerosols, metals, and vehicle emissions.

According to *Environmental Pollution and Water Statistics* (Fourth Digest, 1982), emissions and concentrations of smoke from coal combustion fell by over 80 per cent between 1960 and 1980, due mainly to the introduction of smoke control areas and a general consumer preference for cleaner, more convenient fuels. During the same period, average urban concentrations of SO_2 fell by more than 60 per cent. See *Figure 14*.

National Stream Quality Accounting Network (NASQAN), US: The United States national water quality monitoring network operated by the US Geological Survey. The results over recent years appear to show that the nation's pollution control efforts are having a positive effect. However, since the network tends to emphasize chemical and physical variables, increasing attempts have been made to develop biological monitoring programs using aquatic organisms as a measure of environmental quality. One example of this type of monitoring is the Aquatic Life Survey, jointly designed and managed by the US Environmental Protection Agency and the US Fish and Wildlife Service. The Survey collects information on fish presence, productivity, habitat influences, usability, and population trends.

National Trails System Act, 1968, US: Enacted by the US Congress in 1968, legislation to establish a national trails system. By 1979, 257 national recreation trails had been designated, including 21 trails for those using wheelchairs, and 13 trails for the use of blind people with directional and other signs in braille.

In 1979, the US President announced that: the Forest Service would establish 145 new trails, reaching a goal of two trails for each forest system unit; each federal land management agency would announce by January 1980 a goal for national trails on its lands. By December 1980, 75 new trails would be established on public lands other than national forests; the Secretary of the Interior would assist other agencies in surveying existing trails on federal lands to determine which can be made part of the national trails system; all agencies would encourage states, localities, Indian tribes, and private landholders, to designate trails; legislation would be submitted to Congress to designate the 250 kilometers Natchez Trace National Trail through Tennessee, Alabama, and Mississippi.

The US President reaffirmed his support for the proposed 5,000 kilometer North Country Trail, extending from the State of New York to North Dakota. He also reaffirmed his Administration's commitment to ensure the protection of the existing Appalachian Trail, one of America's best known and most popular recreation trails. The 3,200 kilometer Appalachian Trail winds through 14 states and is readily accessible to nearly half the population of the United States. It had been created by a volunteer movement without parallel in the history of outdoor recreation in

America. Legislation passed in 1978 was designed to protect those threatened portions of the right-of-way located on private lands.

National Urban Policy and New Community Development Act, 1970, US: A measure introduced by the United States Congress to provide for the development of a US national urban policy.

National Wildlife Refuge System Administration Act, 1966, US: An Act which provided for an extension of the US wildlife refuge system, and for more public and multiple uses of refuges. It provided also for wetland habitat for migratory birds. In 1982, an additional 19,400 acres (7,700 hectares) of wildlife habitat were purchased from funds provided by the Land and Water Conservation Fund, a fund established by Congress in 1965 to provide federal dollars for the acquisition of parks and refuges. The Act also promoted the development of habitat evaluation and inventory techniques.

National Wildlife Refuges, US: A national system of refuges and sanctuaries for wildlife developed at a federal and state level in the United States during the twentieth century, assisted by federal funds.

Reservation of federal land as national wildlife refuges was accomplished initially by executive orders of the President, the first refuge being Pelican Island off the east coast of Florida, established by Theodore Roosevelt in 1903. Many early bird sanctuaries, however, were created by the National Audubon society, a citizens' organization, to preserve egret rookeries in the south. The severe decline of waterfowl and shore birds (such as egrets, herons, ibises and pigeons) because of overshooting and drought, and diminishing populations of larger mammals (such as bison, bears, deer, moose, antelopes and beaver), evident during the first decades of the 20th century, stimulated reservation of large federal units such as Malheur and Upper Klamath national wildlife refuges in Oregon, the Pribilof Islands reservation in Alaska, the National Bison range in Montana and the Wichita Mountains wildlife refuge in Oklahoma. By 1929, there were 87 federal refuges in 24 states and the territories. By 1975, however, there were over 280 national wildlife refuges. Through the Wildlife Restoration Act 1937, states were given financial assistance in the acquisition and development of suitable lands for wildlife.

The growth of such a large and varied system of federal refuges, and the concurrrent development of refuges and sanctuaries by the states and private organizations, is evidence of changed attitudes toward wildlife on the part of many American people. Since consumers seldom are dependent on game meat for food, hunting is regulated to preserve species for future generations. During the 20th century there developed a large body of public opinion interested in wildlife for its own sake and as a source of personal enjoyment and scientific study.

Natural Features: The landscape characteristics of a city and its environs, or a geographical area, including forests, meadows, hills, valleys, rivers, and streams which are distinctive, outstanding or prominent.

Natural Parks, Japan: Designated parks in Japan of which there are three categories:

- National parks containing scenery representative of the Japanese landscape and of world importance.
- Quasi-national parks ranking second in scenic attractiveness.
- Prefectural natural parks being representative of localities and regions.

By the end of 1984, there were 27 national parks totalling over 2 million hectares and over 5 per cent of the national land area; 50 quasi-national parks totalling over 1.1 million hectares and nearly 3 per cent of the land area; and 290 prefectural natural parks of nearly 2 million hectares and over 5 per cent of the land area. The whole system embraced over 5.1 million hectares, or 13.6 per cent of the land area of Japan. This was achieved in spite of the human population of 114 million and the need of land for agricultural and other commercial purposes.

Natural parks in Japan are designated without affecting the private ownership of the lands in question. However, the private rights of the land owners are restricted to meet the purposes of nature protection. In more recent years, areas of special significance within designated parks have been progressively brought within public ownership, to permit more effective control and management.

The natural beauty of Japan has represented one of its richest natural resources, and the parks have become points of focal interest for tourists and Japanese alike. Indeed, the pressure by visitors on national parks has created continuing management problems.

A recent development has been the designation of marine parks within national and quasi-national parks; by the end of 1984, some 46 marine parks had been approved totalling almost 1,800 hectares.

Japan possesses more crater lakes, or caldera, than any other country; some of the largest in the world are within the designated national parks. The most magnificent are within the Aso, Akan, and Towada-Hachimantai parks. The Japanese Alps, including the Minami Alps, are noted for gorges, waterfalls, virgin forests and unusual wildlife.

The parks also protect many religious shrines and temples, the most notable being the Grand Shrine of Ise in Ise-Shimo national park. This national park was expanded in 1976. Mt. Fuji, in Fuji-Hakone-Izu national park, is one of the most nearly perfect volcanic cones in the world. It is revered by the Japanese people.

In addition to parks, green zone master plans for towns and cities are being promoted to improve the urban environment. The ideal is considered a green network covering 30 per cent of an urban area. City park improvements are in hand. The Outer Gardens of the Imperial Palace, the Shinjuku Imperial Gardens, and the Kyoto Imperial Gardens, all of which belonged to the Imperial family, have been under the control of the Japanese Environment Agency since 1971 and open for public use.

Natural Resource Conservation: The management of living and non-living resources in such a way as to sustain the maximum benefit for present and future generations; natural resources include soils, rocks, minerals, organic matter, water, air, fauna and flora, the physical and chemical properties of matter, and aesthetic elements. Hence, an attractive natural landscape or mountainous backdrop is a natural resource of value. See *Conservation; Living Resource Conservation; Non-Renewable Resources; Recycling; World Conservation Strategy.*

Natural Resources Data Bank, Chile: A computerized nature resources information service established in Chile in 1980. It provides users with high quality data relating to soils, climate, water, forests, and minerals. Complementary information is also available concerning road and rail infrastructure, population, energy sources, industry, and rural activities.

Natural Resources Inventory: A study of the natural features of a community's environment; the product of such a study should be a series of maps showing the distribution of environmental features coupled with a written explanation of their importance for the community and its future land uses. The inventory should not be solely descriptive but include an interpretation of the data to show how and why it poses some limitation or opportunity for land use. For example, it is important to know where soils are likely to cause problems of erosion, leaky basements, or malfunctioning septic tanks.

Natural resources inventories should indentify areas which pose:

- Hazards to certain uses such as subsidence or landslide prone areas;
- Physical limitations which will make development more expensive or involve major environmental impacts such as soils which are unsuitable for use as disposal fields for septic tank effluent, or steeply sloped areas requiring major earth moving and vegetation removal;
- Opportunities for certain kinds of uses such as a potential for high agricultural production;
- Opportunities for multiple use of an area such as extraction of mineral deposits prior to use for residential or other purposes;
- Opportunities for the enjoyment of unique natural or scenic areas.

Nature and Monuments Conservation Act, 1974, Greenland: An Act passed by the Danish Folketing in 1974 which provides for the conservation of nature, archaeological remains, and buildings, in the dependency of Greenland. Under this Act, the whole of the northeastern part of Greenland has been classified as a national park.

As a hunting, nomadic people, Greenland's Eskimo population has existed for thousands of years in myriads of scattered, tiny communities. It was not until about 250 years ago that European colonization encouraged the population to congregate in rather larger communities. About 1925 climatic changes caused havoc with the traditional industries of sealing, whaling and other forms of hunting. The seal was the first marine mammal of significance to evacuate Greenland waters. Fortunately the disappearance of the seal more or less coincided with an invasion of fish—chiefly cod—but this change meant a swing in the Greenlander's way of life from a subsistence economy to a money economy. These vital modifications in Greenlandic life further promoted the trend toward urban settlements, which was again encouraged by industrialization of the fishing industry after the Second World War.

The physical planning of development in Greenland began around 1950, master plans being prepared at that time for the four more important urban centers. Mining and oil recovery may bring new opportunities to the business community in Greenland, and these must be controlled.

Nature Conservancy, UK: Created by Royal Charter in 1949, a crown body under the supervision of a committee of the Privy Council, with the purpose of advising scientifically on conservation, operating nature reserves, and undertaking scientific research.

The program of national nature reserves includes areas in England, Scotland and Wales ranging in size up to approximately 16,000 hectares and forming the best available examples of natural or seminatural habitats including arctic-alpine vegetation, bogs, woodlands (both deciduous and coniferous), ferns, grasslands, inland waters, dunes, salt marshes, expanses of shingle, sea cliffs and islands.

The majority of the reserves are open to public access but several are closed except to permit holders. Some are primarily selected as "living museums" to conserve selected examples of natural communities of plants and animals or to serve as refuges for rarities. Others are partly or mainly used as open-air laboratories on which experiments are conducted into the effects of climate and drainage, or of different types of management treatment such as grazing, planting, fencing and control of burning. Comprehensive records are kept of the elements and changes in the animal life and vegetation and of the enrichment or impoverishment of soils.

The broad purpose of the Conservancy's nature reserves is to provide suitable and secure arrangements for long-term studies of the natural resources of Great Britain and to preserve for future generations the best possible range of examples of the natural vegetation and animal life. These

210   Environmental Planning

official nature reserves are supplemented by a somewhat larger number of unofficial reserves managed by the Society for the Promotion of Nature Reserves, the Royal Society for the Protection of Birds, the National Trust, the National Trust for Scotland and various regional or country trusts, of which the Norfolk, Yorkshire and Lincolnshire Naturalists' trusts are leading examples. In addition the conservancy has notified the local planning authorities of more than 1,700 sites of special scientific interest, over which the Nature Conservancy must be consulted before permission for development is given. The membership of the Royal Society for the Protection of Birds is now over 330,000.

Nature Conservancy Act, 1975, Sweden: An Act which lays down that the natural environment is a national asset that should be protected and conserved, and that nature conservancy is the business of both the nation and the municipalities. The Act makes it possible to set aside for national parks land belonging to the Crown in order to preserve intact the character of a large area of a certain type of landscape, or for nature reserves in order to protect a region which is of importance for knowledge of the country's natural landscape and/or for open-air activities. Natural objects may be put under the protection of the law as "nature monuments."

By 1984, Sweden had 18 national parks with an aggregate area of more than 625,000 ha; all these national parks occupied land owned exclusively by the nation (or Crown). In addition, there were over 1,100 nature reserves with a combined area of some 875,000 ha. A number of landscape management areas had also been designated, being areas in which scenic beauty can be preserved without infringing on current land-use, usually rural or non-urban.

The original Nature Conservancy Act of 1964 was reviewed by a Royal Commission and in consequence, in 1975, a new and strengthened Nature Conservancy Act was introduced.

Nature Conservation Act, 1976, France: A measure introduced by the French Government in 1976 to strengthen earlier nature conservation measures. The three objectives of policy are to preserve the original wildlife of natural areas and their often splendid landscapes, to offer townspeople opportunities for genuine contact with the countryside, and to help certain farming areas acquire a new momentum for development. The three main tangible expressions of this policy, through which its three objectives are pursued to different degrees, are nature reserves, national parks, and regional nature parks.

Nature Reserves. These are areas, some quite small, where special measures are applied to preserve species of fauna or flora or a natural environment of exceptional scientific interest. The intention is to ensure the survival of species threatened with extinction and the preservation of biotopes of particular interest as relics of primitive environments (such as ice-age peat bogs, or fossil sites) or for biological reasons (such as humid areas, or stopping places on bird migration routes).

National parks. These are large, virtually uninhabited areas requiring stringent measures to preserve wildlife, sites and scenery for the benefit of visitors. They are surrounded by a peripheral zone where the necessary facilities (park services, tourist accomodation, open-air pursuits) are developed in accordance with the needs of the park and also with a view to serving the rural economy.

Regional nature parks. These are inhabited country areas where rural economic development is pursued hand in hand with policies to preserve and enhance natural assets offering opportunities for contact with nature, peaceful recreation and cultural enrichment. The stringent regulations applied in national parks, prohibiting shooting and fishing as well as commercial and industrial activities, are not suited to inhabited areas where they would be impossible to enforce and would merely antagonize the population.

Neighborhood Noise: A term used by the British Noise Advisory Council to cover a great variety of sources of noise (q.v.) which may, and frequently do, cause disturbance and annoyance to the general public in their homes and going about their lawful occasions. They include for example:

- Factory noise;
- Noise from demolition, construction and road works;
- Noise from ventilation and air-conditioning plants in buildings of all kinds;
- Noise from sports, entertainment and advertising;
- Human noise arising from lack of consideration for others (loudspeakers, noisy parties, slamming of car doors, farewell hooting, and the like).

The term does not embrace industrial noise as it affects workers, aircraft noise and traffic noise.

Neighborhood Unit: A concept, dating from the United States in the 1920s, that towns or cities can be subdivided into almost self-contained social units. It has formed part of a search for the ideal size of population which relates to both the provision of services and the retention of identity. Robert Owens (1771-1858) considered 800 to 1,200 persons to be the appropriate size. Sir Ebenezer Howard (1850-1928) divided his garden cities into wards of around 5,000 aiming to provide social services, integration of classes, and a sense of community.

In the United States, two architects, Henry Wright and Clarence Stein, developed the idea of a neighborhood unit even further. They set up the City Housing Corporation in 1924 to build an American Garden City. The initial attempt was at Sunnyside, New York. In 1928, a plan was prepared for the town of Radburn with a projected population of 25,000 to be divided into three neighborhoods. For various reasons, this project came under severe criticism.

The idea of aneighborhood unit persisted in the plans of the early new towns in Britain during the 1950s. The concept has slowly fallen into disregard as a planning approach. Greater personal mobility, the acceptance of the fact that no physical devices create social cohesion, and the thought that people should be able to travel easily into the central business district (q.v.) or the large suburban shopping center have all contributed to this. See *New Towns*.

Neritic Zone: The relatively warm, nutrient-rich, shallow-water zone overlying the continental shelf; the marine counterpart of the littoral zone (q.v.) of a lake. Terminating at the edge of the continental shelf, sunlight normally penetrates to the ocean bottom, permitting photosynthetic activity and promoting the growth of a vast population of floating and anchored plants. The total amount of biomass supported by the neritic zone is greater per unit volume of water than in any other part of the ocean.

Net Present Value (NPV): A value calculated in the following way:

$$NPV = \text{total discounted benefits} - \text{total discounted costs}$$

If the net present value is positive, the benefits exceed the costs and the project is economically acceptable according to this criterion. The net present value does have the merit of demonstrating the magnitude of the benefits. That is of interest where options involve a different capital outlay.

Net Reproduction Rate: The average number of female babies that will be born to a representative newly born female in her lifetime, if existing reproduction and mortality rates continue. If, for example, 1000 girls born in 1978 ultimately produce 1600 girl babies, then the net reproduction rate is 1.6. A net reproduction rate permanently greater than one means ultimate population growth, although this may be deferred if the existing age structure is unfavorable to growth. A net reproduction rate of less than one means an ultimate decline in population. In the United Kingdom, the rate during the period 1935-39 was 0.78, compared with 0.98 in Australia and the United States. In the post-Second World War period all three countries have exhibited rates well above one.

Netherlands, The: See *Acid Rain; Facadism; International Commission for the Protection of the Rhine*.

Netting: A basis for emission trading (q.v.), similar to offsetting (q.v.), with the difference that netting applies to firms that must reduce some pollutants from certain sources to enable them to expand the use of other sources of the same pollutants.

New Brunswick, Canada: See *Federal Environmental Assessment Review Office*.

New Communities Program, US: The systematic development of new towns in the United States of America; over 30 new towns have been constructed with a population range of 17,000 to 400,000. Following the First World War, four new towns were constructed: Radburn, New Jersey; Greenhills, Ohio; Greendale, Wisconsin; and Greenbelt, Maryland. Following the Second World War, construction proceeded on a more ambitious scale. New cities, mostly in the range 50,000 to 100,000 have included: Westlake, Valencia, Irvine, Mission Viejo, Laguna Niguel, Rancho California, Rancho Bernardo, Rancho san Diego, Reston, Columbia, St. Charles, Fort Lincoln, Audubon, Radisson, Roosevelt Island, Riverton, Canada, Flower Mound, San Antonio Ranch, The Woodlands, Clearlake, Park Forest, Jonathan, Cedar-Riverside, and Soul City. About half these towns have had the support of the US federal government. On the whole, the towns have remained largely residential in character unable to attract industry or approach self-sufficiency. The program faltered in the recession of the 1970s and the impact of rising interest rates. See *New Towns Act, 1946, UK*.

New Hampshire, US: See *Regional Planning Commission, US; State Regulatory Controls, US*.

New Jersey, US: See *New Communities Program*.

New Source Performance Standards (NSPS), US: In respect to new emission sources, including coal-fired power plants, standards set by the US federal government; these are known as New Source Performance Standards. The standards represent the best that is technically possible, taking costs, energy, health, and welfare into account. The individual states are responsible for setting emission standards for existing sources of pollution in order to achieve ambient air quality objectives. See *Table 13*.

Table 13: United States of America—New Source Performance Standards for Coal-Fired Power Plants

Pollutant	Allowable Emissions (pounds/million Btu)		
SO_2	1.2	plus	90 percent removal of SO_2 produced from high-sulfur coal
		or	70 percent removal of SO_2 from coal that potentially emits less than 0.6 pounds of SO_2 per million Btu.
TSP	0.3		
NO	0.06 (for bituminous coal) 0.05 (for subbituminous coal)		

Source: *Federal Register* (1979).

New South Wales, Australia: See *Bushfire Prone Areas; Coastal Protection Act, 1979; Commissioner of Inquiry; Department of Environment and Planning; "Do the Right Thing" Campaign; Environmental Planning and Assessment Act, 1979; Heritage Act, 1977; Land and Environment Court; Local Environmental Plan; Regional Environmental Plan; State Environmental Planning Policies; State Pollution Control Commission; Water Classification.*

New Towns: See *National Capital Development Commission, Australia; New Communities Program, US; New Towns Act, 1946, UK; New Towns, France.*

New Towns Act, 1946, UK: The legislative basis of a system of new towns in Britain; the aim was to achieve a measure of urban decentralization, relieving population pressures in the metropolitan areas which in turn would become revitalized. The new towns were to be self-contained, with the provision of local job opportunities; they were not intended simply to become dormitory towns only for commuters to sleep in. Social balance was also intended, at least in the sense of class balance in the community; the towns would however provide for different economic groups and social status.

The first generation of new towns around London were basically in the 50,000 to 100,000 population range; later second generation towns were in the 150,000 to 200,000 population range. The key to the structure of each town was the appointment of a development corportion, the board being appointed by the Environment Secretary. However, the central government has always maintained a strong role. The development corporations have been able to borrow from the UK Treasury for periods of 60 years at current interest rates. The corporations have also been able to buy land needed through a simplified form of compulsory purchase.

The British new towns, over thirty in all, include Bracknell, Harlow, Stevenage, Crawley, Hemel Hempstead, Hatfield, Welwyn Garden City, Telford, Basildon, Corby, Peterlee, Milton Keynes, Petersborough and Northampton. Self-containment has not always been achieved, and life in the new towns for some has been disappointing. A central problem has been the provision of basic infrastructure, such as hospitals, shopping centers in the suburbs, and institutions for higher education and training; some of these could not be provided while the population was relatively small. Life has not proved necessarily better than in the larger cities. However the achievement in the creation of so many new centers has been very great. See *New Communities Program, US.*

New Towns, France: An essential part of a program, conceived in the late 1950s, of reshaping France's urban environment while achieving a better regional balance. Initially, the whole basin of the River Seine around Paris was reorganized on the basis of five new cluster communities: Cergy-Pontoise, St. Quentin-en-Yuelines, Évry, Melun-Sénart, and Marne-La-Vallée. The aim was to reduce the absolute dominance of Paris itself. These new centers combined local employment opportunities with experiments in housing, layout, and traffic, with rapid transit into Paris.

In addition, eight cities, all with more than 100,000 inhabitants and all between 110 and 220 kilometers distance from Paris, have been strengthened and diversified so that they can take more of the industry diverted from Paris. These cities are Amiens, Rouen (with Le Vaudrevil), Caen, Le Mans, Tours, Orléans, Rheims, and Troyes.

Eight further cities have been chosen as "balancing metropolitan centers": Lille, Strasbourg, Nancy and Metz, Nantes and St. Nazaire, Lyon with St. Etienne and Grenoble, Bordeaux, Toulouse, Marseilles with Aix. In addition, many medium cities have been more closely linked to the nation's road network and a range of selective assistance has been given. This assistance has included the restoration of historic town centers, renewal programs for durable but rundown housing, help with pedestrian precincts, parking space, or the addition of an institution of higher education.

New York State, US: See *Acid Rain; Love Canal*.

New York, US: See *Commission on the Height of Buildings, 1913; New York Zoning Ordinance, 1916; Spot Zoning and Rezoning*.

New York Zoning Ordinance, 1916, US: A zoning ordinance (q.v.) upheld by the US Supreme Court in 1916, providing a firm legal footing for the development of zoning (q.v.) in American cities.

New Zealand: See *Commission for the Environment*.

Newfoundland, Canada: See *Federal Environmental Assessment Review Office*.

Niagara River Improvement Program, Ontario: A program formulated in 1981 by the Ministry of the Environment (q.v.), Ontario, Canada, to stimulate corrective action in respect to industrial and municipal effluents discharged into the Niagara and Welland rivers. Some 95 per cent of the pollution entering the Niagara River originates from chemical complexes and disposal sites on the United States side of the river. Comments and suggestions for changes are made to the New York State Department of Environmental Conservation. Waste disposal sites posing a major threat to the Niagara River include the Love Canal (q.v.).

Noise: Sound which is not wanted by the recipient. It is clear that what may be a pleasant and enjoyable sound to one person may not be to another. Beauty, in this case, may be said to be in the ear of the listener. The various sources of environmental noise may be divided into separate categories: road traffic; industry and commerce; domestic and residential; construction and demolition work; road repair and maintenance work; entertaining, advertising and sporting activities; extractive industries; aircraft.

Noise Abatement Zone, UK: A concept introduced into local government planning in Britain in 1974; the Control of Pollution Act 1974 (q.v.) significantly extended the powers of local authorities to deal with noise problems. It made it a statutory duty of every local authority to cause its area to be inspected from time to time to detect noise nuisances, and to assess what action they should take to exercise powers conferred upon them by the Act in relation to noise abatement zones.

The Noise Abatement Act 1960 was repealed by the new Act and replaced by similar provisions. In addition, there were provisions for a local authority to declare a 'noise abatement zone' within its district, a device similar in concept to smoke control areas under the Clean Air Acts. The procedure to be followed in establishing noise abatement zones was also similar to that required for smoke control areas. However, the Act did not aim at outlawing all noise within an abatement zone, for noise is less amenable to control than smoke. Instead it enabled the local authority to take action to prevent noise from specified premises of an intensity greater than the level prevailing at the time the zone was declared; in certain circumstances, the authority could also require those noise levels to be reduced. Special powers were provided to control noise levels on construction sites. However, the problem of aircraft noise was not covered by the Act. By the end of 1984, confirmation had been given by the central government to 39 orders designating noise abatement zones. Overall, however, between 1960 and 1984, the level of noise particularly in urban areas, significantly increased due mainly to traffic. The number of vehicles on the road in Britain increased and lorries became heavier and more powerful. However, there were no statistics to show how noise levels had changed over the country as a whole, as distinct from particular localities. The only national information available on a regular basis was derived from the number of complaints made to environmental health offices; these figures are published in the annual reports of the Institution of Environmental Health Officers.

Noise Control Act, 1972, US: A measure aimed at "the promotion of an environment for all Americans free from noise jeopardizing their health and welfare," the Act authorizes a noise control program encompassing all federal activities and promotes assistance to all communities in the United States for the creation and improvement of noise abatement programs.

Until the 1960s noise control was chiefly handled by state and local governments. In 1968, an Amendment to the Federal Aviation Act gave the Federal Aviation Agency authority to prescribe standards for measuring and controlling civil aircraft noise, including sonic boom. In 1970, an Amendment to the Clean Air Act authorized the establishment of an Office of Noise Abatement and Control within the US Environmental Protection Agency (q.v.). The Amendment also called for public hearings and a special report by the EPA to Congress. The result was the Noise Control Act, 1972.

Programs developed under the Act have been a driving force behind all legislation, state and local, in the United States; a majority of the population is now covered by state legislation and regulations. The EPA has been responsible for programs in noise abatement known as ECHO (Each Community Helps Others); and for the preparation of Model Noise

Control Ordinances in respect to different classes of noise. Many federal agencies now have noise control responsibilities; those undertaken by the EPA include:

- Regulations on the operation of interstate motor and rail carriers;
- Regulations on new products that are major sources of noise, including such controls as anti-tampering, warranty and useful life provisions;
- Labelling of products that produce noise capable of adversely affecting public health or welfare or products that are marketed for their noise attenuation characteristics;
- Providing technical assistance to state and local units of government desiring to develop and enforce noise abatement and control programs;
- Public information dissemination to inform citizens of the hazards of noise to public health and welfare;
- Certification of Low Noise Emission Products.

Noise Exposure Forecast (NEF): A technique for predicting the subjective effect of aircraft noise on the average person, exposure levels being expressed in NEF units. Factors which are taken into consideration are the frequency of aircraft movements and their distribution by day and night; the magnitude and duration of aircraft noise as determined by type, weight and flight profile; and the distribution of the noise energy over the spectrum of audible frequencies. In applying the NEF technique, a pattern of contour lines is drawn on the map of the area surrounding the airport.

Noise Zoning: The statutory classification of areas according to usage; this method allows higher noise levels in areas where their effect will not be noticed but maintains lower levels in more sensitive areas such as suburban residential sites and rural areas. An inherent problem in zoning lies in the difficulty of accurately defining the zones and of setting the boundaries between adjacent zones. Noise zones are not necessarily compatible with established land-use zones.

Non-Attainment Areas, US: A concept introduced in the application of clean air legislation in the United States of America. In the application of legislation, the United States is divided into two general air quality areas:

- Areas in which the national ambient air quality standards (NAAQS) are currently being met; and
- Areas in which air quality exceeds that prescribed by the national ambient air quality standards, known as "non-attainment areas."

In respect to "non-attainment areas," the states must develop regulatory strategies which ensure reasonable further progress in reducing emissions from existing stationary sources, while developing permit requirements for the construction and operation of new sources which would prevent the exacerbation of existing NAAQS violations.

Nonconforming Use: Any legal use of a structure or parcel of land already in existence before the adoption of official controls (q.v.), that would not have been permitted to become established under the current official controls, if those controls had been in effect prior to the date the use was established.

Non-Renewable Resources: Natural resources which, once consumed, cannot be replaced, e.g., a ton of coal once consumed is gone forever in that form. Mineral resources generally are regarded as wasting assets of this kind. However, it is difficult to predict the consequences of exhausting particular resources. For any particular mineral, the exhaustion process is gradual and would be accompanied, other things being the same, by a steady rise in its price. A rising relative price intensifies exploration and ensures treatment of lower grades of ore; recycling and reclamation of scrap and residues are also encouraged. Meanwhile, developments in substitute materials and processes and changes in the pattern of demand could mean that an "indispensable" mineral becomes totally redundant.

Nordic Environmental Protection Convention: An international environmental protection convention signed in 1974 by the Nordic countries, Denmark, Finland, Norway, and Sweden. The contracting countries also signed a protocol containing certain directions for the implementation of the convention. The convention came into being on the initiative of the Nordic Council, to which the four countries belong. At its 20th session in 1972 it requested the Ministers to prepare a draft convention which would give the environmental protection interests of Nordic neighboring countries equal status in the national legislation of each country.

The effect of the convention is that the courts and administrative agencies in each country shall, when deciding on the permissibility of environmentally harmful activities, assess the nuisance such activities may cause in a neighboring Nordic country on the same terms and in the same manner as if the nuisance or potential nuisance were present in their own country. The convention is founded on the assumption that the environmental protection legislation of the countries concerned will develop along much the same lines in the future.

In principle, the convention is only applicable to activities being carried on or that will be carried on in some kind of fixed installation. Dumping or the discharge of oil from ships is not covered by the convention, nor does it contain rules for automobiles or aircraft, or for the dispersion of pesticides.

The Nordic Environmental Protection Convention is an important element in Nordic cooperation in the field of environment protection. The

Convention is in harmony with the Declaration adopted by the United Nations Conference on the Human Environment; by Principles 21 and 22 of that Declaration all countries are called upon to ensure that activities under their jurisdiction or control do not cause damage to the environment of other countries, and to cooperate for the further development of international laws relating to liability and compensation to be paid to victims of pollution and other environmental damage caused by such activities.

North Carolina, US: See *Acid Rain; Environmental Policy Act, 1971*.

Norway: See *Acid Rain; Ministry of Environment; Nordic Environmental Protection Convention*.

Nuclear Hazards: See *International Atomic Energy Agency; International Commission for Radiological Protection; Parker Windscale Inquiry; Ranger Uranium Environmental Inquiry; Three Mile Island Incident*.

Nuclear Regulatory Commission (NRC), US: A US federal commission which has responsibility for the licensing of nuclear plants, and their inspection. It has the power to close down plants which appear to be unsatisfactory in operation, and has done so on several occasions. It was heavily involved in the Three Mile Island incident (q.v.).

O

Ocean Dumping: See *Environment Protection (Sea Dumping) Act, 1981, Australia; Marine Protection, Research, and Sanctuaries Act, 1972, US.*

Offensive or Hazardous Industry: Generally defined as an industry which, by reason of the processes involved or the method of manufacture or the nature of the materials used or produced, requires isolation from other buildings. Such industry is likely to interfere with the amenity (q.v.) of a neighborhood by reason of noise, vibration, risk of explosion or fire, smell, fumes, smoke, vapor, steam, soot, ash, dust, wastewater, waste products, grit, oil, or otherwise.

Office of Environmental Protection, Switzerland: An agency created by the Swiss Government in 1971. Article 2 of the Federal Council decree assigns to the new office the following functions:

- Drafting federal laws and orders concerning the protection of the natural environment and, in particular, the protection of waters, fishing, the preservation of clean air and noise control;
- Enforcing such laws and orders where the cantons are not competent to do so under the federal constitution or the law;
- Supervising the implementation of federal laws and orders by the cantons;
- Cooperating with the federal and cantonal departments which deal with problems relating to the protection of the natural environment, and coordinating the work of these bodies in accordance with a uniform policy;
- Taking emergency measures;
- Publishing technical instructions and formulating principles and guidelines in agreement with the federal services concerned;
- Advising federal services, cantons, municipalities and organizations on questions relating to the protection of the natural environment;
- Informing the public on the need for and significance and organization of protection of the natural environment; and,

- Assisting in the drafting and implementation of international agreements on the protection of the natural environment.

Official Controls, US: Legislatively defined and enacted policies, standards, precise detailed maps, and other criteria, all of which control the physical development of a municipality or a county, and are the means of translating into ordinances all or any part of the general objectives of the comprehensive plan (q.v.). Such official controls may include ordinances establishing zoning (q.v.), subdivision controls, site plan regulations, sanitary codes, building codes, housing codes, and official maps.

Official Map, US: A map adopted by a governing body, such as a municipality or county, which shows existing and proposed future streets, roads, and highways; the area needed for the widening of existing streets, roads, and highways; and existing and future county state aid highways and state trunk highway rights-of-way. An official map may also show the location of existing and future public land and facilities such as parks, playgrounds, schools and other public buildings, civic centers, and travel service facilities. The map may designate the boundaries of areas reserved for the purposes of soil conservation, water supply conservation, flood control and surface water drainage and removal including areas protected against encroachment by buildings or other physical structures or facilities.

Official Plan, Metropolitan Toronto: A comprehensive plan for Metropolitan Toronto, Ontario, Canada, prepared under the State Planning Act by the Metropolitan Toronto Planning Department in 1978 and adopted in 1983. Extensive studies relevant to metropolitan problems have been undertaken in recent years in such subjects as: population projections; employment characteristics; expressways and major arterial roads; regional parks; planning and development controls; the Don Valley transportation corridor; civic design; central waterfront transportation; assisted housing; water supplies and water pollution control; and the disposal of garbage. The Metropolitan Toronto Planning Department also reviewed the findings and recommendations of the 1977 Royal Commission on Metropolitan Toronto. See *Official Plan, Toronto.*

Official Plan, Toronto: A comprehensive plan for the City of Toronto, Ontario, Canada, prepared under the State Planning Act by the Toronto Planning Board and the Toronto Planning and Development Department.

The Planning Board was created in 1943, producing a Master Plan (q.v.) for the City of Toronto and Environs in 1945. An initial Official Plan appeared in 1949, being reviewed in 1955. A new Official Plan was adopted in 1969, undergoing modification in subsequent years; for example, the transportation aspects of the Plan were reviewed in 1976.

A Consolidated Official Plan for the City was aproved by the City Council in 1977, undergoing revision in parts through 1984. A Zoning By-law for the central area and central industrial district was adopted in 1980.

Extensive studies relevant to the Consolidated Plan have been undertaken in recent years in such subjects as: industrial policy and the central industrial district; retailing; buildings of architectural and historical importance; central area traffic management; central area zoning; downtown streetscapes; central area parking and loading; city housing; and the capacity of the existing and committed transportation system serving central Toronto. See *Official Plan, Metropolitan Toronto*.

Offsetting: A basis for emission trading (q.v.), an administrative procedure that provides for inter-firm trading of Emission Reduction Credits (ERC) among activities not in the vicinity of the same plant, and not owned by the same firm. Offsetting was developed prior to the Clean Air Act Amendment of 1977 as an administrative approach to enable growth in the non-attainment regions of the United States.

Ohio, US: See *Litter Laws, US; New Communities Program*.

Oil Spills: A term generally applied to discharges of oil, both voluntary and accidental, from ships, particularly oil tankers, and oil drilling.

During the present century, attempts have been made to control and prevent oil pollution at sea through national legislation in several countries and through international agreements. The Oil in Navigable Waters Act 1922 in Britain, and the Oil Pollution Act 1924 in the United States, are examples of legislation prohibiting the discharge of oil into coastal waters. The International Convention for the Prevention of Pollution of the Sea by Oil, drawn up in 1954, established coastal zones fifty miles offshore into which no waste oil could be discharged. An International Conference on Oil Pollution held in 1962, under the sponsorship of the then Inter-Governmental Maritime Consultative Organization (IMCO), aimed at total prohibition of oil discharges. An agreement came into force on 18 May 1967 that a "total prohibition rule" should apply to new tankers of twenty thousand gross tonnage or more. However, it took about twenty years before tankers built before May 1967 began to go out of service. Zones were also extended under this agreement to include, for example, a large area of the northeast Atlantic and the whole of the North Sea.

A major oil spill occurred in 1967 when the tanker Torrey Canyon went aground on the Seven Stones reef, off southwest England. Most of the cargo of 120,000 tonnes of crude oil landed on the Cornish beaches. In 1974, a tanker belonging to the Miyushima refinery spilled 39,000 tonnes of desulfurized oil off the coast of the inland sea of Seto, Japan. Nearly 9,000 tonnes reached the eastern part of the sea. The direct costs of economic damage to the fishing industry were estimated as amounting to $US 50 million. One of the largest spills of all time, from a tanker, occurred in 1978 off the French coast following the wreck of the tanker Amoco Cadiz. Some 230,000 tonnes of crude oil released polluted 128 kilometers of coastline.

The largest and most prolonged oil spills have arisen not from tankers but from damaged off-shore oil platforms; notably the Santa Barbara "blow-out" of 1969, the massive 1979 spill off the Mexican coast and the 1983 oil spill into the Persian Gulf. See *Figure 15*.

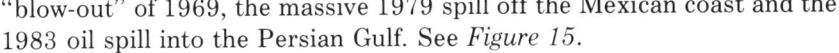

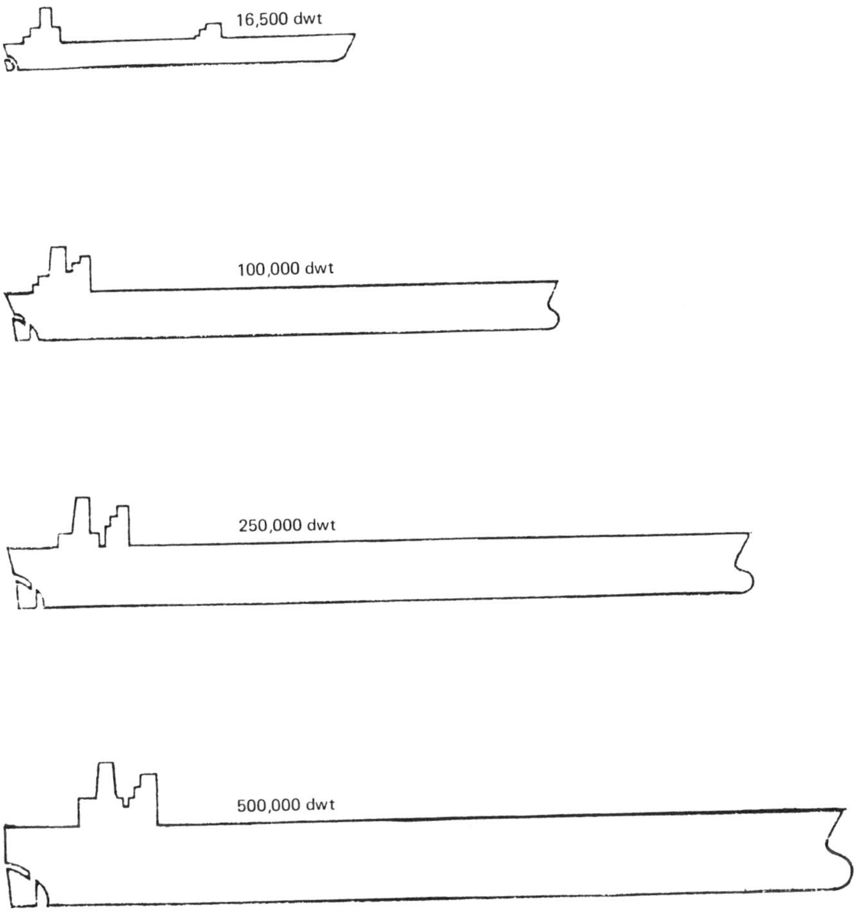

Figure 15: The growth in the size of oil tankers since the second World War. Source: IMO.

Oklahoma, US: See *National Wildlife Refuges*.

One-Hundred-Year Floodplain: The land area adjacent to a water body that is flooded with an average frequency of once in 100 years; besides their function as natural drainage channels, floodplains have important water storage and groundwater recharge capacities. See *Floodplain*.

Ontario, Canada: See *Acid Rain; Environment; Environmental Assessment Act, 1975; Environmental Assessment Board; Environmental Impact Statement; Ministry of the Environment (Environment Ontario); Niagara River Improvement Program; Official Plan, Metropolitan Toronto; Official Plan, Toronto; Ontario Waste Management Corporation; Sudbury, Ontario*.

Ontario Waste Management Corporation: Established as a Crown agency in 1981, a body which reports to the Ontario legislature through the Minister of the Environment; the prime responsibility of the corporation is to design, construct and operate a province-wide system for the treatment and disposal of liquid industrial waste and hazardous waste, and to develop a long-term program to assist in the reduction and recycling of such wastes.

By 1983, the corporation had prepared preliminary planning guidelines on the type of system to be developed, and recommendations on the geological region where the search for suitable disposal sites should be concentrated.

Open Space: Land or water which is not preempted by intensive urban uses and which will be kept free of human-made structures on a permanent basis. Examples of open spaces are parks and recreation areas, water and wetland areas, forests, and institutional holdings with very low building density. An open space system is the network of a city's parks and open spaces, including parkways, that are complementary in scale, function and design. See *Parkway*.

Open Space Plan: Primarily a tool by which a local authority establishes its recreational and open space needs and establishes priorities for open space acquisition and development; an Open Space Plan should incorporate a goal, a set of objectives and an integrated series of policies and programs. An example of a goal might be:

> To provide the community with the widest choice and maximum diversity of recreational opportunities consistent with economic feasibility and the needs of the population.

Examples of objectives are:

- To provide a certain quantity of open space of various types;
- To provide a hierarchy of parks of various sizes and adequate distribution;
- To provide multiple use of open space;

- To locate various types of open space at a certain distance from residential areas;
- To link open space into a linear network;
- Preservation of natural bushland.

Policies and programs will include factors such as:

- Identification of potential recreation and open space areas;
- An appropriate implementation program on a priority basis;
- Site development and management plans for existing areas and those to be acquired.

There are four main steps in developing an Open Space Plan:

- Assessing existing supply,
- Identifying needs,
- Determining how to satisfy unprovided need,
- Establishing programs.

Factors which should be considered when setting standards are:

- Whether they are achievable;
- Whether regional open space and other easily accessible publicly owned land provides for some of the needs of the population;
- The amount of private open space (e.g. back gardens) that exists in the area and whether this compensates for local parks;
- The availability of school facilities and the possibility of these being used by the community;
- The proximity of other alternative recreational facilities, such as city entertainment.

In addition to the actual amount of open space that is supplied, its distribution is also important. Some general principles to be observed in determining whether the distribution of open space is adequate are:

- Playgrounds/small parks should be within walking distance of users;
- Playing fields and other sporting facilities may be concentrated to provide for a catchment that is wider than the immediate locality—they need not necessarily be within walking distance of users;

- Other parks should be easily accessible to the whole area but not necessarily within walking distance.

It is now widely accepted that the application of rigid standards to the provision of open space is not always appropriate. However, a standard of 2.83 ha/1,000 people has been accepted for many years as a yardstick for the adequate provision of open space in urban areas. Table 14 indicates the recreational activities and open space needs for specific age groups.

Oregon, US: See *Litter Laws, US; National Wildlife Refuges.*

Organization for Economic Cooperation and Development (OECD): An international body which came into being in 1961, succeeding the Organization for European Economic Cooperation (OEEC). The role of the OEEC had been to allocate aid received under the European Recovery Program (The "Marshall Plan") to speed recovery in countries whose economies had been seriously impaired by the Second World War.

The members of OECD are today Australia, Austria, Belgium, Canada, Denmark, Finland, France, the Federal Republic of Germany, Greece, Iceland, Ireland, Italy, Japan, Luxembourg, the Netherlands, Norway, Portugal, Spain, Sweden, Switzerland, Turkey, the United Kingdom and the United States of America—a total of 23 countries embracing many of the "developed" and high industrialized areas of the world.

In 1970, the Council of OECD set up an environment committee to advise on patterns of growth and development which would be in harmony with protection of the environment. Through the environment committee, government representatives meet to discuss common problems, to examine possible solutions, and, where appropriate, to develop recommendations and guidelines for national policy formulation. Notably, it seeks to harmonize environmental policies among member countries.

In 1974, the Ministers responsible for environmental policy in the OECD countries adopted unanimously a formal Declaration on Environmental Policy. In essence, the Policy advanced the protection and progressive improvement of the quality of the environment as a major objective of the OECD Member countries. The text tended to mirror the principles embodied in the Declaration on the Human Environment endorsed at the conclusion of the United Nations Conference on the Human Environment held in 1972.

However, a special session of the OECD's Environmental Committee in 1982 concluded that OECD Countries needed to take more account of the environment in their respective policies. The need for the creation of an international institute on the global environment was recognized.

In November 1982, the member nations including the USA adopted procedures to allow the exchange of confidential data, to eliminate the costly and duplicative testing of chemicals by individual countries in respect to the protection of health, safety, and the environment.

A meeting of OECD Environment Ministers in 1985 addressed a

Table 14: Recreational Activities and Open Space Needs for Specific Age Groups

Age	Recreational Activity	Open Space Needs
0-14	Informal play - heavy use of parks, especially those close to home with playground equipment. Informal active sports and organized sports associated with schools and clubs. Participation in family picnics, barbecues, etc.	Provision of small parks with play equipment in walking distance of users. Formal playing fields and areas for informal sport. Fullest possible range of other sporting facilities, e.g., swimming pools. Access to large natural and/or parkland areas for passive pursuits.
15-19	High rate of participation in informal and organized active sports. Medium participation in miscellaneous other recreational activities - "hanging around."	Provision of formal playing fields and areas for informal sport. Wide range of recreational opportunity. Fullest possible range of other sporting facilities.
20-24	Involvement in informal and organized sports continues but much less than earlier ages. Beginning of child rearing, family establishment and associated activities.	Provision of formal playing fields and areas for informal sporting facilities. Access to large natural and/or parkland areas for passive pursuits.
25-29	Participation in active sports continues but slightly less than 20-24 year old group. Highest fertility rates, so family activities important. Medium use of all types of open space and recreation facilities.	Provision of formal playing fields and areas for informal sporting facilities. Versatile areas that provide for a wide range of recreational opportunity. Access to natural and parkland areas.
30-39	Participation in active team sports becoming less important but increased participation in sports such as golf and tennis. Greatly increased participation in passive pursuits such as picnicking, barbecuing, visiting parks and gardens and walking for pleasure.	Provision of demand facilities, e.g., tennis, golf. Access to sporting areas and playing fields and to large natural areas and parkland areas.
40-49 50-59	Watching sport becoming equally or more important than participation (except golf, tennis, etc., and now bowling, etc.). Greatest participation in passive pursuits.	Provision of local parks and demand facilities. Access to large natural and parkland areas and to sporting facilities.
60+	Very little participation in active sports, except bowling, etc. Walking for pleasure and passive use of parks, predominant outdoor recreation activities.	Provision of demand facilities especially bowling. Small parks in vicinity of users. Access to large natural areas and parkland areas.

Source: Department of Environment and Planning, New South Wales, Australia.

variety of issues: hazardous wastes, including movements across national frontiers and the urgent need for tighter controls; air pollution, with special reference to acid rain; exhaust emission standards with particular reference to the differences between Britain and West Germany; a code of conduct for multinational companies operating in the developing world; and better management of water resources.

Oslo Convention: See *Convention for the Prevention of Marine Pollution by Dumping from Ships and Aircraft, 1972.*

Overgrazing: The overstocking of territory with animals for the purpose of short-term gain; the problem occurs in many countries. For example, throughout the Republic of South Africa overgrazing of the veld takes place even during good years with plenty of rain; during dry spells there is no attempt whatever to reduce the numbers of stock to match the carrying capacity.

A consequence of overgrazing is that the most palatable grasses are grazed very short with the result that the root systems degenerate. In grazing these plants, animals tend to remove the entire plant.

Veld fires occur annually throughout the country; some may be caused by accident, others may be part of a grazing management system. However, a large number of planned veld fires are started during the wrong time of the year, with the result that too early grazing takes place with detrimental effects on the root system and growing ability of plants.

The progressive degradation of India's forests compounds the problem of soil erosion also. Overgrazing, for example, has reduced vast forested areas to an unstable low scrub. In addition to the loss of vegetation cover, trampling by cattle hardens the soil, preventing forest regeneration and causing soil erosion in many parts of the Indian subcontinent. Irrigation has tended to exacerbate overgrazing, as the more land there is for cultivation, the less there is for cattle.

Overspill: A surplus of people or activities greater than a city, or part of a city, can reasonably accommodate. Such a surplus may arise from the natural growth of population or from progressive redevelopment in which more land is dedicated for ancillary uses which have become considered socially or functionally essential or desirable. For example, in urban renewal (q.v.) more open space and urban facilities will be incorporated.

If the boundaries of a city can expand, even into adjacent local government areas, overspill from central areas may simply be accommodated further out in the suburban areas. Where restraints are imposed on urban growth as a result of a green belt (q.v.) or a national settlement policy, new towns (q.v.) and growth centers may be established to deal with the overspill.

Ozone (O_3): A triatomic molecule of oxygen; above minimal levels, it is an irritant to human beings and animals. It is a natural constituent of the atmosphere occurring in concentrations of about 0.01 ppm; the toxicity threshold for workers is 0.1 ppm. Ozone is produced by certain high voltage electrical equipment; it is also produced in certain circumstances when photochemical reactions occur in the atmosphere between ultraviolet light (sunlight) and the oxides of nitrogen and hydrocarbons emitted to the atmosphere by motor vehicles. See *Los Angeles Smog.*

Ozonosphere: A layer of the atmosphere between about twenty kilometers and 50 kilometers above the surface of the earth containing ozone (q.v.). The ozone is produced through the splitting of some oxygen molecules, the resulting atomic oxygen recombining with unaffected molecules to yield ozone (O_3).

P

Paris Convention: See *Convention on the Prevention of Marine Pollution from Land-Based Sources, 1974.*

Parker Windscale Inquiry, UK: In early 1978, a British judicial inquiry that reported on a proposed major expansion of the Windscale nuclear waste reprocessing plant which would convert uranium oxide waste into reusable uranium and plutonium. Windscale would thus open the way for greater adoption of plutonium-using fast breeder reactors. The inquiry, conducted by Mr. Justice Parker, heard all sides of the uranium debate. The anti-nuclear lobby argued that a large increase in plutonium output from reprocessing plants such as Windscale would be a threat to world peace. Mr. Justice Parker concluded, however, that there was more likely to be a potentially dangerous proliferation of such plants around the world if Britain did not expand its facilities. The inquiry concluded that there was no sufficient reason to halt the proposed expansion.

In May 1978, the House of Commons endorsed the recommendations of the judicial inquiry that a major reprocessing plant should be built at Windscale. This vote was a rejection of the anti-proliferation policies of the United States Government announced by President Carter in 1977. The United Kingdom was now firmly committed to nuclear waste reprocessing, and a plutonium future. The new plant would take ten years to complete.

The issue remains controversial, however. Earlier in 1976, the UK Royal Commission on Environmental Pollution had concluded:

> "Plutonium appears to offer unique potential for threat and blackmail against society because of its great radioactivity and its fissile properties.
>
> The dangers of the creation of plutonium in large quantities in conditions of increased unrest are genuine and serious. We should not rely for energy supply on a process that produces such a hazardous substance as plutonium unless there is no reasonable alternative.
>
> There should be no commitment to a large program of nuclear fission power until it has been demonstrated beyond reasonable doubt that a method exists to ensure the safe containment of long-lived, highly radioactive waste for the indefinite future.
>
> The abandonment of nuclear fission power would, however, be neither wise nor justified. But a major commitment to fission power and a plutonium economy should be postponed for as long as possible."

The Commission, under the chairmanship of Sir Brian Flowers, expressed confidence that an acceptable solution would be found to the disposal of the "high-level" or most dangerous of the wastes. A solution, both sound and generally acceptable, has yet to emerge. The absence of a solution represents a major stumbling block to the development of the nuclear power industry throughout the world.

Parkway: A road in a linear, landscaped park which is restricted to use by automobiles.

Pedal-Power: Synonymous with the use of the smog-free vehicle, the bicycle. Many organizations now promote the use of the bicycle and the establishment of separate bicycle paths through city suburbs. Canberra, the capital of Australia, can be traversed in a number of directions along bicycle paths which are completely separate from roads; the routes utilize special underpasses.

Pedestrian Precinct: A shopping precinct designed for pedestrians only, with the exclusion of motor traffic. The closing of streets accompanied by conversion to pedestrian precincts has been a common characteristic of cities throughout the world since the 1960s.

Pedway System: A grade separated pedestrian walkway; pedways may ultimately include moving sidewalks. Pedways should be carefully designed and landscaped to maximize their aesthetic appeal. Pedways should connect major activities and create a second level walking network separated from vehicular conflicts, interfacing with people movers and rapid transit stations and with parking garages.

Pennsylvania, US: See *Acid Mine Drainage; Air Pollution; Anthracite and Bituminous Coal Mine Subsidence Insurance Fund, 1961; Bituminous Mine Subsidence and Land Conservation Act, 1966; Environmental Master Plan; Environmental Rights Amendment; Municipalities Planning Code; Three Mile Island Incident.*

People Movers: Auxiliary grade separated transit systems constructed to connect major central city nodes with each other, with rapid transit stations, and with peripheral parking facilities. People movers must be aesthetically and functionally integrated with development.

Performance Standards: Environmental standards offering an alternative to conventional zoning (q.v.); performance standards regulate the levels of nuisance or the effects rather than specifying the permissible uses. For example, performance standards specify the maximum levels of stress to be imposed on natural resources by development. Such standards include:

- The percentage of impervious cover on the site;
- Definition of acceptable levels of increase in stormwater runoff;

- Sedimentation and erosion control;
- Density limitations on sensitive lands, such as areas with a high potential for erosion, landslides, or water pollution.

Performance standards may be applied to building codes (q.v.); the requirement may be specified leaving the methods to the designer. For example, instead of specifying that a window area shall be not less than a specified proportion of the floor areas, it may be enough to specify that a window "shall provide ample light and air." However, subjectivity may displace objectivity; hence both approaches have advantage and disadvantages.

Physical Planning: See *National Physical Planning*.

Planning: The plotting of a course of action involving the allocation of resources; it is a characteristic of most human activity whether individual, family, collective, corporate, or national. Controversy normally ranges around not the desirability of planning but who should undertake it. Those who are against "planning" on the grounds that "things should be left to the market," are simply arguing that human or collective needs will be best met by the planning of entrepreneurs rather than by the planning of government agencies.

Planning, often sophisticated, is the hallmark of all successful corporations and companies; it is also the hallmark of a successful bureaucracy. Public planning involves, or should involve, the reconciliation of social and economic aims, of private and public objectives. Certainly all planning, private or public, should be consumer-oriented.

Nations have judged the proper role of government in a variety of ways. On the one hand are societies with comprehensive planning systems based upon the public ownership of the means of production, distribution, and exchange; and on the other, societies in which government intervention is constrained and the economy rests essentially upon the private ownership of the means of production, distribution, and exchange. Many countries fall somewhere within this range.

However, all countries have accepted the need for some measure of government intervention in the field of environmental planning; at the one extreme it provides an important instrument for public improvement and at the other a form of protection for individual investment. Many countries have also accepted that the government should intervene where the private sector fails, e.g., in the provision of housing for the lower income groups, and in the rehabilitation of extensive blighted areas. In other areas, highways and traffic management need to be undertaken on a scale beyond the normal reach of the private sector; and the supply of electricity to unprofitable rural areas is not normally undertaken by the private sector.

The intervention of government, however, often creates problems of its own, e.g., a lengthening of procedures, the need to have permits and approvals from a dozen different departments, and the inevitable interactions of municipal or local, county, state, and federal levels of government. See *Environmental Planning*.

Planning Blight: Descriptive of a situation when land becomes unsaleable, or is devalued, because of a belief that it will be required for, or affected by, some public project. Blight may result from a town planning scheme which indicates future needs; or as a result of the announcement of a large public undertaking.

In the latter case compensation may be payable to those whose land is acquired for the project. If the blight simply results from a planning scheme, without land acquisition, compensation is not usually paid. A decision to construct an airport at a particular site may result in lengthy public debate, placing the owners of the properties affected under a cloud of uncertainty for a considerable period of time. In the end, the decision may be reversed so that no one receives any compensation even after years of delay.

However, in Britain, even though many of these uncertainties prevail, compensation may be paid in respect to injurious affectation. See *Land Compensation Act, 1973*.

Planning Commission, US: A planning agency created by a municipality or county, under a state enabling act, for the purposes of planning the area within the corporate limits possibly involving the preparation of a comprehensive plan (q.v.) for future development and the establishment of procedures for implementation; the commission acts in an advisory capacity to the government body and is the legal agency through which planning is carried out.

Hartford, Connecticut, established the first planning commission in the United States in 1907. Since then more than 2,000 municipalities and counties have established planning commissions; commissions have also been established for metropolitan areas, regions, and states. Planning tools such as zoning (q.v.) and subdivision regulations have become commonplace while programs now embrace matters of long-range as well as short-range concern.

Planning commissions usually consist of up to a dozen members appointed by the government body; the selected members consist of leaders of the community in business, real estate, industry, labor, and the professions. Their success depends decisively on the selection of persons interested in the welfare of the community, never an easy task. Planning commissions today are concerned with a wide range of matters including:

- Preparation of comprehensive plans, and transportation plans for expressways and parking projects;
- Urban renewal (q.v.) and central business district improvement programs;
- Public housing programs;
- Capital improvement programs;

234 Environmental Planning

- Control of urban sprawl;
- Industrial parks;
- Shopping centers and school sites;
- Cluster subdivisions and planned unit developments to provide more common open space;
- Purchase of easement rights and adoption of other methods to preserve scenic views;
- Open space zoning and acquisition of lands for permanent open space;
- Improvement of living conditions in older neighborhoods.

Planning Gain, UK: In the United Kingdom, a widespread local government practice of acquiring community advantages when determining applications for planning permission; many local authorities bargain with developers to secure community benefits as a condition of planning permission (development consent). An example would be the provision of some needed low-income housing in return for planning permission to build offices.

Planning Inquiry Commission, UK: A concept introduced by the British Town and Country Planning Act, 1968; it envisaged a two-stage process, a policy and need inquiry being followed by a site specific inquiry. It was envisaged that this form of inquiry would be available for specially important and technically taxing development projects. The Town and Country Planning Act, 1971, extended the powers of the Planning Inquiry Commission to allow the spending of money on research.

However, an objection based upon natural justice was eventually recognized to this form of inquiry. An insurmountable difficulty was that one of the Commission's own members would proceed to conduct a public local inquiry into a scheme at a site which the Commission as a whole had designated; local people would not find this fair and would be unlikely to have participated in the policy-and-need inquiry which produced the designated site.

The British Government slowly recognized that where two-tier inquiries were desirable, the public local inquiry or second-stage needed to be conducted by someone totally unconnected with the first stage, precluding an assumption that the site had been pre-selected in any final sense as distinct from simply nominated for further consideration. In special cases, the approval of Parliament for the development would be needed; indeed both Houses would need to approve it under prescribed affirmative resolution procedure. This ensured a third stage, the possibility of debate in either or both houses.

In consequence of further reflection, the planning inquiry commission as originally envisaged by legislation has not been employed in Britain, although special inquiries and public local inquiries have continued to be a feature of the British scene. See *Parker Windscale Inquiry; Project Inquiry; Public Hearings and Inquiries; Roskill Commission.*

Planning Standards: Regulatory or prescriptive standards that relate to individual development projects mainly in respect to the quantifiable aspects of site development; such requirements are imposed by local government in conjunction with zoning (q.v.) restrictions. Standards are applied in many countries in relation to some or all of the following:

- Residential density,
- Lot ratio,
- Daylight and sunlight,
- Car parking,
- Highways,
- Visual privacy,
- Protection from traffic noise,
- Provision of communal play and amenity space,
- Segregation of pedestrians from traffic,
- Restriction on building heights,
- Provision for access and other needs of the disabled in all public buildings,
- Access to public transport,
- Access to recreational facilities,
- Safety and security,
- Building construction.

See *Building Codes.*

Planning System, UK: A national planning system conforming to the following basic principles:

- The planning authorities (i.e. the larger local government administrations) have a duty to prepare a structure plan (q.v.) which expresses publicly the adopted policies for the use of land.
- Local plans are prepared by the smaller local government administrations to develop the policies and general proposals of the structure plan; these provide a detailed basis for development control.
- Structures plans must be approved by the central government (the appropriate Ministry, in this case the Department

236 Environmental Planning

of the Environment (q.v.), while local plans are adopted by the local authority alone.

- Planning permission is required for development.
- When planning permission is refused, the developer has the right of appeal to the Secretary of State.
- In addition, the Secretary of State has powers to call in and determine any application for planning permission where important or controversial issues are at stake.
- In appropriate cases a public local inquiry (q.v.) will be held.
- It is an accepted part of the arrangements for the control of development that the effects of a project on the community and the environment are analyzed before a decision is reached as to whether it can be allowed to take place.
- In appropriate cases, but not generally, an environmental impact statement (q.v.) will be required.
- National Planning Guidelines (q.v.) allow locational and community issues to be thoroughly examined in advance of need so that eventual public debate can take place in an objective setting.
- Community participation is encourged at all stages of plan making.

See *Public Hearings and Inquiries; Town and Country Planning Act, 1947.*

Planning Trilogy: A reference to the three main strands of national domestic policy: economic planning, environmental planning, and social planning. However, some writers on national physical planning, taking a narrower definition of environmental planning, refer to a trilogy comprising land use planning, transportation planning, and environmental planning. See *Corporate Planning; Environmental Planning; National Physical Planning.*

Plat, US: Essentially a map of a piece of land which shows the location, boundaries, area, details of lot boundaries, proposed streets, utilities, public areas, and all other necessary data to demonstrate compliance with subdivision regulations; state statutes provide for the recording of plats, and the selling of lots or parcels of land by reference to the recorded plat.

It is usually unlawful to sell land by referring to an unrecorded plat. The power of municipalities to grant or withhold the recording of a plat is the basis in the United States for the enforcement of subdivision control.

Some US cities have compiled their plats into official city maps, taking in all separate plat areas. The map may be referred to as the "official city plat."

Point Source Pollution: Discharges which occur in a discernible conveyance channel, such as a discharge from a sewage or industrial waste treatment plant. Nonpoint source pollution embraces discharges which are more diffuse in nature, such as urban and agriculture runoffs, soil erosion and sedimentation, and acid mine drainage.

Poland: See *Central Business District (CBD); Council for Mutual Economic Assistance (Comecom).*

Policy Plan: A departure from the more traditional comprehensive plan (q.v.), a form of plan which shifts the emphasis away from a detailed map of future physical development to a statement and definition of development objectives. Rather than relying on a distantly perceived map of an idealized future, policies and standards are written to help guide decisions in a full range of planning areas, including social and economic policies as well as those concerned with physical development.

The policy plan includes a method or system for influencing decisions which are to be implemented immediately or over time. Often such plans may be very general in their long-range view of the entire community, as distinct from the comprehensive plan; however, they may include some detailed short-range plans and a system of implementation for specific districts of the community. Such plans may be incorporated in the capital improvement program (q.v.).

Pollutant Standards Index (PSI), US: An index established by the Council on Environmental Quality (q.v.) in 1976 for the uniform reporting of air quality to the public on a daily basis; the index uses a scale of 0 to 500 to represent daily measured concentrations of the principal air pollutants for which National Ambient Air Quality Standards (NAAQS) (q.v.) have been established. A daily PSI value of 100 for each pollutant is assigned when the daily measured concentration equals the national standard for that pollutant. Thus, a PSI value of less than 100 for all pollutants means that the air quality on a given day is within standards and is said to be "good" or "moderate." A PSI value of greater than 100 means that one or more of the major pollutants exceed the national standard at one or more monitoring sites in a given area on the day in question.

Polluter Pays Principle: A principle which equates the price charged for use of environmental resources with the cost of damage inflicted on society by using them. The price charged may be levied directly, e.g., as taxes on the process which generates pollution or as the purchase price of licenses which entitle the holder to generate specific quantities of pollutants. If the producer or consumer can avoid the additional expense, he or she will attempt to do so; there is, therefore, an incentive to refrain from using the polluting item or to change consumption patterns or production processes in ways which mitigate pollution. Alternatively, in a reversal of principle, measures may take the form of direct payments from the public purse to polluters not to pollute. The difficulty with either procedure is to decide the "right" price to charge, or the "right" subsidy to pay; both involve an assessment of the monetary value of a clean and unimpaired environment.

For this reason, non-market techniques of pollution control are frequently preferred; under the "polluter pays" principle, non-market measures encompass the promulgation of regulations governing in various ways the emission or effects of pollution. Regulations eliminate some of the uncertainty which is inherent in market approaches.

Pollution Control Costs: The internal costs carried by an enterprise to prevent or restrict the emission of potentially harmful pollutants to air, land, and water. Between 1970-1985 expenditure in industrialized countries on pollution control measures amounted to some 0.5 to 1.0 per cent of gross national product (GNP), with Japan spending 2.0 per cent.

Typically, pollution control costs as a percentage of total plant and equipment have been, in Europe and North America:

Iron and steel industry — 20%

Non-ferrous metals — 12%

Electric utilities — 11%

On the other hand, environmental regulations have stimulated the development of clean technologies, new processes and recycling techniques that in many cases have turned out to be quite economic. Planning approvals for new plant and extensions to existing plant have been easier to obtain, and the benefits of pollution control measures have improved significantly the quality of life (q.v) for many people.

Pollution Control Strategy: Or residuals management, a spectrum of measures adopted to combat pollution both at the source and subsequently. Measures include the following:

- The reduction or modification of final demand for goods and services, the production of which generates environmental residuals not readily managed;
- The reduction or elimination of the discharge of residuals through emission standards which require changes in production technology or the adoption of gas cleaning or water treatment equipment;
- The utilization of the assimilative capacity of the natural environment in respect to residuals unavoidably discharged, in a manner unlikely to cause impairment of the environment to any significant degree;
- The recycling of residuals recovered from waste streams or the productive utilization of residuals;
- The management of environmentally sensitive areas should precautionary measures be necessary;
- The minimimization of the potential adverse effects of pollution through land use techniques and buffer zones;

See *Acid Rain; Air Pollution; Best Available Control Technology; Best Practicable Means; Bubble Concept; Buffer Zone; Emission Trading; Land Use Planning; Location of Industry, Optimal; Los Angeles Smog; Noise Abatement Zone; Reasonably Available Control Technology; Recycling; River Thames Clean Up; Smoke Control Area; Sulfur Dioxide; Water Pollution.*

Polychlorinated Biphenyls (PCBs): See *Toxic Substances Control Act, 1976, US.*

Population Differentiation: The increasing heterogeneity of peoples sharing the same geographical locality, and the same economic and political system. Heterogeneity refers to the diversity of population by culture, language, religion, social mores, value systems, ethnicity, and race. The increasing mingling of peoples has been promoted by modern communication and transportation. Historically such mingling occurred through invasion, conquest, and settlement; more recently through regulated immigration policies and through the slave-trade and other compulsory mass movements of people. See *Demography; World Population Plan of Action.*

Population Implosion: The increasing concentration of the people of the world into relatively small portions of the earth's surface; in the history of mankind this is a relatively recent phenomenon. Fixed human settlements came into being in the Neolithic Period, some 10,000 years ago. Human settlements in excess of 100,000 persons occurred only as recently as Greco-Roman civilization. The proliferation of cities of 1 million or more did not occur until the beginning of the 19th century. Since then the rate of urbanization (q.v.) has exceeded the rate of growth of the world's population. See *Demography; World Population Plan of Action.*

Post-Closure Liability Trust Fund, US: See *Comprehensive Environmental Response, Compensation, and Liability Act, 1980, US.*

Prairie Psychology: A general unconcern for the rate at which land is consumed by new development, born of a confidence that the supply of land is virtually unlimited; a view prevalent in many parts of the United States where even first class agricultural land has little national value and can be readily consumed for other purposes. In Britain, there is a marked contrast of attitude, for land is very limited in area in relation to the population and must be carefully husbanded to meet a variety of objectives and purposes. The general result is that many Americans see little need for an overall land-use policy, while in Britain few would deny the necessity of it.

Prevention of Significant Deterioration (PSD), US: A concept of preventing any further deterioration in the quality of air in areas in which that quality is already better than that prescribed by national standards. The US Clean Air Amendment Act, 1977, includes comprehensive rules for the prevention of significant deterioration of air quality in such situations. PSD increments are the ambient concentration limits on industrial activity in each class of PSD area.

For example, in respect to the two most contrasting categories:

Class I areas include national parks, monuments, or wilderness areas, in which only small increases in ambient concentrations over baseline air quality are allowed. Such areas have understandably the fewest (if any) stationary sources. The maximum allowable mean annual increase in concentrations of sulfur dioxide and particulates can be as low as 2 and 5 micrograms per cubic meter respectively.

Class III areas embrace intensive, industrial development and are those in which large increases in ambient levels of pollution may be permitted. However, plant mines still achieve the lowest achievable emission rates (LAER).

Under no circumstances can a new facility located in a PSD area be allowed to raise the concentration of any pollutant in excess of national ambient air quality standards. At present, PSD regulations apply only to the emission of particulate matter and sulfur dioxide. See *Table 15*.

Table 15: United States of America–PSD Increments and Significant Impact Levels

Pollutant and Averaging Period	Class I Increment ($\mu g/m^3$)	Class II Increment ($\mu g/m^3$)	Class III Increment ($\mu g/m^3$)	Significant Impact Level ($\mu g/m^3$)[a]
Sulfur dioxide				
Annual arithmetic	2	20	40	1
24-hour[b]	5	91	182	5
3-hour[b]	25	512	700	25
Particulate matter				
Annual geometric	5	19	37	1
24-hour[b]	10	37	75	5
Nitrogen dioxide				
Annual arithmetic	None	None	None	1
Carbon monoxide				
8-hour	None	None	None	2,000

[a] It is not necessary to analyze the impact of a source beyond the distance at which concentrations due to the source fall below these levels, unless a Class I area will be affected.
[b] Not to be exceeded more than once per year.

In sum, a new plant in a PSD area must not violate national ambient air quality standards, or PSD increments, and must apply the best available control technology (BACT). Coal-fired power plants built in certain parts of the United States can be required to install control technology that exceeds the requirements of new source performance standards. By 1985, most of these federal standards had been delegated to state and local jurisdiction.

Primitive Areas, US: A broad term embracing wilderness areas, wild areas, large natural areas, wild rivers, and wilderness trout streams. Primitive areas provide a complete contrast to metropolitan areas. They provide diversity and a place for people to explore experiences not usually afforded to them in a city environment; they are excellent places to hike, fish and camp. Primitive areas are also important for the protection of environmental qualities; the natural vegetation of primitive areas minimizes soil erosion, assures the protection of excellent stream quality, and facilitates groundwater infiltration. Reservoirs, second homes, highways, public utilities, and other developments eliminate the remote and natural character of primitive areas directly or by stimulating secondary growth and development. See *Wild and Scenic Rivers Act, 1968; Wilderness*.

Program Impact Statements: The adoption of environmental impact assessment procedures in respect to complete programs, or in respect to the broader task of policy formulation. There have been numerous instances under the US National Environmental Policy Act in which impact statements have covered an overall program of works, with additional statements being required for each component project. The process could be extended to evaluate Federal Government policies and planning activities. The notion has emerged of a tiered EIA process, with assessments undertaken in relation to the framing of broad policy objectives, public agency programs and planning decisions, and finally decisions on specific projects.

Project Inquiry, UK: A form of public environmental planning inquiry recommended for Britain in 1979 by a working party comprising representation from the Outer Circle Policy Unit (Justice) and the Council for Science and Society. It proposed a new model of project inquiry to precede the established type of public local inquiry in a small number of cases, i.e., for "major projects involving controversial issues of national policy, especially where these have a high technical content, and more especially where the proponent has close links with government." Such a form of inquiry would be suitable for a Severn Barrage or Channel Tunnel.

This proposal went a long way towards meeting the objection which had arisen to the original planning inquiry commission (q.v.) introduced into the Town and Country Planning Acts but never used. While still a two-stage inquiry, the second part would be carefully separated from the first or policy-and-need part. The first stage would be conducted by a commission or panel comprising a chairman and 6 to 9 others appointed by the

Department of the Environment (q.v.) after wide consultation. The second stage would be handled in the traditional manner with a Ministry inspector sitting with assessors. See *Public Hearings and Inquiries; Public Local Inquiry, UK.*

Protective Covenant: Or restrictive covenant, a legal requirement attached to a deed or statement of ownership imposing duties and restrictions additional to normal municipal requirements. Subdivision regulations are designed to establish minimum requirements and are not necessarily those which might be the most desirable. Protective covenants enable subdividers, developers, and purchasers to adopt a higher set of standards.

Covenants may simply strengthen the existing regulations or they may go much further in terms of type, size, quality, and architectural design of dwellings; they may involve the setting up of an architectural committee. To be of uniform effectiveness, protective covenants should apply to the land and be binding on all property owners in the protected area.

Public Facility: Any use of land, whether publicly or privately owned, for transportation, utilities, or communications, or for the benefit of the general public, including streets, schools, libraries, fire and police stations, county buildings, municipal buildings, power houses, recreational centers, parks, and cemeteries.

Public Hearings and Inquiries: See *Commissioner of Inquiry; Committee on Public Participation in Planning (Skeffington Committee); Examination in Public; Hearing Examiners, Los Angeles; Local Plans; Parker Windscale Inquiry; Planning Inquiry Commission; Planning System, UK; Project Inquiry; Public Hearings and Inquiries, Procedures at; Public Local Inquiry; Public Participation; Roskill Commission; Royal Commission on the Great Barrier Reef; Ranger Uranium Environmental Inquiry; Structure Plan Preparation.*

Public Hearings and Inquiries, Procedures at: A spectrum of proceedings and principles that may be adopted in a public inquiry or hearing into environmental issues and development applications; these procedures may range from court-like procedures with evidence taken on oath and cross-examination permitted, to more informal procedures in which cross-examination is restricted or excluded and in which the normal rules of evidence are not applied. Submissions may be accepted verbally, the whole proceedings being recorded with the subsequent circulation of transcripts, or may be accepted in written form only (thus primary submissions, answers to questions, and closing addresses would all be in writing).

Rules guiding an inquiry should be circulated in advance to all likely participants. The following is an example of a procedure adopted for a public inquiry:

- The hearings connected with the Inquiry will be conducted in public.
- The Inquiry will be informal and not legalistic and persons making submissions may stand or sit as they please. Neither the legal rules of evidence nor formal court procedures will apply.
- Evidence will not be taken on oath.
- If a person relies upon the evidence of some other person, that other person may speak in support of the submission.
- Persons making an oral submission will not be subjected to cross-examination though she or he may be asked questions by the Commissioner or a Solicitor appointed to assist the Inquiry.
- Similarly persons speaking in support of a submission will not be subjected to cross-examination.
- The submission will be recorded on a tape-recorder and a transcript will by typed. A copy of the transcript will be sent to the person making the submission.
- The Inquiry appreciates that persons may wish to comment upon the written or oral submissions made by other parties. They may do so in the course of their own oral evidence. Alternatively, any persons may lodge a further written submission within one month of any other person or body giving oral evidence to the Inquiry.

See *Commissioner of Inquiry; Public Local Inquiry, UK; Public Hearings and Inquiries; Public Participation.*

Public Local Inquiry, UK: The long-established form of public environmental planning inquiry conducted in Britain under housing, development, energy and environmental legislation; many pieces of legislation allow for appeals to the central government or more specifically to the Ministry involved against decisions of local authorities in respect to a wide range of matters. It is customary for a public local inquiry to be conducted by one or more of Her Majesty's Inspectors specially appointed for the conduct of inquiries. Objections are carefully heard in a non-intimidatory atmosphere, the inspector reporting to his Minister on his findings, decision, or recommendations. In some instances, the inspector(s) will be assisted by technical assessors with expertise in the area under consideration.

The subjects of objection include power stations, transmission lines, highways, housing development, redevelopments schemes, slum clearance schemes, smoke control areas, zoning and rezoning, airports and airport extensions, industrial installations of many kinds, pipelines, storage facilities and docks.

However, an inspector even sitting with technical assessors lacks an investigative arm to aid with major and technically demanding public inquiries; in some cases therefore the evidence of major bodies cannot always be as searchingly examined as might be desirable. This is particularly the case where the proponent has a virtual monopoly of expertise in the area of controversy, e.g., the Central Electricity Generating Board in the area of gaseous diffusion from high stacks.

HM Inspectors enjoy a high reputation in Britain as a body of independence, integrity, and objectivity. See *Planning Inquiry Commission; Project Inquiry; Public Hearings and Inquiries.*

Public Participation: The participation of members of the public in the deliberations leading to important environmental planning and development decisions; such participation requires an interested and educated contribution by individuals, groups, and voluntary organizations utilizing opportunities provided in law for such contributions. There may of course be active demonstration by members of the public outside of this framework in relation to such controversial issues as the mining, export, and utilization of uranium; nuclear tests; rainforests and hydroelectric dams.

Public participation is a cornerstone of the National Environmental Policy Act, 1969 (q.v.) in the United States. Under the NEPA regulations, agencies are required to make "diligent" efforts to involve the public in the various procedures and provide notice of hearings, public meetings, and the availability of environmental documents.

When an agency is considering a project whose anticipated effects are of national concern, public notices must include advertisement in the *Federal Register* and direct written notice to national organizations reasonably expected to be interested in the matter.

An agency must arrange for public hearings or public meetings whenever: environmental controversy concerning the proposed actions exists; substantial interest in the holding of a hearing is expressed by members of the public; or there is a request for a hearing by another agency with jurisdiction over the action or project.

When a draft EIS is to be considered at a public hearing, the agency must make the statement available to the public at least 15 days in advance, together with any relevant supporting documents. Scoping (q.v.) provides an additional opportunity for public participation at an early stage of the decision-making process. The NEPA process has shown that not only can the public be heard, but will be listened to.

Administrative inquiries or hearings are one of the means adopted by governments to enable ministers to achieve public participation and to ensure that the rights of individuals are protected before decisions on certain matters are made. Normally such inquiries or hearings arise because of a general concern felt by the community on some issue, or as a result of a right to a hearing given by an Act of Parliament before a decision is made.

In some cases ad hoc organizations are set up for the purpose; in others a minister may appoint a person to hold an inquiry or hearing. There is no difference in practice between an inquiry and a hearing. Very often it simply means that certain persons are entitled to a hearing at an inquiry.

One important objective of the administrative inquiry system is to provide a reasonably informal forum so that the "individual in the street," whether he or she is simply interested in the issue being investigated or is directly affected by a proposal, is not daunted by the prospect of giving his of her views. To facilitate this, the person conducting the inquiry ensures that the venue for the inquiry is generally convenient and that all parties know the approximate time they will be called. These inquiries are not bound by the rules of evidence, cross-examination is not normally needed and evidence is not on oath.

Interested parties may if they wish be represented by counsel, attorney or agent, but also they may put forth their views themselves. The person conducting the inquiry should ensure that all parties are given a fair hearing in a congenial atmosphere.

A low level of consciousness of the possible adverse environmental effects of projects, a climate of "development at any cost," a lack of tradition of active public interest in affairs, a lack of education in the public at large, and the authoritarian character of many governments are all great obstacles to public participation in the environmental impact assessment process. See *Public Hearings and Inquiries.*

Public Policies, Components of: The segments of national domestic policies that impinge most sharply on the quality of life (q.v.). These components and aims are;

- Economic Policy. To fully deploy the available factors of production (land, labor, and capital) in such combinations and in such places as to produce the goods and services demanded by society at a minimum of cost. The result may be a mixed economy (an economy with substantial public and private sectors) with varying degrees of national planning. It may encompass a resources and energy policy.

- Social Policy. To modify the effects of economic policy, where necessary, to achieve equity, e.g., a more equitable distribution of after-tax incomes, supportive services for those unable to participate in the economic process, satisfactory housing for low-income groups, education and fairness of opportunity for self-help and advancement, and a high and stable level of employment for all social groups.

- Environment and Conservation Policy. To protect land and other resources (including the free goods of air and water) from degradation due to overuse and misuse. To some extent, environment policy supports economic and social policies insofar as it ensures self-sustaining yields, or protects areas of natural beauty. It will interrelate with any resources and energy policy.

- National Settlement Policy. To combine and blend the three strands of economic, social and environment policies to achieve an acceptable national balance in population distribution; in effect to strike a balance between efficiency, equity and ecological considerations within a spatial context.

Public Trust Doctrine: A view that common property resources such as rivers, coastal beaches, clean air, and wilderness areas, are held "in trust" by government for the benefit and use of the public. It follows from this view that before a government ratifies a decision about the use of public lands or resources, it must be shown that some observable public benefit will ensue from the proposal, and that that benefit will outweigh the facilities lost to the public if the project goes head.

Q

Quality of Life: In current usage, a concept embracing a miscellany of desirable things not recognized, or not adequately recognized, in the market place. Some qualities of life of a community which cannot readily be valued or measured include such matters as civil liberties, compassion, justice, freedom, and fair play. There are also such things as observance of law, heath and education, clean air and water, recreation, wildlife, and enjoyment of wilderness—desirable "goods" which are partly or wholly outside the market economy. The morale of the nation is also relevant, the attitude to its history and achievements in literature and the arts, conduct in war, and contributions to international peace. The list does not end there, for the individual's life and its quality are strongly influenced by personal and family relationships, community relationships, personal safety, security of employment, job satisfaction, travelling time, housing conditions, working conditions, diet and general stress, qualities in respect to temper and temperament, and religious and social influences. However, current usage, while emphasizing a range of matters of crucial importance, shifts attention almost entirely from the determinants of the basic standard of living of development, production, and productivity. Indeed, it has been fashionable in some circles to regard development as inimical to the quality of life rather than its supporting arch, and for developers to be regarded with disfavor. This view is as unbalanced as any earlier view that development was invariably "progress," and progress was the sum of all good things.

Quebec, Canada: See *Federal Environmental Assessment Review Office.*

R

Rain Forest: A dense, luxuriant, closed, mesomorphic community; a global vegetation type containing many tree species associated with high rainfall and humidity, and a relative absence of frosts. There are many types of rain forest, though it is usual to consider three major divisions—the tropical, subtropical, and temperate.

The degradation and destruction of rain forests in tropical and subtropical regions has proceeded apace. Yet according to the International Union for Conservation of Nature and Natural Resources (IUCN) (1980) in the *World Conservation Strategy* the available estimates and projections of the global rate of tropical forest degradation and destruction are extremely unreliable and unsatisfactory. The most recent, and the only one based on a country by country analysis, was prepared for the timber industry in order to assess the area of forest likely to be available to it. The estimate was not concerned with other uses (such as the preservation of genetic diversity) or the impact of logging operations on those uses.

Ranger Uranium Environmental Inquiry, Australia: A public inquiry set up by the Australian federal government in 1975, under the provisions of the Environment Protection (Impact of Proposals) Act, 1974-75 (q.v.). The presiding commissioner was Mr. Justice Fox. The Commission was required by the Minister for Environment and Conservation to inquire into the environmental aspects of: the formulation of proposals; the carrying out of works and other projects; the negotiation, operation and enforcement of agreements and arrangements; the making of or the participation in the making of decisions and recommendations; and the incurring of expenditure by or on behalf of the Australian Government and the Australian Atomic Energy Commission and other authorities for and in relation to the development by Ranger Uranium Mines Pty. Ltd. of uranium deposits in the Northern Territory of Australia.

The first report of the Inquiry was presented on October 28, 1976, dealing with the generic issue of nuclear development and the basic question of whether any uranium mining should be allowed to proceed anywhere in Australia.

The second report was presented on May 17, 1977. This latter report dealt in detail with the Ranger proposal itself. The Fox Commission sat, therefore, for over two years, hearing evidence from the proponent, the Australian Atomic Energy Commission, environmental organizations, overseas experts, and the Aborigines of the Northern Territory. The Commission heard a total of 303 witnesses, the evidence resulting in 13,525 pages of transcript and 419 exhibits.

The response of the Australian Government was the acceptance of almost all of the Inquiry recommendations. Where the government decided on a course of action different to that recommended, it did so on the basis of achieving the same purposes and satisfying the same principles. The key elements of the policy are that the Australian Government retains its right to be selective in choosing the countries to which uranium exports will be permitted. In the case of non-nuclear weapon states, sales will be made only to subscribers to the Non-Proliferation Treaty. Nuclear weapons countries receiving Australian uranium must not use it for explosive or military purposes.

Reasonably Available Control Technology: Equipment, devices, methods, or techniques which will prevent, reduce, or control emissions of air pollutants based heavily on a cost effectiveness approach. See *Best Available Control Technology; Best Practicable Means.*

Reclamation Reform Act, 1982, US: An Act which called for the development of water conservation plans by all recipients of federally developed water supplies; it also increased the basic acreage eligible for federal irrigation water. The Act introduced "full cost" pricing of water from federal projects; new provisions also increased the recovery of federal costs in the delivery of water to large western farming operations. The Act is administered by the US Bureau of Reclamation. It was the first comprehensive revision of reclamation law in the United States in over 50 years.

Recycling: The return of discarded "waste" materials to the production system for utilization in the manufacture of goods, with a view to the conservation, as far as practicable, of non-renewable and scarce resources. Recycling goes beyond the reuse of a product (such as glass bottles and jars) and involves the return of salvaged materials, such as paper, metals, or broken glass, to an early stage of the manufacturing process. Some recycling has always been profitable to certain industries, e.g., the return of steel scrap to the steel industry and glass to the glass industry. The capacity of an industry to recycle is in many cases limited by technical as well as economic considerations. In recent years, a body of opinion has emerged favoring an increased tempo and scale of recycling in order to conserve resources for the future of mankind beyond what may be profitable in the shorter term. In addition, many people favor positive collection and recycling programs as a means of reducing indiscriminate tipping, littering of the streets, despoilation of beauty spots, dumping on waste land, and unsightly domestic accumulations; items for recycling include small items such as drink cans and bottles, and larger items such as discarded consumer durables and automobiles.

About 160 million tonnes of mining and quarry waste, and ash and clinker from power stations, is generated in Britain each year. Manufacturing industries produce roughly 25 million tonnes and there are also large quantities of agricultural and forestry wastes. In addition, households, shops and offices generate an increasing amount of waste, about 17 million tonnes in 1978. It was estimated that household waste in Great Britain was, in 1978, about 27 per cent paper, 29 per cent vegetable and putrescible matter, 10 per cent glass, 7 per cent metal and 5 per cent plastics, the remainder being fine matter (mostly dirt, ash and grit) together with a small percentage of rags and unclassified material. Some of this waste is reused immediately. For example, some mining and quarry waste is used in land reclamation and much agricultural waste is immediately used on the land. Industry reuses or recycles a large proportion of its process waste, either in-house or with the help of a well-established reclamation industry. In the late 1970s, between 65 and 70 per cent of the ferrous metals consumed by manufacturing industry was produced by recycling, as was about 60 per cent of lead, about 30 per cent of paper and board and roughly 25 per cent of glass. Very little household waste is recycled—for example, only 1 per cent of waste paper from households is reclaimed. A Waste Managmeent Advisory Council advises on a national program for waste reclamation.

Recycling is a term also applied to a process in which existing buildings are adapted for reuse. This process often infuses new life into old buildings and utilitarian structures which take on a role completely different to their original function. Thus old warehouses may be converted into home units or shopping centers. Recycling may effect significant economies compared with demolition and new construction, while retaining a heritage atmosphere. Much recycling has occurred at The Rocks, Sydney, Australia.

Redevelopment of Central Business District Areas Act, 1976, Kansas, US: An Act intended to assist in the development and redevelopment of the central business district areas in the State of Kansas, thus promoting the general welfare of the citizens. The Act authorizes cities to acquire certain property and to issue special obligation bonds for the financing of redevelopment; in the public interest, the power of eminent domain (q.v.) may be exercised.

To utilize the Act, a city governing body must adopt a resolution finding that the area to be redeveloped is blighted, and that redevelopment is necessary to promote the economic welfare of the city. Feasibility studies must show that the benefits will exceed the costs and that the income will be sufficient to pay for the project. A public hearing on the prepared redevelopment plan must be held. A two-thirds vote of the governming body is required to adopt the plan.

The Act provides for relocation assistance. No persons may be displaced until there is suitable housing available at rents within their ability to pay. Payment must be made to a retailer who suffers damages by reason of the liquidation of his/her inventories necessitated by relocation. Payments must be made to persons or businesses displaced as a result of the acquisition of real property. See *Business Improvement District Act, 1981.*

Regional Administrative Centers: A concept of carefully identified centers within a region suitable, for example, for the co-location of federal and state offices and private sector regional administrative offices. The infrastructure of such towns is deliberately strengthened, bolstering selected rural centers. This may have the effect of inhibiting or slowing down migration from the rural areas to the largest centers of population in the state, creating with other forms of encouragement a series of regional "capitals" offering a reasonable range of higher order services and opportunities. A regional administrative center may interlock with a decentralization (q.v.) policy and program.

Regional Development Commission, US: An agency encompassing a large geographical area and many jurisdictions, created to prepare area-wide development plans. The Minnesota Regional Development Act, 1969, for example, permits the creation of multi-county development commissions for this purpose. Some state planning bodies now attempt to tackle problems through the preparation of comprehensive state-wide development plans. See *Southern California Association of Governments.*

Regional Environmental Plan, New South Wales: A plan formulated within the terms of the Environmental Planning and Assessment Act, 1979 (q.v.) of New South Wales, Australia, relating to a region within the state, e.g., the Hunter Valley and the Illawarra Regions. A regional environmental plan is preceded by an environmental study and a draft regional environmental plan. In turn, both the study and the draft plan must be published and publicly exhibited, the public being given every opportunity to express opinions and make submissions. A public inquiry headed by a commissioner of inquiry (q.v.) may be held. Any regional plan must be consistent with relevant state environmental planning policies (q.v.); local environmental plans must in turn be consistent with a regional environmental plan. See *Local Environment Plan.*

Regional Planning Commission, US: A planning commission (q.v.) enabling the representatives of several municipalities and counties to jointly prepare a coordinated comprehensive plan (q.v.) for the development of a region with a view to encouraging the most appropriate use of land; the economic location of public utilities and services; the development of the most advantageous transportation and communications systems; the development of adequate recreational areas and regional parks; the promotion of good civic design; and the wise and effective expenditure of public funds.

As an example, the State of New Hampshire has eight regional planning commissions embracing some 234 separate municipalities. The boundaries of the regions were delineated by the Office of State Planning. The powers of the regional planning commissions are strictly advisory. Each commission has a duty to prepare a comprehensive master plan for the development of the region; the planning boards of the individual municipalities may adopt any part or all of a master plan. The finances of the commission are met by the member municipalities and counties. Another example is the Southern California Association of Governments (q.v.).

Regional Seas Program, UNEP: A program organized by the United Nations Environment Program (q.v.) to protect the marine environment and its living resources against pollution and over-exploitation through the development of conservation policies. By 1984, action plans had been developed for the protection and management of the marine environment in respect to ten regional seas. International and other regional conventions for the protection of the marine environment had been brought into effect, including the Convention for the Prevention of Marine Pollution by Dumping from Ships and Aircraft (Oslo, 1972), the Convention on the Prevention of Pollution from Ships (London, 1973), and the Convention on the Prevention of Pollution from Land-based Sources (Paris, 1974).

The regional seas examined have included the Baltic Sea, the North Sea, the Mediterranean Sea, the West African coastal waters, the Gulf of Mexico, the North American coastal waters (Atlantic, Pacific and Arctic), the South-West Atlantic (Brazil and Argentine Basins), South-East Pacific Coast (Ecuador, Peru, and Chile), the Indian Ocean, Australian and New Zealand coastal waters, the Persian Gulf and the Red Sea. See *Table 16*.

Table 16: Marine Pollutants and Nature of Concern

Input or Activity	Concern
1. CO_2	Climate shift, temperature change, sea-level change, etc.
2. Metals	Potential toxic effects
3. Microorganisms	Public health risk
4. Radioactive waste disposal	Public health risk
5. New chemicals	Toxic effects on people and organisms
6. New energy production	Alteration or disturbance of habitats
7. Deep-sea mining	Increased turbidity, sea-bed disturbance

Source: United Nations Environment Program GESAMP: The Health of the Oceans UNEP Regional Seas Reports and Studies, No. 16, 1982, p. 6.

Regulatory Impact Analysis (RIA), US: An analysis which US federal agencies must perform on all new "major" regulations to ensure that the potential benefits to society outweigh the potential costs of the proposed action; this requirement was introduced under the US President's Executive Order 12291 of 1980. Analyses are to be submitted to the Office of Management and Budget (OMB) for review. The measure applies to all proposed regulations likely to have an annual effect on the economy of $100 million or more; to result in a major increase in costs or prices for consumers, individual industries, federal, state, or local government agencies, or for geographic regions; or likely to have significant adverse effects on competition, employment, investment, productivity, innovation, or the ability of US-based enterprises to compete with foreign-based enterprises in domestic or export markets. Such benefit-cost calculations are still not the final determining factor in the making of regulations (for political considerations are not taken into account) but they have become an indispensable tool in the analysis of the likely effects of government regulations. See *Benefit-Cost Analysis.*

Reith Report on New Towns, 1946, UK: The report of a committee chaired by Lord Reith; it recommended that garden cities on Ebenezer Howard lines should be built by public corporations drawing finance directly from the Treasury; such corporations would be autonomous and free from the existing structure of local government. See *New Towns Act, 1946.*

Residential Development Standards: The statutory requirements imposed by municipalities in respect to the construction of dwellings in residential zones (q.v.). The objectives of setting development standards include:

- Protecting the investor against the less attractive features of the free market, e.g., the danger that lots within the same subdivision will be sold for the construction of inferior or different classes of dwelling considered to threaten ultimate resale values;
- The achievement of attractive residential environments promoting also safety, health and general welfare;
- Safeguarding privacy, and access to light, sun, and air;
- Ensuring the minimization of future maintenance costs;
- Ensuring the road system efficiently serves each property.

Standards embrace such matters as:

- Allotment size;

- Frontage width;
- Front, side and rear boundary setbacks:
- Services;
- Aesthetic standards;
- Dwelling size;
- Density;
- Height;
- Car parking:
- Fencing;
- Landscaping

Minimum standards protect investors and minimize uncertainties. Excessive standards increase costs unduly both to the individual and the community, e.g., lower residential densities encourage the peripheral expansion or urban areas which in turn entails higher commitments for physical and social services. See *Development Standards; Subdivision; Subdivision, Factors to be Considered.*

Residential Zones: Parts of local government areas or environmental precincts (q.v.) zoned or reserved for housing within a local environmental, structure, or development plan. Each type of housing (detached, attached or group, and flats or units) may be confined to specific areas, though there has been an increasing trend in many parts of the world to accept a mixture of housing forms where there is no great variation of scale. Widely different housing forms continue to be considered by many people to be incompatible.

Families in single-story detached houses do not want high-rise flats erected on an adjacent site. Some of them object to any change at all (be it high-rise flats or any other multiple dwelling use), but it appears that the main objection is to development which is of vastly greater intensity and height than detached houses, while other types of dwellings would be acceptable to the majority of people if height and intensity of land use would be reasonably compatible with the existing dwellings. The real objective of separate residential zones should therefore be not to segregate the types of dwellings but rather to group the intensity of residential development. If residential areas were zoned mainly by the intensity (i.e., density) and height of development, the physical form of dwellings (in rows, clusters, intertwining, overlapping or stepped) could be left to the housing preferences of the public.

Density zoning allows not only more variety in dwelling types but also more refinement in the technique of zoning. Many more than three residential zones may be created. Low, medium and high density zones may be expressed at various density levels and may be connected by transition zones. Even at detached housing densities a great deal of variety could be achieved. A planning authority may wish to preserve trees or other landscape characteristics of land which is not in public ownership and where some residential development has to be accepted. If site areas per dwelling of the order of half a hectare (or one acre) were specified, this would ensure that the area would virtually retain its open, undeveloped character. At medium densities different scales and intensities of development may also be achieved. For example a "medium" density zone where one dwelling per 200 square meters (about 2,200 square feet) of site area is allowed will probably have attached dwellings of two stories, whereas a "low to medium" density zone where one dwelling per 320 square meters (about 3,500 square feet) is allowed, will probably encourage single-story courtyard houses—although other types will be permissible. Similarly, areas intended for higher densities may be divided into those where low-rise buildings with a great deal of private open space are desired, those where development may be more intense and those where high-density buildings are appropriate.

Density zoning would also bring more certainty to the field of residential development. There would be more certainty from the viewpoint of the planning authority: future populations of particular areas could be more accurately forecast than at present and services and facilities be planned accordingly. There would be more certainty also from the viewpoint of developers who would be aware of the development capacity of each site. This would reduce the speculation and delays that now surround the purchase of sites for development, and eliminate the need for long options while development consents are obtained.

Resource Areas: Areas that are essential for maintaining the quality and diversity of the natural environment and that are therefore of long-range environmental, economic, social and cultural value.

Resource Conservation and Recovery Act, 1976, US: An Act with historical roots in the Solid Waste Disposal Act, 1965 (q.v.); its primary object was to promote and protect the public health and the environment, and to conserve valuable material and energy sources. These objectives were to be accomplished by providing technical and financial assistance to state and local governments and interstate agencies for the development of solid waste management plans; prohibiting future uncontrolled open dumping on land, and requiring the conversion of existing open dumps to controlled and managed nonhazardous facilities; regulating the treatment, storage, transportation, and disposal of hazardous wastes which have adverse effects on public health and the environment; providing for the promulgation of guidelines for solid waste collection, transport, separation, recovery, and disposal practice and systems; and establishing a cooperative effort among the federal, state and local governments and private enterprise in order to recover valuable materials and energy from solid waste.

The RCRA established the statutory framework for comprehensive federal and state regulation of hazardous wastes. The Act required the identification and listing of hazardous wastes, taking into account such factors as toxicity, persistence, and degradability in nature, the potential for accumulation in tissue, and other characteristics. It directed promulgation of such standards for generators of hazardous waste as might be necessary to protect human health and the environment. These standards were to include requirements for record keeping, labels for containers, disclosure of chemical components, use of a manifest system, and reporting to the Environmental Protection Agency (q.v.). Similar standards were to be developed for transporters of hazardous wastes in cooperation with the US Department of Transportation.

The Act also provided for the establishment of a permitting system to control the treatment, storage, and disposal of hazardous wastes, thus ensuring that all hazardous waste facilities would be operating under conditions specified in a RCRA permit. The Act also directed the Environmental Protection Agency (q.v.) to promulgate guidelines to assist States in the development of their own hazardous wastes programs; the EPA was also given a wide range of enforcement tools for ensuring compliance with the regulations. Regulations were promulgated in May, 1980, and in January, 1981 (amended in July, 1982). There were now in effect detailed regulations governing every aspect of hazardous waste management.

In November, 1984, the US President signed into law a bill reauthorizing and amending the Resource Conservation and Recovery Act. The new legislation broadened government restrictions on land disposal of hazardous waste and greatly increased the number of waste generators subject to EPA regulation. Other provisions significantly improved the quality of landfills and surface impoundments, and placed underground storage tanks under EPA regulation. By specified dates, the EPA must decide whether it is safe to continue land disposal of a large variety of hazardous wastes. Should the EPA fail to meet these deadlines, so-called "hammer clauses" will go into effect prohibiting such disposal. The specified dates range from November, 1986, to May, 1990, for various categories of waste.

In addition to ruling on various types of land disposal, the EPA must promulgate regulations specifying methods of treatment capable of substantially reducing the toxicity of the waste or its likelihood of migration from a disposal unit or injection zone. Wastes which are so treated will be exempt from the ban on land disposal.

The 1984 RCRA requires that permits for hazardous waste facilities be renewed every ten years; land disposal permits, however, are subject to review every five years. Applications for permit renewal are subject to all the requirements that pertain to the issuance of new permits. The RCRA also specifies that these applications must reflect improvements in control and measurement technology that have occurred since the previous permit was issued. In certain cases, private citizens are authorized to bring legal action where past or present hazardous waste management practices pose an imminent danger. They can bring this action against companies, governmental entities, or individual citizens. See *Comprehensive Environmental Response, Compensation, and Liabiity Act, 1980, US ("Superfund")*.

Resource Recovery: The extraction and utilization of materials of value from a waste stream; materials recovered, for example, may include metals and minerals which are used as raw materials in the manufacture of new products. Other forms of recovery include energy recovered by utilizing components of waste as a fuel, production of compost using solid waste as a medium, and reclamation of land through landfill (q.v.). See *Recycling*.

Restriction of Ribbon Development Act, 1935, UK: A measure introduced into Britain intended to control linear development along main roads through the machinery of the Town and Country Planning Act, 1932, (q.v.).

Ribbon Development: The practice of building along both sides of a highway; such development may be only one building deep, with open country behind. Such development gives the visual impression of each town being connected directly with its neighbor, the country never being reached. At best the edges of town and country are blurred; at worst they are eliminated. Frequently, large areas of the backland become incapable of development due to lack of access.

Right of Common Access, Nordic: In principle, a right of entry by the public which includes passage on foot over all types of land. However, growing crops or the right to privacy of a houseowner are by no means unprotected. As regards passage on foot over most land, the rights are similar to the Scottish ones, also of Norse origin, that make much of their countryside accessible, in contrast to the more restricted situation in England.

Care is taken in Swedish planning legislation to protect the Right of Common Access. This is particularly true of the current work on National Physical Planning, which calls upon municipalities to implement the general guidelines for physical planning that the Riksdag (Parliament) laid down in 1972. One important criterion for the further protection of natural areas and for planning decisions in the municipalities is the wish to keep open such areas that are now open to everybody. This means that valuable areas will be protected by various planning restrictions to prevent land use—urban development, afforestation with dense plantations, etc.— which is in conflict with free accessibility to the land.

One planning measure to ensure the accessibility of valuable land is the general ban on building within 100 meters of the shore. Similar regulations have been in operation in some coastal areas and around certain lakes for a long time. The reasons for these regulations are to prevent valuable and beautiful shores from being spoiled by development and to keep them accessible to everybody.

Landowners have sometimes tried to get around the Right of Common Access by putting up fences to exclude people. Sometimes this is perfectly legitimate and necessary to protect the privacy of a house, but in several cases the aim has been to exclude people, e.g., from a beach which is not in the immediate vicinity of a house. In such cases the county administrations have the power to force the landowner to make passage on foot possible, e.g., by putting in a stile or gate.

In Finland, the recreational use of nature is based on the Right to Common Access. By virtue of the Right to Common Access everybody can, without asking for the landowner's permission, stay, stop for temporary camping, and pick berries, mushrooms and flowers. The Right to Common Access does not exist in the neighborhood of someone's house. Since seashores and the shores of lakes especially in southernparts of Finland are to a great extent occupied by holiday housing, possibilities for recreation on the basis of the Right to Common Access have been remarkably reduced in areas most valuable from a recreational point of view. Another phenomenon which reduces essentially the possibility to enjoy the Right to Common Access is the intensified forestry in the form of clear cutting, with the accompanying radical disruption of the soil (especially by plowing). It has proved necessary to establish a network of special recreational areas easily attainable from urban settlements.

Risk and Hazard Assessment: An essential component of the environmental impact statement (q.v.) relating to any major project; it embraces those potentially adverse effects of a project involving fire, heat, blast, explosion, or flood, arising from the fixed plant or from the transportation system used in importing raw materials and exporting finished products. An assessment should reveal the hazards of life and limb, and to property, expressed in the form of a risk probability.

Overpressure damage arising from explosions may give rise to damage ranging from glass breakage up to complete demolition; heat flux damage may range from ignition of timber to the structural failure of steel. A potential hazard with the storage of inflammable gases and other combustibles is the boiling liquid expanding vapor explosion (BLEVE). All these events may be associated with human shock, injury and death. The risk to life and limb of heavy trucks and road tankers moving through towns must not be overlooked, in terms of traffic accidents as well as the risks of leakage and explosion. The movement of inflammable and radioactive wastes requires special study.

The central problem in environmental impact assessment (q.v.) is determining what the risks are, and what the consequences of various failures may be; then a level of acceptability has to be determined. No project is without some risk, and some may carry considerable risks. Risks may be reduced by a variety of safety precautions and back-up measures. Gas storage vessels, for example, may be sand-mounded.

One of the keys to an acceptable level of risk is choice of location, with particular reference to the proximity of dwellings. Thus safety may rest in part on physical design and procedures, and in part on distance. The routing of vehicles along their own haulage roads, or other prescribed routes, may reduce the risks and hazards to pedestrians, residents, and other vehicles in towns and villages. The buffer zone (q.v.) becomes most important in some instances, in separating hazardous industries from residential areas, schools, and hospitals.

Safety criteria in the context of land-use planning greatly depend on local conditions.

River Thames Clean-Up, UK: A program launched in the 1960s to clean-up the River Thames, London, England; pollution of the Thames had been a longstanding problem. In the 19th century, it was not unknown for both Houses of Parliament to have to adjourn because of the smell from the Thames. The Committee on Pollution of the Tidal Thames (the Pippard Committee) reported in 1961 that the Thames estuary down to Gravesend was in a badly polluted and frequently offensive state. Hydrogen sulfide was evolved through the activity of sulfate-reducing bacteria when the water was completely de-oxygenated. The Water Pollution Research Laboratory had cooperated with the Committee in investigating these conditions.

It was found that the sewerage works at Barking and Crossness, Mogden, Acton and Dartford were responsible for about nine-tenths of the sewerage effluent discharged directly into the tidal Thames, the Barking and Crossness works together contributing roughly half the polluting load from all sources. The direct discharge from Acton, while comparatively small, was found to be a significant source of pollution.

Direct industrial discharges accounted for about 3 per cent of the total polluting load. Sulfide discharged from the sulfur dioxide scrubbing plant at the Battersea and Bankside power stations, while it had no material effect on the rest of the estuary, caused an appreciable loss of oxygen from the river water near the power stations. The Report recommended that the aeration of discharges should be increased. While the Upper Thames and the tributaries contributed dissolved oxygen and nitrate to the estuary, the lower reaches were found to contribute polluting matter only. Of the tributaries, the Wandle was found to have the greatest polluting effect. The introduction of synthetic detergents in and after 1949 was followed by a marked lengthening of the zone devoid of oxygen.

The Committee recommended the pursuit of measures to prevent offensive conditions and the creation of a margin of safety. It was noted, however, that to raise the dissolved oxygen enough for salmon to pass through the estuary would require extremely high standards of effluent purity, and imply an expense out of proportion to any likely gain. The Committee recommended that new polluting discharges to the estuary above Lower Hope Reach should be avoided if at all possible; even below this point, proposals for new discharges should be carefully considered. The Committee thought that the imposition of uniform standards of quality was not suitable save for minor effluents; standards for effluents must control the quantity of polluting matter discharged, hence the volume as well as the strength of each effluent must be known. There can be no doubt that improvements at sewage works and in the treatment of other effluents had led to a marked improvement in the condition of the Thamses.

By 1969, the sand goby, flounder and smelt were being found on the screens at Fulham power station which is situation 16 kilometers above London Bridge and 64 kilometers from the sea. Other fish found at Fulham have included eel, pike, bream, roach, perch and trout; salmon have yet to return. Undoubtedly, the Thames has not been cleaner for more than a century. In 1985, a seal reached central London in search of fish.

River Water Quality Classifications, UK: Classifications used in the national surveys of rivers in England and Wales. Four quality standards are used according to the following criteria:

Class 1. Unpolluted or recovering from pollution. BOD (biochemical oxygen demand) of less than 3 mg/l and receiving no toxic effluents.

Class 2. Rivers of doubtful quality and needing improvement.

Class 3. Rivers of poor quality requiring urgent improvement.

Class 4. Grossly polluted rivers with a BOD of more than 12 mg/l and incapable of supporting fishlife.

Recent results have shown that out of 90,255 kilometers of inland waterways, 85 per cent were unpolluted (Class 1) and only 2 per cent were grossly polluted (Class 4), the remainder lying intermediate between these extremes.

Water quality is also being measured at 240 stations near the tidal limits of rivers in the United Kingdom with the objective of keeping a record of the total quantities of pollution that are being released into coastal waters and originating from sources inland.

Roads, Hierarchy of: A hierarchy of importance in respect to which a fourfold classification is often adopted by traffic authorities:

- arterial roads
- distributor roads
- collector roads
- local roads

Arterial roads predominantly carry through traffic from one region to another. Distributor roads connect the arterial roads to areas of development or carry traffic directly from one part of a region to another; they may also relieve traffic on arterial roads in exceptional circumstances. Collector roads connect the distributor roads to the local road system in areas of development. Local roads are the subdivisional roads within an area of development; these are used solely for local access purposes to residential, commercial, industrial, and recreational districts and activities.

The heirarchy can be likened to a tree. There is the trunk (arterial roads), the limbs (the distributor roads), the branches (the collector roads) and the twigs (the local roads). See *Freeways*. See *Figure 16*.

Roskill Commission, UK: An environmental planning inquiry conducted in Britain with the task of identifying the best site for a Third London Airport; the Commission comprised six commissioners under the chairmanship of Lord Roskill, a High Court Judge. The Commission reported in January, 1971, in favor of Cublington, Bukinghamshire, with a note of dissent by Professor Buchanan in favor of Foulness on environmental grounds.

The analytical methods and conclusions of the Commission leaned heavily on benefit-cost analysis (q.v.). The inquiry was broad in the sense that the Commission was not faced with a specific proposal. The inquiry occupied 258 sitting days over 22 months.

Royal Commission on Environmental Pollution, UK: A commission set up in 1970 as a permanent or standing body to advise the British Government as a whole in relation to environmental matters. The terms of reference have been: "To advise on matters, both national and international, concerning the pollution of the environment; on the adequacy of research in this field; and the future possibilities of danger to the environment." The Royal Commission was intended to serve as a general watchdog on pollution and environment protection.

A number of substantial reports have been published. In 1975, in its Fifth Report ("Air Pollution Control: An Integrated Approach"), the commission urged the creation of a central inspectorate to control all forms of pollution. It recommended also the creation of a diploma in pollution control; this would be awarded to engineers and scientists suitably qualified and experienced in all aspects of pollution control to meet the needs of policy-makers.

In 1983, the Royal Commission on Environmental Pollution released its report *Lead in the Environment*. The Royal Commission could find "no compelling arguments for the retention of leaded petrol [gasoline], except as aninterim measure to enable the majority to existing cars to be phased out." It was recommended that the United Kingdom initiate negotiations with the European Economic Community to secure the removal of the minimum level of lead in gasoline of 0.15 g/l contained in Article 2,

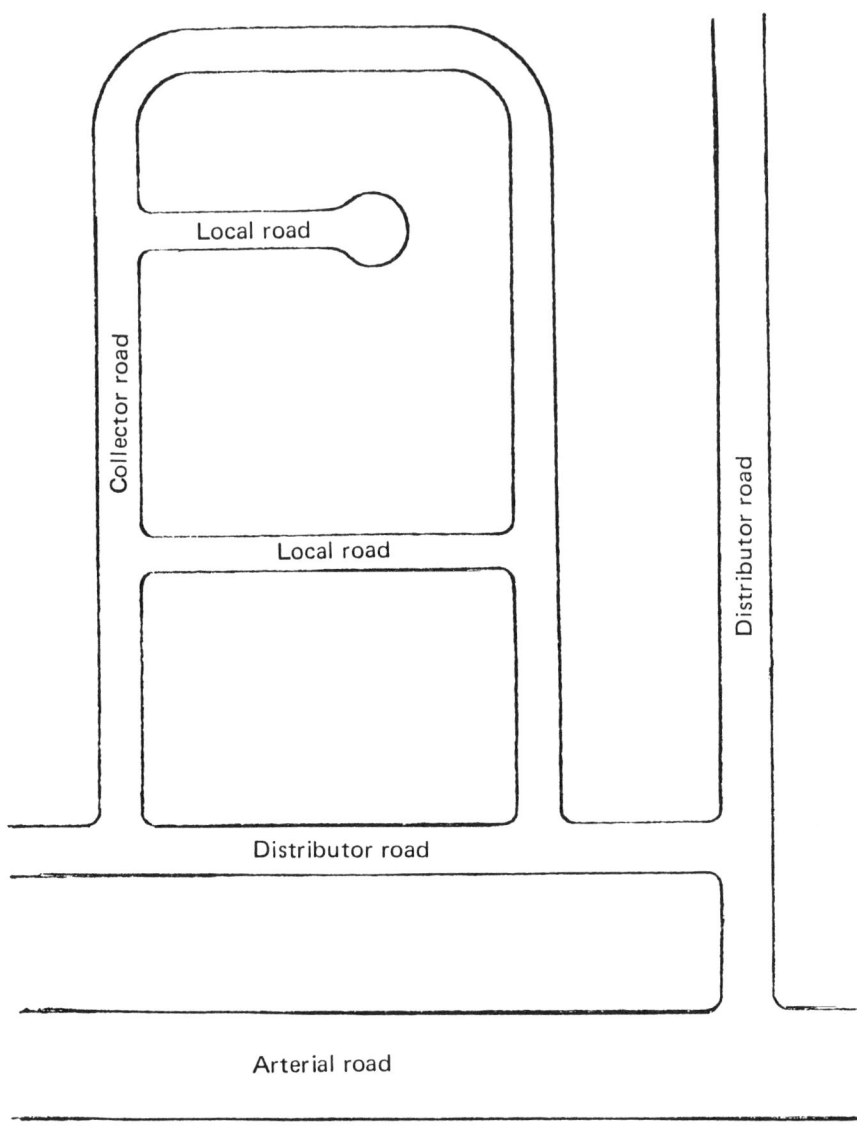

Figure 16: Functional classification of roads.

Directive 78/611/EEC. The objective was to achieve unleaded gasoline as soon as possible.

The Environment Secretary announced to the House of Commons on 18 April, 1983, that he accepted the Royal Commission's recommendation and would pursue formal negotiations in respect of a timetable for the introduction of unleaded gasoline. The aim was to have all new cars running on lead-free gasoline by 1990.

Royal Commission on the Great Barrier Reef, Australia: A commission set up by the Australian federal and Queenland State governments to examine the implications of oil drilling on the Great Barrier Reef. The Reef lies off the eastern Queensland coast of Australia and is widely regarded as the largest assemblage of living corals in the world. The Commission concluded in 1975 that if petroleum drilling occurred, some spills ranging from small to substantial would occur. Two commissioners found there was a risk of uncertain magnitude to marine life and recommended that drilling should not be permitted in large areas of the Reef. The chairman (the remaining member of the commission) was not convinced that drilling could be conducted anywhere on the Reef unless more research provided some clear evidence that no harm would result.

Following the report of the Royal Commission, the Australian federal government decided that no further exploratory drilling would be permitted until research into the effects of oil on coral reefs had been undertaken. However, subsequently large areas of the Reef were proclaimed as a marine park under the control of a Great Barrier Reef Marine Park Authority, thus precluding oil drilling.

With the declaration in 1983 of five additional sections of the Great Barrier Reef Marine Park, some 98.5 per cent of the Great Barrier Reef became protected. The Great Barrier Reef Marine Park has an area of about 344,000 square kilometers, and is by far the largest marine park in the world.

Under section 38 of the Great Barrier Reef Marine Park Act, 1975, operations for the recovery of minerals including petroleum within the marine park are prohibited. The Great Barrier Reef Marine Park (Prohibition of Drilling for Petroleum) Regulations prohibited, from September 1983, the drilling for petroleum in any part of the Great Barrier Reef Region not included as part of the marine park.

Royal Society for Nature Conservation, UK: A voluntary body in the United Kingdom which in 1984 had a membership of 140,000. In 1981, it published *Towards 2000*, a report which reveals that twice as many trees are destroyed by farmers in Britain as by Dutch Elm disease, and that a quarter of all hedgerows had been removed since the Second World War. The report emphasized that nearly half of all the semi-natural and ancient woodland in Britain had been lost since the mid-1940s and half the ponds in East Anglia; while Dorset's Wild Heathland had been trimmed from

17,600 hectares in 1931 to under 6,000 hectares in 1981. Some 31 of the 54 species of butterflies resident in 1950 were now hard to find; in 1979, the large blue butterfly had been declared distinct.

The report went on to discuss agricultural practices and planning controls on farming operations, including stop-orders, and the concept of heritage farms (relieved from capital taxes while subject to an approved conservation plan).

S

Saemaul Undong (New Community Movement), Korea: A pan-national drive to achieve a better life through the improvement of living surroundings and real income inspired by the spirit of diligence, self-help and cooperation. This movement, which originated from the traditional Korean spirit of mutual assistance, was initiated in 1970 and contributed much to the protection of the environment. The protection of nature, improvement of housing, construction and upgrading of rural roads, and energy savings at the rural community level are regarded as the major achievements of the Saemaul Undong. Though the Government provides partial assistance in procuring material and technical guidance, the movement is a voluntary drive of the Korean people. This movement has spread to cities, factories and schools and is facing a new era of substantiation in the 1980s. See *Environmental Conservation Law (ECL), 1977, Korea; Han River Basin Environmental Master Plan, Korea.*

Safe Drinking Water Act, 1974, US: An Act which provides for the safety of drinking water supplies throughout the United States of America by establishing and enforcing national drinking water quality standards. Under the Act, the Environmental Protection Agency (q.v.) has the primary responsibility to establish national drinking water quality standards, and to supervise public water supply systems and other sources of drinking water. Regarding the enforcement of drinking water quality standards, a state can qualify for "primary enforcement responsibility" if it adopts regulations at least equal to the federal regulations in protecting public health, and can provide adequate surveillance and enforcement procedures.

In 1984, the Environmental Protection Agency recommended maximum contaminant levels (MCLs) for a group of nine chemical compounds that might cause health problems if found in drinking water at significant levels; these were volatile synthetic organic chemicals found in industrial solvents, degreasing agents, and dry-cleaning fluids.

By 1985, most states and territories had accepted primary enforcement responsibility. Two classes of regulations have been introduced: primary regulations are concerned with the protection of public health, while secondary regulations are concerned with the taste, odor, and appearance of drinking water.

The Act also includes provision for the control of the underground injection of water and other substances which might endanger drinking water sources. It also provides for the designation of particular aquifers as the sole source of a drinking water supply.

Safe Growth Plan, Tennessee, US: Environmental measures introduced by the State of Tennessee in 1981 for dealing with aesthetic issues including protecting valuable natural areas and scenic rivers, picking up roadside junk and litter, and removing unsightly billboards, as well as improving garbage collection systems and the ways of dealing with hazardous wastes. Under the plan, prisoners have been used to pick up trash (bottles, paper, old mattresses and tires) alongside county state, and federal roads. Penalties have been increased for persons caught littering; pull-tabs and rings on cans have been banned. A variety of other measures have been adopted to improve the environment.

By 1984, the state's 40-year old Safe Drinking Water Act had been completely rewritten. Water supply systems and wastewater treatment plants were being improved. A major reorganization of the state environmental agencies had taken place and many functions brought under the newly-named Department of Health and Environment. Surface mining controls were being improved. A Superfund had been established for the identification, investigation, and clean-up of abandoned hazardous waste dump sites. Funding was also made available to encourage communitites to investigate the benefits of having a hazardous waste treatment and disposal facility.

The state also began to acquire unique biological, historical, archaeological and cultural areas of statewide significance. A model Protection Plan for the Collins State Scenic River had been adopted, and a Protection Handbook for the Scenic Rivers Program prepared. A Scenic Parkway System of some 2,300 miles had been established. An inventory of wetlands was being prepared. Conservation and environmental education curricular materials had been introduced into classrooms through Project CENTS— Conservation Education Now for Tennessee Students.

Sanitary Epidemiological Service, Soviet Union: A national system of observation and control in the Soviet Union relating to air and water pollution, noise, and environmental problems generally. The service is also responsible for measures to provide adquate health conditions at work and in the home, and to protect the population from infectious diseases. It is also responsible for determining standards for foodstuffs.

Sanitary inspectors have broad powers; they take part in the formulation of town plans and in discussions relating to the siting of plants, factories, shops, and child-care institutions. Officially, they have the right to stop production at a plant that is polluting the environment. No enterprise or residential building can be handed over for use without the permission of the sanitary inspector. Since 1974, the USSR five-year and long-term development plans have laid down general targets for water, air and forest protection.

It is claimed that in Moscow, between 1968 and 1978, more than ten thousand boiler houses were closed and replaced by district heating plants using natural gas. Virtually all Moscow's flats are heated in this way from central sources. The more polluting industries are being located outside the city, while those remaining in the city are being fitted with pollution control equipment. In the period 1963-78, it is claimed that air pollution from dust and sulfur oxides in Moscow fell by 65 per cent. Some 22 stations monitor the air. Strategic plans involve moving more industrial enterprises out of the city.

Moscow has become noted as one of those rare cities where a white collar will remain white throughout the day. It is said that fifteen years ago in Moscow, fresh snow falling in the city would stay clean for only two or three days, whereas now it stays white for up to two weeks.

Saskatchewan, Canada: See *Federal Environmental Assessment Review Office.*

Scoping, US: A significant innovation in the National Environmental Policy Act, 1969 (q.v.); when a federal agency determines that a proposed action requires the preparation of an environmental impact statement (q.v.), it must take prompt action at the outset of its planning process to identify those issues that require full analysis, and separate them from less significant matters that do not require detailed study. This sifting out of issues is known as "scoping."

Further, the scoping process helps the agency identify any environmental review and consultation requirements imposed by laws other than NEPA and to allocate responsibilities among lead and cooperating agencies. Thus, where an EIS is required for a coal-fired power plant, the scoping process would afford the opportunity to coordinate the preconstruction environmental reviews required by the Clean Air Act, Clean Water Act, and other environmental laws into the overall EIS process. To ensure such coordination, affected federal, state, and local agencies and interested members of the public must be invited to participate in this scoping process.

Scott Report, 1942, UK: The report of the Committee on Land Utilization in Rural Areas; the report advanced the basic argument that agriculture should occupy a privileged place in the national economy, to be protected against competition for the land it used. A developer would have to show that a proposed change of land use from agriculture was positively in the national interest. The post-war years witnessed continuing government support for agriculture although land, some of it agricultural, continued to be needed for inevitable urban expansion. The committee also recommended the creation of national parks and nature reserves.

Selective Decentralization: See *Decentralization.*

Service Activities: Useful functions performed by persons or organizations within the framework of the economic system; service activities may be classified as follows:

- The transport and distribution of food, energy, water, clothing, and other goods;
- The transport of people;
- The provision of managerial, administrative, organizational, and educational services;
- The meeting of various personal and collective needs relating to security, health, communication, entertainment, and finance.

Within these groups, activities may vary greatly in scale. For example, retail services may vary from a small isolated corner shop to a large supermarket; postal services may vary from a rural post box to a General Post Office; medical services may vary from a single general practitioner in a village to specialized services in a large public or private hospital. In other words, there is a *hierarchy* of service activities ranging from "low-order" services found in every center whether city, town or village, to "high order" services found only in the major centers. Each service activity has a *threshold population* and *market range*.

The threshold population is the minimum population necessary to support the service activity. If the population falls below the threshold level, the activity will operate at a loss and eventually fail. Vastly different threshold populations apply to a corner shop and a theater or orchestra. The market range of a service activity is that distance which people are willing to travel to reach the service, time and cost being taken into consideration. People may travel some distance to a supermarket or major store, tending to ignore local shops save in emergency. See *Central Place Theory; Growth Pole Theory*.

Setbacks: Requirements that new buildings shall not be nearer than a minimum prescribed distance from boundaries and roads. Generally, residential buildings are required to set back from front and rear boundaries of their sites and detached houses and flat buildings are required to set back also from the side boundaries. The objectives of setbacks from the street alignment are to achieve a satisfactory street appearance and to allow for landscaping between street and buildings. Although it is usual to fix standard building lines for streets, there may in some cases be good reason to vary these to create interest in the street scene. It is certainly logical to relate building lines to the height and continuous facade of buildings, so that high and bulky buildings are set back further than low and small ones.

The purpose of setbacks from side boundaries is to provide access, light, air circulation and, most importantly, sunshine and privacy. Formula-type setback requirements related to height of buildings satisfy the first three of these, but are inadequately related to sunshine and privacy. Moreover, rigid setback requirements are often criticized by designers as too restrictive, leading to environments where every building is positioned in the center of its site. Some councils provide flexibility in their regulations by specifying a minimum total setback from both boundaries which is more than twice the minimum setback from one boundary. This allows limited freedom since a reduction in the setback from one side can be offset by an increase in the other. Theoretically it would be possible to devise controls which provide more flexibility than this, but in practice this is difficult.

Of the two main purposes of setbacks from boundaries, ensuring sunshine and privacy for adjacent sites, the first is easier to express as a direct requirement. It may be stated simply in terms of periods of midwinter sunshine over the face of adjacent buildings. The requirement of visual privacy—or protection from being overlooked—cannot, however be so simply translated into regulations. It can be related to the type and orientation of windows and the type of rooms behind windows, but it must be expressed as a distance. An effective formulation of the privacy requirement would be complex and difficult to understand as well as to administer.

An alternative to formula-type setbacks is the English system of "daylight Codes" designed to ensure sufficient daylight within buildings.

Severance: A term often used when a new road operates as a barrier to movement; the communities on either side are severed from each other. The severance may be physical or it may be psychological. The severance is physical when the barrier is impassable or operates to inhibit trips. For example, freeway standards demand that there should be no access by pedestrians, and controlled access by vehicles. Person-proof fencing may be constructed along the route; the communities on either side will lose cohesion.

Psychological severance may arise despite the provision of crossings, including bridges and underpasses. People may still feel cut-off from areas with which they previously had easy contact. Elevated roads in particular contribute to this sense of separation.

Sewage Treatment: The modification of sewage to make it more acceptable to the environment. Sewage treatment may be divided into four main stages:

> Primary Treatment. The removal of suspended matter by physical and mechanical means, e.g., screening, grinding, flocculation or sedimentation.
>
> Secondary Treatment. The removal of finely suspended solids and colloidal matter, and the stabilization and oxidation of these substances and the dissolved organic matter by means of air and the activity of living organisms.
>
> Tertiary Treatment. The attainment of higher effluent standards for many purposes.
>
> Sludge Disposal. The disposal of the suspended matter removed.

A decision as to the stages of treatment to be adopted depends on what is to become of the final effluent. A town situated close to the sea may discharge its sewage with suitable pre-treatment at a suitable distance out to sea; this approach is known as disposal by dilution. Where conditions are not satisfactory for this method of disposal, further pretreatment becomes necessary. With regard to inland towns, the final effluent which will be discharged into a watercourse undergoes at least two stages of treatment, and probably three.

The safe collection and disposal of domestic sewage remains one of the major tasks facing the developing countries. In many areas of the world, human excreta is still being disposed of in shallow latrines, into watercourses, or on the surface of the ground. In these areas, water pollution is still biological in the main, whereas in developed areas chemical pollution becomes much more prominent. The significance of effective sewerage systems in promoting health cannot be underestimated. As Barbara Ward has commented in *The Home of Man*, "Nothing... so reduced the death rate of the 19th century poor as the sewerage systems built at the instance of the 19th century rich" (Ward, 1976, p. 10).

In Egypt, the population of 43 million is entirely reliant upon the Nile. The existing sewerage system in Cairo was laid down by the British at the beginning of the 20th century to serve 2 million people. By 1981, the population had grown to 12 million. Hence burst sewage pipes have been a regular feature of life in Cairo.

In December, 1982, a main pipe in the Nile-side suburb of Giza broke causing a torrent of sewage to flow for 10 days. Earlier in the year, erupting sewage in a southern slum of Old Cairo led to riots. In 1981, an aid and cooperation agreement was signed between the United Kingdom and Egyptian governments to help finance massive renovation and extension works to Cairo's sewerage system, with completion of works by 1990 at the earliest.

Shift and Share Analysis: A technique using a comparison of the growth rate of some element, such as a region or city, within a system, with the growth rate of some relevant larger aggregate, such as the national economy. Differences in growth rates may be explained in terms of two components:

> Structural component, which describes the tendency for (say) a metropolitan area to grow slowly or quickly because of its own population composition or economic structure. For example, it may possess fast growing industries.

> Regional component, which describes the tendency for an area's growth rate to be influenced by factors not accounted for in its population or economic composition.

A region may enjoy a comparative advantage by virtue of its special location with respect to markets, raw materials, labor and other factors such as inputs from nearby firms.

Singapore: See *Keep Singapore Clean Campaign.*

Sinkholes: Depressions in a limestone region formed by the solution of surficial limestone or by the collapse of underlying caves.

Site Planning: The art of arranging buildings and other structures on land to achieve function and efficiency and, as far as practicable, harmony. Site plans are necessary at an early stage whenever substantial groups of buildings or structures may be erected, embracing factories, mine developments, shopping centers, cultural centers, housing estates, and redevelopment schemes. Site planning lies along the boundaries of architecture, engineering, landscape architecture, economics, and city or urban planning. Site plans locate structures and activities in three-dimensional space and, when appropriate, in time.

Site Rehabilitation Objectives: Objectives often defined for works sites or exploration sites that have been cleared of vegetation or otherwise disturbed to ensure satisfactory rehabilitation; these objectives should have been approved before the commencement of operations. Rehabilitation objectives include:

- Reconstruction or approximation of the original topography or creation of a final topography which blends with the adjoining undisturbed landscape,
- Prevention of undesirable water ponding by creation of a suitable surface drainage pattern which also inhibits erosion,
- Restoration of a soil condition capable of supporting plant life similar to that of the surrounding landscape,
- Replanting with local endemic plant species, with particular regard to erosion control and visual reinstatement,
- Maintenance of the rehabilitated area until such time as the vegetation cover is self-sustaining and the surface is stabilized against erosion,
- Removal from the site of all unnecessary plant, equipment and construction materials and debris.

Skyways, Skywalks, Pedways: Different names for elevated, enclosed walkways connecting buildings in a city core or central business district (q.v.); crossings above congested streets offering pedestrians a safer, weather-protected, alternative to city sidewalks. The earliest systems were developed in the 1960s in the northern states of America and Canada. The largest system in the United States is in Minneapolis-St. Paul; the twin cities have over 50 skywalks. Major systems are also to be found in

Louisville, Charlotte, Des Moines, Rochester, Dallas, and Edmonton. Most cities in the world have skyway systems, though much reliance is still placed on pedestrian precincts to separate pedestrians from traffic.

Slum Clearance program, UK: The massive nationally organized and locally-implemented program for the clearance of dwellings unfit for human habitation and the rehousing of the occupants conducted in Britain between the two World Wars, and again from 1954 onwards. The local authorities were required to review the condition of their older houses and estimate the number of unfit and substandard dwellings in their areas; they were then to prepare programs for clearance and programs of new construction to provide for all displaced residents requiring accommodation. Between the two World Wars some one-and-one-quarter million people were so rehoused. Between 1956 and 1982 well over one million houses were cleared and some 4 million people rehoused.

Compensation (q.v.) to the owners of unfit houses acquired for demoliton was generally restricted to the market value of the cleared site, although additional payments could be made in some instances. Exchequer subsidies were payable to the local authorities in respect of every dwelling built to rehouse a family from an unfit house. The acquired sites were frequently redeveloped by the local authority for its own housing schemes.

Smoke Control Area, UK: Introduced under the British Clean Air Act 1956, a concept of a geographical area proclaimed smoke-free.

The introduction of smoke control areas has been at the discretion of each local authority, and this accounts for the very uneven progress in introducing smoke control areas even in the most densely populated and industrialized areas. Many local authorities have completed, or nearly completed, their programs, while adjacent local authorities may be as much as 20 years behind. The first smoke control order was introduced by the County Borough of West Bromwich, which encountered the first organized opposition by the public to the introduction of these areas. However, opposition tended to subside once the benefits were realized. By 1970, about 75 per cent of all premises in Greater London were covered by smoke control orders; however, for England as a whole, only about 50 per cent of premises were so covered. While solid progress was commendable, there was little doubt that the whole of the "black areas" could have been covered earlier by smoke control orders had local authorities all pulled together and enthusiastically implemented the recommendations of the government's Committee on Air Pollution (under the chairmanship of Sir Hugh Beaver) which reported in 1954. By 1985, about 7.7 million premises were covered by some 5,240 smoke control orders. The land area involved amounted to about 739,000 hectares mainly in the conurbation areas. It is thought that local authorities will eventually cover around 12 million premises.

As regards pollution levels in residential districts, smoke has decreased as a whole by between 70 and 80 per cent since 1956 and the amount of sulfur dioxide in the air from coal has been more than halved. This is a considerable achievement especially as over four-fifths of the total smoke was emitted from the chimneys of domestic dwellings. In London over 95 per cent of the smoke control program has been completed. As a result December sunshine has increased by 70 per cent, and visibility on an average winter day has improved from 2½ to 6 kilometers. Sheffield, which quickly completed its smoke control program, is now reckoned to be the cleanest industrial city in Europe, instead of one of the dirtiest.

Social Forestry: The growing of trees for fuel as well as raw materials. In 1979, the World Bank financed a social forestry project in the Uttar Pradesh state of India; the project will make possible an increase of at least 20 percent in the fuelwood supply for some 6 million people. Most of the three plantings are along roads, railways and canals, eliminating the need to walk long distances for fuel. A greater abundance of fodder will result in greater milk production; organic wastes now used for fuel will be feed for agricultural use. Forest-based cottage industries will be provided with more raw materials.

Social Impact Assessment (SIA): An assessment of the impact on people and society of major development projects; social impact assessment is often a weak point in environmental impact assessments. Social impacts may be defined as those changes in social relations between members of a community, society, or institution, resulting from external change. The changes in social relationships may lie in the areas of:

- The consequences of severance (q.v.), both physical and psychological;
- General life-style;
- Group relationships;
- Cultural life (language, rituals, dress);
- Attitudes and values;
- Obligations to kin, marriage patterns, and visiting patterns;
- Social tranquillity (disrupted, for example, by the arrival of a large all-male work-force);
- Relocation of large populations.

In the past, many proposed major projects and plans have not been assessed for social impacts; in consequence, adverse social impacts have not been ameliorated by appropriate measures adopted and prepared in advance. The results have been most notable in respect to hydroelectric dams where it has been necessary to relocate large populations into alien environments. The results have been social discontent, unhappiness, increased illness, and a loss of productivity resulting in loss of income. The closer proximity of different social groups with contrasting cultural characteristics may intensify the adverse effects. On the other hand, some of the social changes may be beneficial to some groups as with successful slum clearance programs.

Social Indicator: An attempt to measure by means of an index the degree of human welfare or quality of life in a given area, national or regional. The need for such an index has arisen from widespread discontent with the use of the concept of gross national product as a measure of human well-being. An integrated social indicator reflects a set of social and environmental indices, as well as conventional economic indices. It is influenced by such factors as real income; employment, housing, educational and cultural characteristics; ease of access to social, community and transport services; recreational opportunities; depletion of non-renewable resources; air, water and noise pollution. In regional terms, a social indicator becomes a basis of comparison, and an important element in the development of urban and regional policies and the determining of priorities within and between programs. Subjective social indicators deal with the degree of satisfaction felt by people with various aspects of their lives. Work on social indicators, both objective and subjective, has been conducted in several countries.

Socioeconomic Elements: Matters or elements which have a sociological and economic relationship; such as people's way of life, their housing, welfare and leisure in relation to business, the trades and crafts, employment and the resources of the area.

Sociosphere: The area of study of the social scientist; it is analogous to the hydrosphere as the area of study of the oceanographer, the biosphere (q.v.) as the area of study of the biologist, among many others, and the lithosphere as the area of study of the geologist. The sociosphere embraces people, their roles and patterns of behavior, their organizations and groups, and social interactions. See *Environment*.

Soil Erosion: The loss of soil as a result of natural and human activities. Natural erosion is the starting point of pedogenesis (creation of soils) which is indispensable to sustain human life; on the other hand, accelerated erosion due to bad soil management, deforestation, brush and forest fires, overgrazing and poor agricultural practices destroys the soil, with far-reaching consequences. Among the most important measures for combating soil erosion are re-afforestation, reduction of overgrazing, return of manure to the land, terracing of mountain slopes and contour plowing.

Solid Waste Disposal Act, 1965, US: An Act authorizing a research and development program with respect to solid waste disposal in the United States. It sprang from a recognition that continuing technological progress and improvement in the methods of manufacturing, packaging, and marketing of consumer products had resulted in an ever-mounting increase in, and a change in the characteristics of the materials discarded by the purchasers of such products. In addition, increasing urbanization had presented many communities with serious financial, management, intergovernmental, and technical problems in the disposal of solid wastes. Improper methods of disposal has resulted in scenic blights and created serious hazards to public health. A failure to salvage and reuse material had also resulted in unnecessary waste and depletion of natural resources.

The aim of the Act was to initiate national research and development programs for new and improved methods of disposal, with provision for recovery and recycling. Technical and financial assistance was to be provided to state and local governments in the development of programs.

The law was reauthorized in 1970, a special report on the problems posed by hazardous wastes being requested. Following this report, the US Congress responded by enacting the Resource Conservation and Recovery Act, 1976 (q.v.). The RCRA established the statutory framework for comprehensive federal and state regulation of hazardous wastes. However, the Act included provision for developing methods for the disposal of solid wastes which are environmentally sound and which conserve valuable resources. Federal grants have encouraged each state to develop its own solid waste management plan. In order to receive approval, each state plan must meet certain minimum requirements which emphasize the closing or upgrading of all open land dumps and a prohibition on the formation of new ones.

Under the RCRA amendments of 1984, the Environmental Protection Agency gained enforcement authority; the EPA is to step in if the states fail to meet deadlines for developing programs to ensure that their solid waste management facilities comply with the RCRA's existing and added criteria.

South Africa: See *Overgrazing*.

Southern California Association of Governments, US: A regional planning agency which prepares and coordinates regional plans, and forecasts regional population growth and growth locations. It comprises representation from six counties embracing 105 cities.

Spatial Allocation Models: A fashionable tool in planning studies; models which simulate the impact upon land use of changes in population, employment, or in the transport network.

Special Environmental Agency (SEMA), Brazil: The Secretaria Especial do Meio Ambiente (SEMA) or Special Environmental Agency established as a federal agency in 1973 within the portfolio of the Ministry of the Interior. Its task was to identify environmental problems and provide solutions, recommend standards or objectives in respect of pollution, and cooperate with other organizations to preserve fauna and flora threatened with extinction.

The Secretary of the Agency has powers to approve plans and projects, delegate functions, propose to the Minister of the Interior legal and regulatory measures thought necessary to control pollution, prevent contamination, and protect the environment. The Minister has issued a number of regulations since 1973 relating to air and water pollution, and solid waste disposal.

For administrative and management purposes, in 1975 Brazil was divided into 13 regions; the metropolitan areas of Sao Paulo, Rio de Janeiro, Belo Horizonte, Recife, Salvador, Porto Alegro, Curitiba; the others were the catchments of Cubatao, Volta Redonda, Medio e Baixo Tiete, Paraiba do Sul, Rio Jacui e Estuario do Guaiba, and Pemambuco.

The carrying out of a Program of Ecological Stations, by the Special Environmental Agency (SEMA), was linked with the Second Plan for National Development and to the Second Plan for Scientific and Technological National Development of Brazil.

The objective of this program was to protect natural environments, representative of the chid ecosystems, and promote comparative ecological studies between the protected areas and those nearby inhabited by people. Basic ecological data has been obtained from this program, data important to the development of national policies and technologies for using, controlling, and preserving the environment, and using natural resources sensibly. There are sixteen established ecological stations so far, with other prospective areas being studied.

In addition to the federal agency, there are also environmental protection agencies at state level, for example: *Companhia de Saneamento do Parana*, for Parana, *Companhia Catarinense de Aguas E Saneamento*, for Santa Catrina, and the *Companhia Rio Grandense de Saneamento*, for Rio Grande do Sul.

Special Use Corridors: Areas of land set aside for future use by public authorities and private enterprise in providing for such things as major roads, pipelines, transmission lines and other public utilities and for meeting regional open space requirements. Corridors also serve as physical and visual boundaries between or adjacent to existing and future urban areas. They are essentially long term proposals to cater for demands arising from the expected population of the region and linkage requirements between cities and towns.

Provision for corridors is necessary because experience demonstrates that personal hardship and high social and economic costs result if land for future public purposes is not reserved early, and needs to be acquired later when the population density is much higher.

The size and location of the service corridors will depend upon the area demands of various known uses and likely future uses; in some cases they will need to be several kilometers wide and in others perhaps as little as 100 meters to allow for a limited number of linear service lines to pass through an urban area or along its boundary. They must also provide some flexibility and room for future unforeseen requirements.

Spot Zoning and Rezoning: Zoning (q.v.) which affects only a particular piece of property or a small group of adjoining properties. In some instances a spot zoning for a particular purpose may serve the good of the community as a whole; in other cases it may merely benefit or profit a particular owner. Generally, zoning and rezoning should take place within the framework of a comprehensive plan (q.v.) intended for the benefit of the community as a whole; but from time to time zonings and rezonings appear to be arbitrary or capricious. In New York, in 1968, an appellate court struck down a small area rezoning because it was not based on any underlying policy or comprehensive plan; the rezoning authority failed to show that theproposed amendment was "for the benefit of the community as a whole, following a calm and deliberate consideration of the alternatives" (Udell v. Haas (1968) 21 N.Y. 2d 463).

Star Town: A form of town development designed to deal with growth and contain urban sprawl; compact satellite towns are developed within easy access of the main town center and in a peripheral ring around it.

State and Local Air Monitoring System (SLAMS), US: A nationwide uniform monitoring network for air pollutants established by the US Environmental Protection Agency (q.v.) under the Clean Air Act, 1977. Information from the system has been collected and stored in the EPA system for Storage and Retrieval of Aerometric Data (SAROAD). A subset of the system known as the National Air Monitoring System (NAMS) has been established for use at the national level for analysis of the status and trends of air quality throughout the nation. See *Pollutant Standards Index (PSI)*.

State Development and Redevelopment Plan, New Jersey, US: A plan prepared by the State Planning Commission (q.v.), initially in 1985 with updating every three years thereafter. It is essentially an integration of state and local plans, recommending land uses and infrastructure development. It is not a state imposed policy, nor legally binding on state agencies. It is designed to help municipalities to meet their Mt. Laurel decision (q.v.) obligations.

The plan provides a guide for state-level capital investments, a context for state functional plans and regulations, a basis for reviewing proposed federally-funded projects, a basis for impact analyses of large-scale subdivisions and projects, a context for municipal planning (as provided in the Municipal Land Use Law), a framework for county and regional plans, and a means for fostering coordinated land use planning between all levels of government.

State Development Plan, Maryland, US: A plan for the State of Maryland, based on studies of social, economic, physical and environmental conditions and trends, aimed at the coordinated development of the State. The preparation and amendment of the plan has been the responsibility of the Department of State Planning.

The plan contains recommendations for the most desirable general patterns of land-use within the state based on the best information concerning topography, climate, soil and underground conditions, water sources and bodies of water, and other natural and environmental factors, as well as in the light of information concerning the present and prospective economic bases of the state. It also contains recommendations relating to water and sewerage facilities and the relationship of land-use within the state to land-use in adjoining areas. It notes trends in industrial and other developments, analyses changes in respect of demography, reviews the habits and standard of life of the people of the state, and identifies areas of critical state concern.

In respect to circulation patterns, the plan makes recommendations for transportation and communication facilities, and major routes and terminals. It recommends the best general locations for major public and private works and facilities such as utilities, flood control works, water reservoirs, pollution control facilities, and military and defense installations.

The plan also contains a comprehensive analysis and evaluation of the capital plans and programs of the State departments, agencies, commissions, and instrumentalities; and reviews all federal grants, loans and services available to Maryland. It contains also a policy plan for the Patuxent River watershed.

State Environmental Planning Policies, New South Wales: State planning policies declared under the Environmental Planning and Assessment Act, 1979 (q.v.), of New South Wales, Australia. Such policies relate to matters of significance for the environmental planning of the state or a major segment of it. Policies introduced have related to development standards, minimum standards for residential flat development, medium density housing, strata subdivision of buildings for residential purposes, housing for aged and disabled persons, group homes, surplus public lands, and traffic generating developments.

State Implementation Plans (SIP), US: See *National Ambient Air Quality Standards (NAAQS), US.*

State Plan, Hawaii, US: Adopted by the Hawaii legislature in 1978, a comprehensive state plan setting forth goals, objectives and policies as well as implementation procedures to guide future development; the first comprehensive state plan enacted by any US state. The plan addresses policy areas such as population, the economy, physical environment, facilities systems, and sociocultural advancement. Its stated goals are to achieve:

- A strong viable economy, characterized by stability, diversity, and growth, that enables the fulfillment of the needs and expectations of Hawaii's present and future generations;

- A desired physical environment, characterized by beauty, cleanliness, quiet, stable natural systems, and uniqueness, that enhances the mental and physical well-being of the people; and

- Physical, social, and economic well-being, for individuals and families in Hawaii, that nourishes a sense of community responsibility, of caring, and of participation in community life.

Twelve "functional plans" have also been prepared to implement the goals, objectives and policies of the state plan. In addition, county general plans and development plans describe the desired population and physical development pattern for each county. Figure 17 illustrates the relationships between the state plan, functional plans, and the county plans.

State Planning Commission, New Jersey, US: An agency created in 1984 with the task of preparing a State Development and Redevelopment Plan (q.v.) and to assess long-term capital improvement needs; it also has the duty to provide technical planning assistance to local government. The agency has a membership of 21 including the State Treasurer and four other cabinet members, four legislators, six public members, and appointees from county and municipal government. The Office of State Planning is established in the Treasury. See *Mt. Laurel Decision, 1983; Planning Commission*.

State Pollution Control Commission, New South Wales: A state agency responsible for pollution control in New South Wales, Australia. The State Pollution Control Commission began operations in 1971. Under the State Pollution Control Act, measures are taken to control pollution and the disposal of wastes and to protect the environment from harm. Environmental standards are set and practical programs formulated. The commission also undertakes some environmental studies and initiates research. Formal inquiries have been held into controversial matters. The commission administers the Clean Air Act, the Clean Waters Act, and the Noise Control Act.

State Regulatory Controls, US: Legal controls enacted by a state of the United States for general application. In the State of New Hampshire, for example, state regulatory controls deal with wetlands protection, water supply, sewage disposal, erosion control, and air quality among many other matters. Public hearings are part of the permitting process for any major state project in these areas.

While Table 17 is related specifically to the State of New Hampshire, it represents a fair impression of the distribution of duties in most US states. It will be noted that there is no reference to city and regional planning; generally, responsibility for the making and enforcement of planning and land use regulations is vested by state enabling acts (q.v.) in the municipalities and counties. Within the framework of statutory guidelines and court interpretations, each municipality may shape and direct its own affairs.

Stormwater Runoff: That portion of rainfall which flows over the land surface directly into streams and lakes; it is distinguished from the portions of rainfall that evaporate or seep into the soil to recharge groundwater

280   Environmental Planning

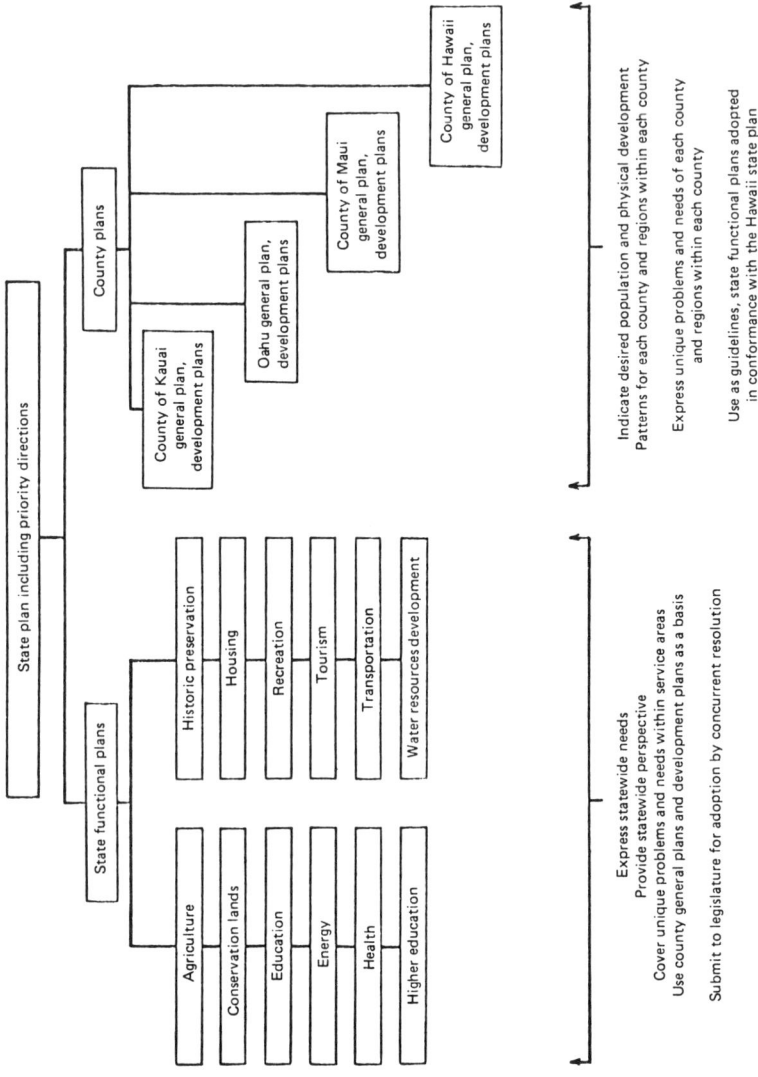

Figure 17: Hawaii: Relationship of state plan, state functional plans and county general plans.

Table 17: State of New Hampshire, US—State Agency Environmental Planning Activities, 1985

Regulated Activity	Description	Administrative Agency
Air Pollution	Construction and operation of all major, new sources of air pollution or modification of existing sources	Air Resources Agency
Airports	Siting and construction of airports	Aeronautics Commission
Archeological Excavations	Field excavations on State lands and on the bottom of State waters	Department of Resources and Economic Development
Boat Moorings	Boating and mooring sites within tidal waters or harbors of the State	Port Authority
Dams and Reservoirs	Construction or reconstruction of dams and reservoirs	Water Resources Board
Dredge and Fill	Dredging, excavating, mining, filling, transporting of forest products or undertaking construction in or on the border of surface waters of the State, or altering the characteristic of the terrain	Water Supply and Pollution Control Commission
Dredge and Fill	Construction, filling, excavation or dredging of surficial or sub-surface materials in areas adjacent to State waters	Wetlands Board
Driveways	Construction of roads and driveways connecting to public ways	Department of Public Works and Highways
Drive-In Theaters	Construction of outdoor theaters	Department of Public Works and Highways
Electric Power Facilities	Location, operation and construction of power plants	Bulk Power Evaluation Committee
Energy Facilities	Construction of energy facilities	Energy Facility Evaluation Committee
Hazardous Waste Facilities/Transportation	Construction, alteration or operation of hazardous waste treatment, disposal or storage facilities; transportation of hazardous materials within the State	Office of Waste Management

(continued)

Table 17: (Continued)

Regulated Activity	Description	Administrative Agency
Hydropower Facilities	Construction or reconstruction of dams and reservoirs for hydroelectric facilities	Water Resources Board
Junkyards	Operation and location of junkyards adjacent to Federal-aid highways	Department of Public Works and Highways
Mining	All mining of minerals and metal deposits on State lands	Department of Resources and Economic Development
Outdoor Advertising	Installation of advertising devices adjacent to Federal-aid highways	Department of Public Works and Highways
Pesticides	Commercial or private application of pesticides	Pesticides Control Board
Public Swimming Pools	Installation, operation or maintenance of a public swimming pool or bathing place	Water Supply and Pollution Control Commission
Road Construction Across Public Waters	Construction of public highways, access roads or private ways across a watershed tributary to a lake, pond or reservoir used for public drinking water	Water Supply and Pollution Control Commission
Sewage Treatment Facilities	Construction of any new public sewage installation or sewage treatment facility or repair of existing one	Water Supply and Pollution Control Commission
Solid Waste Facilities	Solid waste disposal, storage, treatment and processing sites	Office of Waste Management
Subdivision/ Waste Disposal Systems	Design and installation of subsurface sewage or waste disposal systems; subdivision of land	Water Supply and Pollution Control Commission
Timber Harvesting	Cutting of more than 50 percent of timber in areas adjacent to great ponds, streams, rivers, brooks and public highways	Department of Resources and Economic Development
Transmission Lines	Construction and maintenance of transmission lines on State owned lands and across public waters	Public Utilities Commission

(continued)

Table 17: (Continued)

Regulated Activity	Description	Administrative Agency
Underground Utilities	Installation and repair of underground utilities within limits of State highway rights-of-way	Department of Public Works and Highway
Waste Disposal	Discharge and disposal of sewage or waste into surface and ground waters of the State	Water Supply and Pollution Control Commission
Water Supply Systems	Construction or modification of any public water supply system	Water Supply and Pollution Control Commission
Wood Processing	Operation of any mill processing primary forest products	Department of Resources and Economic Development
Youth Camps	Any camp or vacation resort operating longer than 30 days	Water Supply and Pollution Control Commission

Source: New Hampshire, 1985.

supplies. Development increases the amount of impervious cover, such as roofs, roads, or parking lots, on a site; this reduces the amount of water that can be absorbed by the soil and increases the amount of stormwater flowing to streams and rivers.

Drainage systems for individual sites have been designed to remove stormwater as quickly and efficiently as possible. Pipes and concrete channels have been used extensively to speed the removal of water from the entire site, and hence more quickly to surface waters. The effects of many developments have been: more frequent and higher flood peaks downstream; more nuisance-type flooding of local roads and properties; lower stream flows during dry periods; lower groundwater levels; and lower water quality downstream because the fast moving stormwater flushes more salts, oils, and other pollutants from urban areas into the receiving waters.

Planning agencies should ensure that drainage plans for each individual site and development are compatible with plans for drainage throughout the basin; encourage the use of natural swales, retention ponds, and porous pavement to slow runoff from the site if possible to a predevelopment rate; restrict the amount of impervious cover allowed on a site; minimize the removal of vegetation during construction; and minimize runoff problems during construction. See *Floodplain*.

Strategy Plan: A diagrammatic plan with text illustrating broad principles and a general concept for population distribution and land use. It represents a policy statement and is based on surveys and certain assumptions on a regional scale, such as population growth. See *District Structure Plan; Structure Plan, UK; Structure Plan Preparation; Structure Plan Principles; Urban Structure Plan*.

Strategy Plan for Southeast England: A development plan approved by the Department of the Environment (q.v.) in 1971. Five major growth centers were planned away from the capital, London, while other areas were to be protected. London itself was envisaged as having a declining population. The region then sustained a third of the work-force of the United Kingdom with real incomes some 16 per cent above the national average.

Stream Valleys: The bottomland and slope areas along streams and ravines.

Street Furniture: Descriptive of such items as railings, lamp posts, litter bins, seats, signs, bollards, and many other free-standing items in any street scene.

Street Improvement Schemes: Schemes, usually voluntary, involving the collective redecoration of property, the removal of unnecessary clutter, the replacement of ill-designed or confusing signs, the improvement of lighting and street furniture (q.v.), and perhaps a revision of traffic flow. Such schemes require leadership, the creation of a representative committee to speak for the occupants and coordinate plans, perhaps the appointment of an architect, financial arrangements and supervision. The results can be striking and rewarding to all concerned.

Streetscape: The combination of roadway, sidewalk, driveway, building characteristics, and landscaping or other appurtenances which are located in or along a public street producing a general impression or appearance encompassing all or part of such characteristics.

Structure Plan: A plan which illustrates the basic land use distribution and communications network (such as main roads and railways) of the urban areas identified in terms of the strategy plan (q.v.). The structure plan provides a framework for planning at the district level. See *District Structure Plan; Structure Plan, UK; Structure Plan Preparation; Structure Plan Principles; Urban Structure Plan*.

Structure Plan, UK: An environmental plan formulated in Britain at the regional level, incorporating major policies. Only structure plans need to be approved by the central government; local plans are made and adopted by local government, subject only to conformity with the overall policies of the relevant structure plan. This separation of regional from local issues, introduced in 1968, represented a fundamental departure from the old planning system.

The change was in response to objectives relating mainly to three issues: the length of existing planning procedures, excessive central government involvement in local planning, and insufficient public participation in planning decisions. The role of structure plans is to set out, for the area covered, land use, transportation and environmental policies and proposals which should be formulated against the background of economic and social considerations. Structure plans are also the means of giving effect to regional strategies and for establishing a broad basis for detailed local planning. They are written documents, illustrated by diagrams.

Structure planning poses severe problems to British planners in relation to techniques. Structure plans must link the economic and social concerns of urban spatial form and land development in a broad sense; they must also be based on the preparation and elaboration of alternatives which need to be evaluated comparatively against stated aims. Structure planning then, has become more comprehensive in its procedures, involving problem appraisal, policy studies, forecasting, design of alternatives, evaluation, testing, implementation, monitoring and review. This has increased the complexity of the planning task. Planners have therefore been searching for ways to cope with complexity and been attracted by the many quantitative techniques, such as models and cost-benefit analyses, which have been evolved mainly to deal with the evaluation of alternatives.

Structure Plan Preparation: The stages of evolution of a structure plan, the main steps of which may be summarized as follows:

- Identifying the land which should be conserved,
- Identifying the land which is suitable for urban development,
- Assessing the influence which existing and committed development will have on expansion proposals,
- Estimating the type and range of facilities required in the area under consideration and examining the interactions between them,
- Determining the most suitable urban form and structures,
- Defining the principles on which to base the structure plan,
- Formulating objectives and policies,
- Allowing public participation at each planning stage,
- Estimating the costs and the benefits of alternative structures.

See *Public Hearings and Inquiries; Strategy Plan; Structure Plan; Structure Plan, UK; Structure Plan Principles.*

Structure Plan Principles: Principles which may be divided into plan principles which would apply no matter what form of structure plan is ultimately adopted, and physical or technical matters related to the land itself. An example of the former is "provide an efficient and effective transportation system appropriate to the total structure;" an example of the latter would be "development within flood liable areas is prohibited." Structure plan principles may be defined as follows:

- Ensure the efficient and convenient distribution of all land uses;
- Provide a reasonable choice of residential accommodation in future urban development;
- Avoid exclusionary zoning;
- Designate land for the widest range of public facilities appropriate to the structure plan scale;
- Provide an efficient and effective transportation system appropriate to the density and distribution of uses;
- Prohibit development within permanently flood liable areas;
- Relate the future urban design and form to the drainage capabilities of the area to ensure acceptable limits on flooding;
- Achieve a high degree of economic and social self-containment for the sector;
- Relate surface development to existence of economically viable mineral resources, to avoid sterilization;
- Relate surface development to areas of expected mine subsidence;
- Ensure that the location of use zones does not contribute to pollution;
- Ensure that the form and scale of urban development does not give rise to pollution of existing waterways and water bodies;
- Prevent development in areas of high landscape value, and ensure a compatible range and intensity of use within areas of lesser landscape where some development may be allowed.
- Retain sufficient flexibility in the design of future urban areas to allow for unforeseen intensification of any land use or communication link;
- Relate the structure plan to planning in contiguous areas;

- Prevent urban development in areas where adequate sewerage and drainage systems are not available;
- Ensure that location can be coordinated with the provision of services;
- Ensure that the proposed land uses result in economies in the provision of infrastructure (q.v.);
- Prevent development of areas which could be required for water storage purposes for the population envisaged;
- Formulate appropriate land use controls to restrict development in catchment areas;
- Wherever possible restrict urban development to areas which are not required for agriculture or primary production;
- Protect the recreational and tourist use of the coastal areas.

Subdivision: Simply a division of land into parts for residential purposes; the basis of a housing estate. The term may be applied also to the subdivision of a building into individual ownerships, i.e., strata titles. Subdivision regulations or controls define the standards for subdivision development; subdivision regulations and ordinances in many United States cities have specified that new streets conform to the overall city plan and that new lots be properly laid out for building sites. Developers have been required to provide land for streets, playgrounds, and school sites. The duties of developers have progressively increased over the years, and now embrace most and often all the requirements indicated in Table 18.

Most countries have adopted regulations requiring the approval of plans for new subdivisions before the owner or developer is allowed to proceed with construction. Such plans prescribe the layout of streets and open spaces, access for traffic and protection from through traffic, the location and types of utilities, houses, shopping centers and public buildings. As in the case of zoning, subdivision control gives effective direction to city growth only where it is combined with overall planning.

While developers tend to accept basic requirements readily enough, opposition arises when the requirements are perceived to be serving the needs of a larger community or other development, e.g., when the roads are to be heavily paved or the water mains of excessive size. Furthermore, excessive standards may tend to push the price per lot beyond the reach of the market which may lie in the lower-income brackets. See *Zoning*.

Subdivision Application, Factors to be Considered: Factors to be taken into account by local government, in considering an application for the subdivision of land; these factors include:

- The size and shape of each separate parcel;

Table 18: Subdivisions—Contemporary Requirements for Developers

Roads, paved or sealed

Suitable intersections with main roads

Curbing and guttering

Sidewalks

Parking strips

Turning circles

Drainage for stormwater

Street lighting

Street signs

Noise bunds

Buffer zones

Rights of way

Easements

Landscaping and tree planting; preservation of special characteristics

Playgrounds

Water mains

Fire hydrants

Water supply

Electric power

Open-space and parks

School-sites

Financial contribution towards nonland capital costs such as community facilities, or water supply or sewerage headworks

- The length of road frontage of each separate parcel;
- The situation and planning of the separate parcels in relation to public convenience, present and prospective;
- Whether the district is or probablly will be suitable as a residential district;
- The draining of the land, the drains proposed to be constructed and the drainage reserves and drainage easements to be provided;
- Whether the land is subject to flooding or tidal inundation;

- Whether the land is or probably will be subject to subsidence or slip;
- The standard number of houses to the hectare fixed by the council;
- The amount of land to be provided as a public reserve out of the land to be subdivided;
- The provisions of any environmental planning instrument, including the use to which the land is proposed to be, or may be, put following the subdivision, in accordance with or consistently with that instrument; and
- Whether any trees on the land should be preserved.

Sudan, The: See *Infant Mortality Rate.*

Sudbury, Ontario: The site of a major smelting operation; in August, 1980, the Ministry of the Environment, Ontario, Canada, issued a directive to Inco Limited in Sudbury limiting its sulfur dioxide (q.v.) emission to 2,500 tons per day, effective immediately, and to 1,950 tons per day by December, 1982. These reductions were accomplished on schedule.

Sulfur Dioxide: A colorless, pungent gas formed when sulfur burns in air. It is considered to be one of the most important air pollutants either alone or in combination with other gases and substances. Most of the sulfur dioxide in the general atmosphere of cities comes from the combustion of the sulfur present in most fuels. All the sulfur in oil, and from 80 to 90 per cent of that in coal and coke is emitted from the stack or chimney as sulfur dioxide, with a small proportion being already converted to sulfur trioxide.

The discharge into the atmosphere of large quantities of oxides of sulfur has led to a problem known as "acid rain" (q.v.). Every year, some 15 million tons of sulfur dioxide are emitted into the atmosphere from sources in the northeast of the United States of America, and some 2 million tons from sources in Ontario, Canada. In addition, many millions of tons of oxides of nitrogen are also emitted. These substances tend to oxidize to various sulfates and nitrates, becoming dissolved in rain and snow. See *Table 19.*

Studies by the Organization for Economic Cooperation and Development (OECD) (q.v.) and the Ministry of the Environment, Ontario, Canada, have confirmed that sulfur pollution can originate from very distant sources. Studies reveal that in Europe, sulfur is not only a local but a transfrontier pollution problem of common concern to all European countries.

Some countries, such as Norway, import much more pollution than they cause themselves. It has been shown that precipitation over Scandanavia, eastern and northeastern United States, and Ontario, has become more acid over time. The Scots complain of pollution from the English.

Table 19: Sulfur Dioxide Emissions—Northeastern North America

Source	SO$_2$ Emissions x 10^3 Tons/Yr Urban	Utilities
Canada		
Ontario	1,741.0	191.0
United States		
Ohio	578.4	2,338.3
Indiana	433.5	1,666.4
Kentucky	74.2	1,387.7
Illinois	332.4	1,255.6
Michigan	189.4	1,158.8
Pennsylvania	535.5	1,119.6
West Virginia	149.7	1,020.9
Tennessee		721.4
Missouri	226.4	499.0
District of Columbia		216.5
New York	636.8	209.5
Wisconsin		192.1
Maryland	129.5	
Massachusetts	139.9	
Virginia	141.6	
U.S. Total	3,567.3	11,785.8

Source: U.S. Environmental Protection Agency, 1984.

The long-range transport of air pollutants is not limited to sulfur and nitrogen compounds. For instance, an analysis of Greenland ice layers has revealed that tetraethyl lead concentrations from gasoline have multiplied by 50 during the period of 1925-75. See *Convention on Long-Range Trans-Boundary Air Pollution*.

Superfund: See *Comprehensive Environmental Response, Compensation, and Liability Act, 1980*.

Surface Mining Control and Reclamation Act, 1977, US: A measure passed by the US Congress in 1977 establishing the first uniform federal controls over surface or strip mining, which had previously been a state matter.

Briefly, the law imposes requirements in five areas:

- Under *performance standards* operators are required to: restore stripped land to a condition capable of supporting the use that it was capable of supporting before mining; restore land to original contours; minimize disturbances to hydrologic balance; and, permanently revegetate the stripped land.

- Under the *permitting* and *bonding* sections, operators must demonstrate their financial ability to conduct reclamation activities and purchase insurance covering personal injury and property damage.

- Under the *inspection* and *enforcement* provisions, one complete inspection every three months and one partial inspection every month, without advance notice to the operator, is required.

- Under the *citizen rights* section, public participation is allowed in nearly all phases of the law's implementation. This participation includes the right to be involved in administrative and judicial proceedings, to inspect surface mining activities, and to challenge any regulations in court.

- Under the *designation* of *unsuitable lands* provisions most surface mining is precluded if such operations: are not compatible with existing landuse plans; affect fragile or historic lands; affect renewable resources lands; or, affect *natural hazard* lands. Natural hazard lands include areas subject to landslides, cave-ins, severe wind or soil erosion, avalanches, or frequent flooding. Surface mining is now almost always precluded near national parks, wildlife refuge systems, and wilderness preservation systems.

The cumulative costs for strip-mine reclamation were expected to total more than $15 billion between 1979 and 1988 (Council on Environmental Quality, 1980).

The Surface Mining Control and Reclamation Act 1977 is administered by the Department of the Interior. The Department may delegate enforcement to individual states whose reclamation plans meet minimum standards.

Surface or Strip Mining: Or open-cut mining; a technique of mining employed wherever coal is relatively near the surface; the overlying earth and rock are mechanically stripped to expose the coal, which is then removed with or without blasting. In many countries, including the United States of America, Canada, and Australia, strip mining has demonstrated great economic advantages while presenting graver environmental problems than underground mining.

However, in the United States strip-mined coal has one environmental advantage. Much of the coal in the east or Appalachian Region traditionally won by underground mining tends to have a relatively high sulfur content. Much of the coal in the west, however, is close to the surface and can be readily strip-mined; this coal tends to have a relatively low sulfur content. Coal from the western States such as Montana and Wyoming usually has a sulfur content of less than 1.0 per cent; it has therefore a distinct advantage for use in areas where pollution by sulfur dioxide is significant. Furthermore, only a small portion of the strip-mined coal needs to be cleaned before combustion in contrast to much deep-mined coal. There has been a rapid increase in surface or strip-mining in the United States, reaching a 1984 production of 800 million short tons.

The largest amount of thermal coal produced in Canada originates from strip mines, accounting for over 80 per cent of Canada's total 1984 thermal coal production used for power generation. It is anticipated that this technique will continue to be the major form of thermal coal production in Canada. By the year 2000 about 75 per cent of Canada's thermal coal production intended for domestic power generations will be obtained from strip mining operations.

In the prairies region, the mining of coal often takes place in major agricultural areas. In Alberta, for example, two-thirds of the coal fields in the central and southern districts are on arable land, most if not all of which is presently cultivated, and to a large degree privately owned. Thus a preplanned coal mining and reclamation program is necessary for each new development. Planning needs to consider the land, its present and future use, and the social impact on farming communities. To ensure that environmental disturbances are minimized, all the provincial governments in this part of the country have developed land reclamation regulations.

It seems likely that land which has been subjected to strip mining will be out of normal production for about five to nine years; but there is still some uncertainty about the extent to which the reclaimed land will return to the level of productivity of its premined state.

Strip mining presents a number of environmental and pollution problems; these may be countered by a variety of methods, not all of them necessarily wholly successful. Figure 18 presents a scheme for the control of water flows in an open-cut mine.

Swales: Slight depressions, sometimes swampy, in the midst of generally level land.

Sweden: See *Acid Rain; Car Scrapping Act; Environment Protection Act, 1969; Greater Stockholm Finger Plan; National Environment Protection Board; National Franchise Board for Environment Protection; National Parks; National Physical Planning; Nature Conservancy Act, 1975; Nordic Environment Protection Convention; Right of Common Access; Water Act, 1984.*

Switzerland: See *International Commission for the Protection of the Rhine; Office of Environmental Protection.*

Sydney Metropolitan Waste Disposal Authority: An authority established by the New South Wales Government, Australia, in 1971 to introduce in the metropolitan area of Sydney an environmentally acceptable

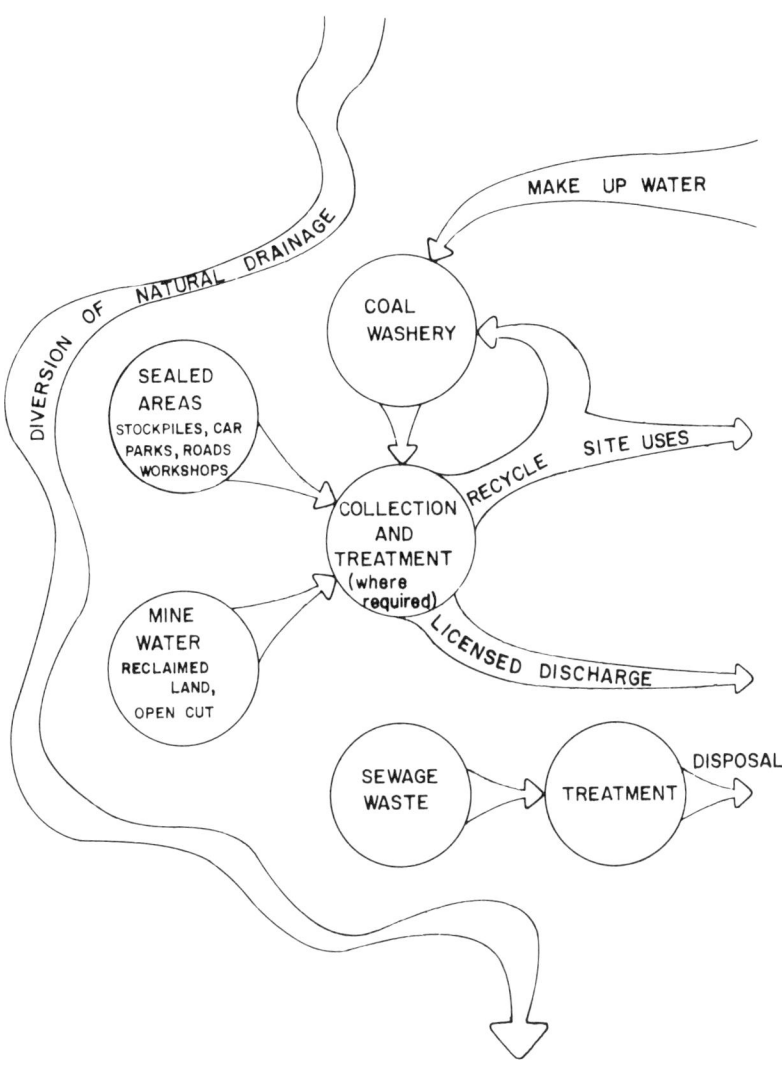

Figure 18: Planned water flows in an open-cut mine. Source: State Pollution Control Commission, New South Wales, Australia.

solid and industrial liquid waste management system. The system was to embrace the collection, transportation, and disposal of wastes. In 1985, the system embraced seven major landfill sites, eight regional transfer systems, an incinerator and an aqueous waste treatment plant. The authority has been high successful.

T

Taking Issue, US: An argument that may arise when zoning (q.v.) creates an excessive gap between the value of land for the use to which it is restricted and its value for more intensive development; an owner may claim the zoning is unconstitutional because it confiscates the increased value that might be realized if the land were put to its most profitable use. This is commonly called the "taking issue."

The argument is based on guarantees of the Fifth Amendment to the United States Constitution and certain state constitutions, e.g., Article 1, Section 10 of the Pennsylvania Constitution which states: "... nor shall private property be taken or applied to public use, without authority of law and without just compensation being first made or secured." However, well-conceived and precise land use regulations are generally upheld by the courts. It is in particular instances only that the courts have found that the cost of a particular land control was so burdensome on a landowner that it amounted to an illegal taking. Land use controls should withstand a taking challenge where the owner of the land is left with a reasonable, profitable, use, even though it is substantially less than the most profitable use, and the restriction is necessary for the public welfare.

Tennessee, US: See *Safe Growth Plan; Tennessee Valley Authority.*

Tennessee Valley Authority, US: An authority created by the US Tennessee Valley Authority Act, 1933, to promote the use of electric power for agricultural, domestic and industrial purposes in the Tennessee Valley; it formed part of the New Deal and was the first major piece of regional planning on a truly large scale attempted in the United States. As an incidental to its main purposes, the Authority achieved much in the area of water management and improved agricultural practices.

Texas, US: See *Acid Rain; Great South Western Industrial Estates.*

Thailand: See *National Environment Board.*

Three Mile Island Incident, Pennsylvania, US: An incident at the Three Mile Island nuclear power station, near Harrisburg, Pennsylvania, in March, 1979. The station comprised two plants, each of about 800 megawatts, owned by the Metropolitan Edison Utility. For more than 20 years, leading nuclear scientists had been assuring critics and skeptics that a major disaster at a nuclear power station, though theoretically possible, could hardly occur in practice due to the stringent security precautions against all conceivable mishaps.

At 4 A.M. on Wednesday 28 March 1979 an incident occurred which shook this confidence, and resulted in a general emergency being declared in the area of the Three Mile Island plant. It arose from an unexpected combination of circumstances at the No. 2 plant. While the initial failure occurred at 4 A.M. it appears that it took until 7 A.M. for the plant operators to appreciate the nature and scale of the accident and report it to the state authorities.

The initial reaction by state authorities was to declare a state of emergency at 8 A.M.; however, this was cancelled by 11:30 A.M. The following Friday, following a deliberate discharge of radioactive material to the atmosphere, the State Governor (Richard Thornburgh) advised residents within a 16 kilometer radius to stay indoors. In addition, pregnant women and children within an 8 kilometer radius of the site were advised to leave the area. By Saturday this phase of evacuation had been completed. Subsequently, plans were made to evacuate population within an 8 kilometer radius of the plant and up to 24 kilometers in the down-wind quarter. It is thought that about 200,000 people left the area voluntarily.

The radioactive material released to the atmosphere in the incident was detected 16 to 33 kilometers away from the site, although no excessive amounts of radioactivity were detected, and no one suffered physical injury. The gravest risks presented by the incident were those of a massive gas explosion, or a reactor core meltdown. The incident at Three Mile Island demonstrated an immense potential hazard. In the actual event, the gas bubble subsided and the damaged reactor slowly cooled. President Carter, who visited the plant, ordered a full inquiry into the accident.

The Presidential commission of inquiry reported that the company operating the reactor lacked knowledge and expertise, and was highly critical of certain design features. It urged fundamental changes in organization, procedures and practices, and in the approach of the Nuclear Regulatory Commission (q.v.). The Nuclear Regulatory Commission subsequently imposed fines of $155,000 for safety maintenance, procedural and training violations. In November, 1979, the Pennsylvania Public Utility Commission ordered the Metropolitan Edison Company to show cause why its license to sell electric power should not be revoked.

Since the incident, which proved bad news for the nuclear power industry worldwide, governmental confidence has slowly returned. In October, 1981, the Reagan Administration announced that it was proposed to allow spent fuel from nuclear power stations to be reprocessed; the plutonium recovered could be used as fuel in future fast breeder reactors or, mixed with uranium in present day light water reactors. The Clinch River fast-breeder prototype would also proceed.

Threshold Costs: The costs of overcoming a physical, quantitative, or structural limitation in the growth of towns. Physical limitations may consist of marshland, steep slopes, or river restrictions, some of which may be overcome in whole or part by filling, draining, redirecting, or new bridge construction. Quantitative limitations may be set by the existing maximum capacities of public utilities such as sewage plants, drainage systems, water works and storage systems, and road systems; major investments are needed to meet the needs of expansion. Structural limitations may appear when a population outstrips the capacity of existing shopping centers; if space is limited new multi-deck shopping centers may be necessary perhaps involving an expensive redevelopment of the surrounding area.

Threshold costs imply exceptionally high injections of capital investment; the average costs of development should then decline until the next threshold is reached.

Tiering: The coverage of general matters in a broader environmental impact statement (q.v.), such as a national program or policy statement, with subsequent narrower statements of environmental analyses, such as regional or basinwide program statements or site-specific statements. The narrower statements would incorporate by reference only the general discussion and concentrate solely on the issues now arising.

Times Beach, US: An example of the consequences of environmentally hazardous chemicals (q.v.), in this case dioxin. In February 1983, the Reagan Administration offered to buy-out all 2400 residents of Times Beach, Missouri, after confirming that the town was too contaminated with the toxic compound dioxin to be safe for habitation. The dioxin levels had been found by the Environmental Protection Agency to be 100 times what was considered safe.

Times Beach was one of at least 100 communities in Missouri where dioxin mixed with waste oils was sprayed on dirt roads in the 1970s to keep down dust.

Town and Country Amenities Act, 1974, UK: An Act which imposes a duty upon local authorities to draw up proposals for the preservation and enhancement of defined urban conservation areas, and to submit such proposals to a public meeting. It introduced important new provisions for the basis of compensation when listed buildings are acquired compulsorily by a local authority for preservation. In addition, buildings could no longer be demolished in conservations areas without appropriate consents.

Town and Country Planning: In Britain, physical planning as it extends to the whole country; in the United States the usual term is "city and regional planning." Planning activities in Britain have included:

- Schemes for the redistribution of industry and the building of new towns,
- Slum clearance and development schemes,
- Urban renewal (q.v.),

- The creation of green belts to restrain the outward spread of towns and cities,
- The creation of national parks and the preservation of national monuments,
- The protection of heritage buildings,
- The provision of urban amenities in the countryside for the agricultural population,
- The preparation and approval of structure plans for county areas,
- The preparation and approval of local plans for municipal or local authority areas.

Town and country planning is also referred to simply as town planning, physical planning, spatial planning, territorial, urban and regional, and environmental planning (q.v.). See *Local Plans; Planning System UK; Slum Clearance; Structure Plan; Town and Country Planning Act, 1947; Town and Country Amenities Act, 1974.*

Town and Country Planning Act, 1932, UK: A British measure aimed at broadening the planning process by allowing local authorities (the local elected municipal councils) to prepare planning schemes for almost any kind of land. Comprehensive plans were to be drawn up within a three-year period and then approved by Parliament. A developer could, within this period, apply for planning permission under the clauses relating to "interim development control." By 1942, 73 percent of England and 36 per cent of Wales were subject to "interim development control;" but only 5 per cent of land in England and 1 per cent in Wales were under a fully operative scheme. At that time, there were 1,400 planning authorities often linked through joint committees. The Second World War inflicted considerable damage on British towns, and new construction ceased. Post-war hopes surrounded the concepts of reconstruction, finding expression in the Town and Country Planning Act, 1947 (q.v.).

Town and Country Planning Act, 1947, UK: A British measure that embodied the principle that all development rights belonged to the nation; it created machinery for the comprehensive planning of all land in England and Wales (there was parallel legislation for Scotland). The system created in 1947 obliged planning authorities to draw up "development plans," a combination of strategy and projection covering a period of twenty years. Development plans consisted of a written statement and a number of maps. Between them, these expressed intentions for development and intentions for preservation in the different areas under the planning authority's control. Plans were to encompass urban and rural areas.

Plans became the responsibility of the local planning authorities, primarily the larger units of local government, the counties and county boroughs. There were drawn up by planning departments, professional and expert administrations which the authorities were obliged to employ. After approval by the planning authority, the plans were submitted to the minister in Whitehall (the central government) who could accept them, reject them, or require their modification. Development plans had to be revised every 5 years. Between 1947 and 1951, the central government department was the Ministry of Town and Country Planning; from 1951 to 1970 the Ministry of Housing and Local Government; and from 1970, the Department of the Environment (q.v.).

Individuals and organizations were directly affected by the system in that they required "planning permission" for any development, subject to a number of exemptions. For large projects, planning permission could be divided into two halves, "outline" and "detailed" permission, in order that architectural effort should not be wasted on detailed plans for a non-starter. Anyone who was refused planning permission by a local authority could appeal to the minister. This could be done by written representation, and could result in a public local inquiry (q.v.). At such inquiries, an inspector from the ministry heard the case put by the developer and any objections to the development which other individuals or organizations might wish to put. The inspector would then report in writing to the minister who finally refused or allowed development. In the 1970s some 10,000 inquiries were held each year, with the inspector's recommendations being accepted in 95 per cent of cases. See *Town and Country Planning Act, 1968*.

Town and Country Planning Act, 1968, UK: A British measure aimed at speeding up the planning process by a distinction between "structure plans" which could be ordered, and must be approved, by central government, and local plans which do not need to go through the same process. A later document outlined seven functions for the structure plan:

"1. Interpreting national and regional policies.

2. Establishing aims, policies and general proposals.

3. Providing framework for local plans.

4. Indicating action areas.

5. Providing guidance for development control.

6. Providing a basis for coordinating decisions.

7. Bringing the main planning issues before the minister and public."

The public was to be consulted on both structure and local plans, enlarging the area for public participation (q.v.). See *Planning System, UK; Town and Country Planning Act, 1947*.

Town and Country Planning Act, 1971, UK: An Act which consolidated and reenacted most of the current planning legislation in Britain, including the Town and Country Planning Act, 1968 (q.v.) and the Civic

Amenities Act. It was augmented by the Town and Country Planning (Amendment) Act, 1972.

Town Development Act, 1952, UK: An Act which made provision for the relief of congestion in large industrial towns in England and Wales by encouraging the transfer of population from these areas to places suitable for expansion where employment was available. Centers brought within the scope of the Act included the already well established towns in the South of England of Basingstoke, Andover, and Haverhill.

Townscape: The complex of built and unbuilt spaces that comprise the urban landscape or urban environment; the townscape presents a variety of scenes and settings, e.g., elegant, distinctive, monotonous, friendly, forbidding, garish, bleak, and crude. It has been described as the visible expression of the collective activities and attainments of the inhabitants; a reflection of interactive influences.

Toxic Substances Control Act, 1976, US: A measure introduced by the US Congress to regulate environmentally hazardous chemicals (q.v.).

Under the Toxic Substances Control Act of 1976, the Environmental Protection Agency was authorized to regulate the manufacture, distribution, use, or disposal of chemical substances. The Agency was required to adopt rules requiring the testing of chemicals that presented a risk of injury, will be produced in substantial quantities, and reasonably be expected to result in extensive human or environmental exposures. The cost for such testing would be borne by the manufacturers. Chemical substances specifically controlled under existing legislation (pesticides, tobacco, food, food additives, drugs, cosmetics, nuclear materials, firearms and ammunition) were not included in the Act.

By November, 1977, the Administrator of the EPA was required to publish an inventory of existing chemicals. Any substance not on the inventory would then be considered a new chemical substance for which manufacturers must give 90-day notice of intent to manufacture. This procedure gave the Administrator the opportunity to evaluate the hazard-causing potential of the new chemical substance or significant new use and control of its introduction into commerce if deemed harmful. The 1985 inventory comprised over 60,000 chemicals.

Polychlorinated biphenyls (PCBs) were the only chemical substances specifically mentioned in the Act. The use, distribution, manufacture and processing of PCBs were prohibited after January, 1978, except for enclosed uses or uses approved by the Administrator. Manufacture was totally prohibited after January, 1979. Since "manufacture" was defined to include importation into the US, imports were also prohibited. Distribution was prohibited after July, 1979, except for PCBs being recycled for use in a manner not prohibited by the Act.

Subject to certain exceptions, the Act did not apply to a chemical substance manufactured, processed, or distributed in commerce for export from the United States, provided it was so labeled. An exception to this would occur if the Administrator of the EPA found that the substance to be exported presented an unreasonable risk to health within, or to the environment of, the United States.

Trade Waste: Any solid, gaseous, or liquid matter which is refuse from any industrial, chemical, trade or business process or operation, including any building or demolition work.

Traffic Management: The organization of a more efficient movement of traffic within a street system or city center by rearranging the flows, controlling the intersections, and regulating the times and places for parking.

Traffic Segregation: The separation of different types of traffic; a term most commonly used with reference to the separation of pedestrians from vehicles. Such segregation may be achieved by arranging the pattern of vehicle roads and pedestrian ways that they never cross, or by physical separation using overpasses, underpasses and skyways (q.v.).

Transfer of Development Rights (TDR), US: When all or a portion of the rights to use or develop a given piece of land are removed from it and transferred to another piece of land; in the process, the first piece of land has lost development rights, and the second has gained development rights that it did not have before. Usually in such a transfer, the second property owner pays the first for those development rights at a price which reflects the approximate value the rights would have had if the first owner chose to develop. In this way, the first owner makes nearly as much money on his land by not developing as would have been made by developing. The second landowner will consider the price paid for the development rights worth it if some development may then be undertaken which allows a recovery of the cost of development rights.

The unique feature of development rights transfer is that it does not rely soly on government severely restricting property rights through regulation nor on purchasing developing rights with public funds. Instead, the real estate market becomes the vehicle. It is however necessary for government to permit changes in the use of the property to which the development rights are transferred.

The main advantage of allowing the transfer of development rights appears to be in making agricultural zoning politically acceptable. Local governments sometimes hesitate to preserve large areas of agricultural land by zoning (q.v.) because of the risk of giving ground for a "taking issue" (q.v.) and the unpopularity among farmers of such measures. TDR has proved a way for farmers and other restricted landowners to have their cake and eat it too—to be paid for the development value of their land while allowing development to occur elsewhere. Once farmers have sold their development rights, agricultural zoning ordinances may be imposed without opposition.

The customary procedure is for a local government to create certificates of development rights which are made available to landowners in the agricultural district. Landowners who purchase these certificates are entitled to increased densities and increased floor areas in designated areas.

Transition Areas: Areas identified as extreme breaks in physical scale, character or use, and which require special design attention to ameliorate any adverse effects on adjoining areas.

Transport Modal Split: The proportional division between different methods or modes of travel of the total number of person trips within a study area. See *Transport Models; Transportation Study.*

Transport Models: Theoretical models which seek to describe and predict the travel patterns of large numbers of people in the light of existing information regarding land use activities, movement between them, and current trends within the study area. The model analyzes the reasons behind the decisions to make journeys (trip generation), the probable destinations of those journeys (trip distribution), the routes they will take (traffic assignment), and the means of transport that will be used (modal split). See *Land Use Prediction Models.*

Transport Plan: A compilation of policy statements, goals, standards, maps and action programs for guiding the future development of the various modes of transportation of municipality, county, or regions and its environs such as streets and highways, mass transit, railroads, air transportation, trucking and water transportation, and includes a major thoroughfare plan.

The circulation element of the General Plan (q.v.) for the City of Los Angeles sets forth objectives, policies and programs to guide the location and development of the city's transportation facilities linking parts of Los Angeles with each other and other parts of the metropolitan area. It is recognized that the transportation network will significantly shape the development of the city and the region although the private automobile, modified to conform to required air quality standards, will continue to provide the principal means of transportation in the foreseeable future.

However, it is conceded that traffic congestion remains a serious problem during peak hours. Public transportation is provided primarily by buses, which compete with the automobile for space, offer limited routes and time schedules and attract insufficient revenues; public transit facilities remain inadequate and the non-driver has limited mobility. However, a rapid transit system is now being developed to improve the situation and efforts are being made to meet the special needs of various communities.

The objectives of the Los Angeles Transport Plan are stated as follows:

- To provide an integrated transportation system coordinated

with land use which adequately accommodates the total travel needs of the community;
- To minimize the use of air polluting motor vehicles;
- To improve the safety and efficiency of all transportation modes, particularly during peak travel periods, giving priority to public transportation;
- To achieve within an exclusive right-of-way a rapid transit system as an effective alternative to the private automobile for trips between centers and also between the regional core and outlying suburban areas;
- To locate, operate and maintain transportation facilities to be compatible with adjacent areas;
- To promote improved design and appearance of transportation facilities;
- To minimize conflict between vehicular and pedestrian traffic;
- To utilize the transportation system as a tool in developing planned land use patterns so as to minimize detrimental effects upon urban life.
- To provide adequate local bus transportation throughout the metropolitan region, as a part of the integrated transportation system;
- To provide for the efficient movement of freight;
- To encourage the development of air rights over publicly owned rights-of-way in areas of high intensity where appropriate and consistent with public health, safety and welfare;
- To improve the city's highway and freeway system as a major component of the city's integrated transportation system;
- To encourage the use of bicycles as a viable means of transportation.

See *Los Angeles Smog.*

Transport Planning, Objectives of: The aims of transport planning as they have evolved during the twentieth century. They include the following:

- The reducing of the length of the journey to work;
- The reducing of journey time;

- The provision of reasonable access to opportunities within a region for the population of that region;
- Roads and public transportation should work in combination to handle the transport task of the region, rather than in competition;
- The improvement of the urban arterial road network to safely and efficiently carry all major traffic movement;
- The reducing of existing congestion on arterial roads;
- The reduction of through traffic on residential streets, especially truck traffic;
- The reduction of the conflicts caused by traffic passing through shopping centers;
- The improvement of accessibility to employment, shopping and recreation centers by private and public transportation;
- The enhancement of public transport operations to meet the needs of public transport users and reduce the demands for parking spaces in the city center;
- Provision for future traffic growth;
- The provision of routes for vehicles and tankers carrying inflammable, explosive, or hazardous loads, to avoid residential streets and densely populated districts;
- The progressive reduction of all heavy vehicle traffic through city, town, and village centers;
- To coordinate the movement of road, rail, water, and air public transport for the convenience of the public;
- To promote the safe movement of vehicles, and also promote the safety of passengers using public transport at all hours;
- To undertake research and the application of the results of research to improve the quality of roads;
- The achievement of cost-effectiveness in road provision and improvements.

Transport Policies and Programs (TPPs), UK: Required under the Local Government Act, 1972, annual programs that must be prepared by county councils which set out the policies and objectives for transport planning for the county for a period of 15 years ahead, together with a budget for both roads and public transport.

The TPP must be in accord with the board policies of the structure plan (q.v.); it is very significant in terms of physical planning in setting out the policies for both highway development and the infrastructure of public transport, as well as influencing the level of public transport through a revenue support subsidy.

The Transport Act, 1978, provided for the preparation of annual Public Transport Plans (PTPs) for the shire counties which take th epolicies of the TPPs a stage further and provide details of the financial arrangements with the public transport operators in the county.

Transportation Study: A study designed to show how much travel may be expected within the study areas in the future and how it is likely to be concentrated. It is conducted within a three part framework:

- A survey and analysis which assesses the present demand for movement, how it is met, and what relationships exist between the characteristics of the study area and the demand for movement;
- An extrapolation of such relationships to produce estimates of future travel demands, together with an outline of the possible ways of meeting those demands;
- An evaluation to assess which of the transportation proposals would provide maximum benefit to the community at minimum cost.

The base data may be collected by home interviews; traffic surveys, and special observations; from land use maps and socioeconomic data sources. Models are often complex and sophisticated, although the assumptions on which they are based are often weak particularly those relating to land use and economic forecasting.

Costing has become more sophisticated. At one time, investment costs were measured generally in terms of capital cost of construction plus the market prices for the land required for the right of way. Today, costs arising from the displacement of people and businesses, the opportunity cost of the land consumed, and the environmental effects, are also taken into account. The environmental effects include visual intrusiveness, noise and vibration, delays at pedestrian crossing points and danger to pedestrians.

The most intricate and involved stage in the transporation planning process is the construction and simulation of possible networks or systems which meet future movement forecasts within the broad objectives of the development plan. To assist in the design of these networks a large number of computer-aided transport models have been developed.

Undoubtedly, the transportation plan which emerges from a study will play a major role (perhaps a dominant one) in determining the nature and function of urban development. However, because of the uncertainties inherent in the transportation study (which some consider better regarded as an exploratory device) reviews of the plan become necessary as each major stage is reached.

The Los Angeles Regional Transportation Study (LARTS) embraces current and future traffic demands; the information is provided to local jurisdictions in an effort to coordinate highway and freeway developments. Studies were originally primarily concerned with automotive vehicles, but now extend to the mass and rapid transit needs of the region.

Trophic: Concerned with nutrition. An ecosystem has two basic trophic components: An autotrophic (self-nourishing) component in which fixation of energy predominates (e.g., plants), and a heterotrophic (other-nourishing) component in which utilization and decomposition of complex materials predominate (e.g., most animals).

Tropical Cyclone Warning System: A warning system developed by the World Meteorological Organization (q.v.) to assist in mitigating the effects of tropical cyclones in all regions. The meteorological satellite has marked a turning point in this development. Supplemented by ground radar stations the new system ensures that no major storm can now remain undetected for more than a few hours or even minutes, and its subsequent movement can be closely watched. Of course, advance warning may not be enough to avoid staggering losses.

To take but one, albeit a very tragic example, the existence of the particularly devastating tropical cyclone in the Bay of Bengal, which in November 1970 shocked the world by the unprecedented death toll of nearly a quarter of a million people and by the enormous damage to property which it caused, was clearly visible on satellite photographs 3 days before its onset, and its subsequent movement was closely monitored by satellite and radar. In respect to natural disasters, almost all the countries which suffer most are developing countries. These countries need assistance in facing up to these storms and their consequences.

Truck Routing: A concept in which trucks and road tankers are confined to a particular designated road system. The aim is often to exclude heavy trucks from residential districts and local roads, save for the purposes of delivery or collection. It may also be employed to exclude heavy trucks from the centers of towns, villages, and shopping areas. In respect to mines and quarries, it is sometimes possible to construct separate haul roads.

The acceptance of truck routing is becoming increasingly a condition of approval or consent for industrial developments such as oil refineries, gas terminals, waste disposal or processing plants, quarries, mines, sand extraction operations, logging, servicing of container terminals, enrichment plants, and port traffic. The encouragement of all heavy traffic to use designated roads is consistent with the concept of a functional road hierarchy.

Heavy trucks are often regarded as a major hazard and source of discomfort by the public, often being involved in the more serious traffic accidents, and carrying loads of flammable or explosive liquids and materials. They are a source of noise and vibration, and psychologically overwhelm the ordinary motorist and pedestrian. The enforcement of approved truck routes remains, however, a problem particularly when many outside contractors transport material to a particular plant. Surveillance is essential.

Turkey: See *Infant Mortality Rate.*

Twilight Zone: Or blighted area, usually a ring of older property rapidly decaying around the central area of any large town or city. The cause of such decay is often partly physical and partly social. See *Urban Renewal.*

U

Union of Soviet Socialist Republics (USSR): See *Council for Mutual Economic Assistance (COMECOM); Environmental Administration, Soviet Union; Joint Committee for Environmental Protection, US-USSR; Lake Baikal; National Parks; Sanitary Epidemiological Service.*

United Kingdom (England, Wales, Scotland, and Northern Ireland): See *Abercrombie Plans for London; Acid Rain; Advertisement Control; Amenity; Barlow Report, 1940; Buchanan Report, 1963; Civic Amenities Act, 1967; Clean Air Acts, 1956 and 1968; Committee on Public Participation in Planning, 1969; Compulsory Purchase; Conservation and Development Program; Control of Pollution Act, 1974; Corporate Planning; Country Parks; Countryside Commission; Department of the Environment; Distribution of Industry Act, 1945; Dobry Report; Enterprize Zones; Environmental Recovery Areas; Examination in Public; General Improvement Areas; Glasgow Eastern Area Renewal Project, 1976-87; Greater London Development Plan; Health and Safety at Work Act, 1974; Herbert Report, 1960; Heritage Building and Conservation Areas; Heritage Coasts; Housing Improvement Grants; Industrial Air Pollution Inspectorate; Inner Urban Areas Act, 1978; Land Commission; Land Compensation Act, 1973; Landscape Evaluation; Litter Control; Local Government Act, 1972; Local Plans; London Smog Incidents; National Planning Guidelines; National Heritage Memorial Fund; National Parks; National Survey of Air Pollution; Nature Conservancy; Noise Abatement Zone; Parker Windscale Inquiry; Planning Gain; Planning Inquiry Commission; Planning System; Public Local Inquiry; Public Project Inquiry; Reith Report on New Towns, 1946; Restriction on Ribbon Development Act, 1935; River Thames Clean-Up; River Water Quality Classifications; Roskill Commission; Royal Commission on Environmental Pollution; Royal Society for Nature Conservation; Scott Report, 1942; Slum Clearance Program; Smoke Control Area; Strategy Plan for South-East England; Structure Plan; Town and Country Amenities Act, 1974; Town and Country Planning Act, 1932; Town and Country Planning Act, 1947; Town and Country Planning Act, 1968; Town and Country Planning Act, 1971; Town Development Act, 1952; Transport Policies and Programs; Twilight Zone; Urban Development Corporations; Urban Renewal; Uthwatt Report, 1942; Water Act, 1973.*

United Nations Conference on Desertification (UNCOD): Held in Nairobi, Kenya, in 1977 the first worldwide effort to consider the global problem of advancing deserts. Some 95 countries participated together with 50 United Nations bodies, 8 intergovernmental organizations, and 65 non-governmental organizations. The outcome of the conference was a World-Wide Plan of Action to Combat Desertification, with 26 specific recommendations. In addition, the Executive Director of the United Nations Environment Program (q.v.) was asked to convene a Consultative Group for Desertification Control

The core members of the Consultative Group have comprised Australia, Bolivia, West Germany, India, Iran, Iraq, Japan, Libya, Mexico, Niger, Senegal, Sudan, Sweden, the United States of America, Upper Volta, and Uruguay; also seven agencies including the World Bank. The purpose of the group is to assist in mobilizing resources for activities forming part of the Plan of Action.

An Inter-Agency Working Group on Desertification was established in September, 1978, to coordinate the activities of the specialized agencies and other United Nations organizations and programs.

In all these activities, particular emphasis is placed on the arid and semi-arid regions of Africa and to a lesser degree on similar regions in southwest Asia and Latin America.

A number of projects and programs aimed at assisting countries with desertification problems were not originally conceived as anti-desertification measures. However, through redesign and reorientation, they now serve this end. They usually involve better planning and management of soil resources.

There are, however, some formidable obstacles to the implementation of the Plan of Action, apart from the massive scale of the problem. The governments of countries faced with desertification problems are confronted with conflicting demands on limited financial and human resources. They appear generally, to be unable to assign a sufficiently high priority to desertification prevention or control, and have only to a limited degree included such measures in their national development plans. Insufficient financing has plagued the program. See *Desertification Control.*

United Nations Conference on New and Renewable Sources of Energy: A United Nations Conference held in 1981; it adopted a program of action for increased use of new and renewable sources of energy in a socially equitable, economically and technically viable, and environmentally sustainable manner. The plan particularly emphasized the need to consider the environmental aspects of programs for the exploration, development, and utilization of new and renewable sources of energy.

United Nations Conference on the Human Environment, 1972: Held in Stockholm, from 5 to 16 June 1972, the first United Nations Conference on the Human Environment; the main purpose of the Conference was defined as being: "to serve as a practical means to encourage, and to provide guidelines for, action by Government and international organizations

310   Environmental Planning

designed to protect and improve the human environment and to remedy and prevent its impairment, by means of international cooperation, bearing in mind the particular importance of enabling developing countries to forestall occurrence of such problems." The Conference was attended by 113 delegations.The principal achievements of the Conference were the agreements reached on:

1. A Declaration on the Human Environment,
2. An extensive program of international action (the Action Plan),
3. A permanent environment secretariat (now based in Nairobi, Kenya),
4. An environment fund of $US 100 m for expenditure in the first five years to support new environmental initiatives.

The major disappointment of the Conference was the failure of the Soviet Union and countries from eastern Europe, with the exception of Yugoslavia and Rumania, to attend. The Conference was attended, however, by China.

In 1982, a session "of a special character" of the Governing Council of the United Nations Environment Program (q.v.) was convened to review the major achievements of the Stockholm Action Plan for the Human Environment and to make recommendations for future action over the forthcoming 10 years.

United Nations Conference on the Law of the Sea (UNCLOS): International discussions which have continued since the early 1970s, aimed at establishing a new legal regime for the oceans and their resources. The discussions have embraced the concepts of 200-mile economic zones and an International Seabed Authority, and have been concerned also with the management and conservation of mineral and food supplies as yet unexploited. The need to prevent pollution and over-exploitation has been recognized. All nations would have a general obligation to protect and preserve the marine environment as a whole.

The UNCLOS negotiations have benefited from earlier agreements adopted for the purpose of protecting the marine environment. Early international agreements focused on pollution from ships and, in particular, oil pollution, e.g., the Convention for the Prevention of Pollution of the Seas by Oil, 1954; the Convention on the High Seas, 1958, with two articles concerned with control of pollution; the Convention on the Prevention of Marine Pollution by Dumping of Waste and other Matter, 1972; and the Convention for the Prevention of Pollution from Ships (MARPOL), 1973.

While much progress was made in the negotiations over a number of years, there was in the end a failure of consensus which precluded, for the time being, a new legal agreement.

United Nations Conference on Trade and Development (UNCTAD): A Standing Conference of the United Nations set up in 1964 to assist the less developed nations towards improved rates of economic growth. The Conference endeavors to do this by arranging aid and finance, and by promoting trade.

The Conference collaborates with UNEP in providing governments with information on the impact of environmental policies and measures on international trade and development; it seeks to ensure that environmental measures do not needlessly create obstacles to trade, especially to the disadvantage of developing countries. Particular attention has been paid recently to the environmental policies of the United States of America and their effects on foreign trade.

United Nations Conferences: See *Habitat; United Nations Conference on Human Settlements, 1976; United Nations Conference on Desertification, 1977; United Nations Conference on New and Renewable Sources of Energy, 1981; United Nations Conference on the Human Environment, 1972; United Nations Conference on the Law of the Sea; United Nations Conference on Trade and Development; United Nations Water Conference, 1977; World Conference on Agrarian Reform and Rural Development, 1979; World Food Conference, 1974; World Population Conference, 1974.*

United Nations Educational, Scientific and Cultural Organization (UNESCO): A United Nations agency which came into being in 1946; its objective is the expansion of education as a tool of progress in the world's developing countries. A major activity has been the Man and Biosphere Program which was launched in 1970. The objective of the Program has been to develop the scientific basis for the rational use and conservation of the resources of the biosphere, and for the improvement of the global relationship between humankind and the environment. It concentrates on the technical, scientific and educational aspects of conservation.

The Belgrade International Workshop on Environmental Education was convened in October 1975, under the UNESCO-UNEP program, with the purpose of reviewing and discussing the trends and emerging issues in this area, and to formulate guidelines and recommendations which could be applied internationally. The result was the Belgrade Charter, and the Belgrade Recommendations.

The Intergovernmental Oceanographic Commission (IOC) established within UNESCO in 1962, is the sole intergovernmental marine science body. Its purpose is to promote scientific investigation with a view to a better understanding of the nature and resources of the oceans through the concerted action of its members. The preservation of the marine environment is an important element in its program.

In 1972, the International Convention for the Protection of the World Cultural and Natural Heritage was adopted; by 1979 over 30 UNESCO member-states had become parties to this Convention. The Convention established a World Heritage Committee which met for the first time in June 1977. It created also a World Heritage Fund to enable financing of preservation projects; the Fund receives voluntary contributions from the member nations.

The Committee decided that the first action to be taken would be to call nominations for the 'World Heritage List' in the form of detailed case studies. Arrangements were made to assist countries which lacked the resources to prepare their own case studies.

United Nations Environment Program (UNEP): A Program and organization created by the United Nations Conference on Human Environment, 1972 (q.v.); based in Nairobi, Kenya, it has a Governing Council of 54 members charged with implementing the recommendations of that Conference. Activities are financed from an Environment Fund to which nations contribute. Between 1972-82, the United States of America provided about 40 per cent of the agency's total financing. The secretariat is headed by an Executive Director.

The Governing Council has instituted a three-level approach for the development of the Program:

> Level One (the Review) seeks to identify new issues requiring the attention of governments, to consider issues in relation to the activities being undertaken or planned to deal with them, and to assist in the establishment or revision of objectives and priorities.
>
> Level Two (the Program) is based on the objectives and priorities established by governments and defines a program of action to meet those objectives. Basic tools in this regard are the "program frameworks," which outline the conceptual approaches followed in programming, point out the principal actions to be taken and the principal "actors" involved, and indicate realistic time-scales. Within each program framework, specific projects may be identified.
>
> Level Three (the Fund) consists of those actions and projects identified within the program frameworks at Level Two which are selected for support by the Environment Fund. Such support will normally consist of only a portion of costs involved and is based on the degree to which the provision of such support is likely to contribute to the implementation of the program activity concerned. Funding in full is normally considered only in respect to relatively small projects undertaken as the initial stages of program development, when preliminary work is required in the course of elaborating the program in particular areas which may then lead to a larger project.

A special session of the Governing Council of the United Nations Environment Program was held in Nairobi from 10 to 18 May 1982; the session was held to mark the 10th anniversary of the 1972 Stockholm Conference. It was attended by ministers in charge of environmental affairs from 105 of the 157 UN member countries and a number of other delegates from various countries and UN organizations. The session adopted a 'Nairobi Declaration' which assessed the present state of the global environment together with a document, *The Environment in 1982: Retrospect and Prospect*, which sought to evaluate the principal results of the Action Plan adopted at Stockholm, and set the areas of priority activity for the next 10 years.

The Nairobi gathering was not, however, a grand affair on the Stockholm scale. While all heads of state were invited, only three turned up: the Presidents of Zaire, Sudan, and Kenya (the host country). Delegations were small; Australia with 17 delegates at Stockhom, provided two for Nairobi. Furthermore, the documents presented suggested a mixed record of achievement over the decade. UNEP itself claimed "fair-to-good progress" in respect to some of the elements of the action plan, while conceding "very slow" progress in other areas.

However, UNEP claimed that "despite deficiencies in environmental data, the scientific knowledge, technology and organizational means of avoiding or solving many of the environmental problems identified in the 1960s and 1970s are now at hand, to a significant extent as a result of activities associated with the implementation of the action plan." Several specific achievements were named:

- The Global Environmental Monitoring System (GEMS) focussing on pollution, health, climate and renewable natural resources was operating but still incomplete;
- INFOTERRA, a worldwide information network on the environment was functioning but was not being used as frequently or as widely as it should be;
- The International Register of Potentially Toxic Chemicals was established, providing information on about 45,000 known chemicals in use in the world;
- A World Conservation Strategy to protect plants and animals had been well received;
- The Regional Seas Program to protect the world's seas had been considered a major success, particularly in respect to the Mediterranean.

On the other hand, progress in the implementation of action plans adopted to combat desertification, to improve water supply and management, and to improve human settlements, had been "distressingly slow." In addition, governments had been unwilling to provide the data needed to establish a register of radioactive releases to the biosphere; and Principle 26 of the Stockholm Declaration calling upon countries to reach prompt agreement on the elimination of nuclear weapons had gone unheeded.

The arms race, preoccupation with conflict and survival, economic recession and an energy crisis, had combined to weaken the will of nations while reducing the resources available for matters of international concern.

United Nations Industrial Development Organization (UNIDO): Established in 1967, a United Nations agency, headed by an Industrial Development Board, which attempts to promote industrialization and meet the urgent needs of developing countries. In accordance with the Lima Declaration and Plan of Action, the aim is to increase the share of total world production of the developing countries to 25 per cent by the year 2000.

Global industrial secor studies are undertaken at the International Center for Industrial Studies. One of the purposes of these studies is to assess the implications of the Lima Declaration, including the environmental aspects. A number of case studies on the environmental aspects of various industries have been supported by UNEP. Five case studies were published in 1976 relating to an iron and steel plant in Brazil, a cement plant in Iran, a chemical plant in India, a textile plant in Thailand, and a chemical plant in Turkey. Studies have also been undertaken into fertilizers, leather and leather products, vegetable oils and fats, pharmaceuticals, and the agro-industries.

United Nations Organizations: See *Economic Commission for Europe; Economic Commissions for Africa, Latin America, Asia and the Far-East; Food and Agriculture Organization; United Nations Educational, Scientific and Cultural Organization; United Nations Environment Program; United Nations Industrial Development Organization; World Bank; World Health Organization; World Meteorological Organization.*

United Nations Water Conference (UNWC): A United Nations Conference held in Mar del Plata, Argentina, in March, 1977, which attempted to focus the attention of policymakers on the water needs that are likely to arise between now and the year 2000; the steps that could be taken to meet them; and the difficulties likely to be experienced by those who failed to make adequate provision. The proposal to hold this Conference was first made by the Committee on Natural Resources in New York in 1971; following endorsement by the United Nations General Assembly in 1975, this Committee then undertook the necessary preparatory work for the Conference. Some 216 thematic papers were contributed by 51 countries for consideration at the Conference. The great majority of the papers dealt with questions relating to policy formulation and planning objectives, including legal and institutional problems. However, a good number of papers dealt with the environmental impact of water development, both in the developing and the developed countries.

The Conference was deeply concerned with the simple fact that less than one-fifth of the world's population can get water simply by turning a tap; for the remaining four-fifths (among them Indian villagers, nomads in the African Sahel, and the inhabitants agglomerated around the great cities of Latin America and other regions) the getting of water is part of the daily struggle for existence. On the other hand, a great deal of water is wasted under current agricultural practices; large production increases could be achieved with the more efficient use of water through improved water management and better farming practices. Since the total amount of water in the world is fixed, additional supplies can only come from better distribution, greater efficiency in use, a more economical application in various uses, and good water management at all levels, local, regional, and national.

The Conference revealed that countries have a great deal to learn from each other in respect to good water practices. China was clearly strong in small water works, such as dams and canals; India and Egypt had acquired great knowledge about irrigation; the Soviet Union and the United States were strong in river basin management; the Philippines, Afghanistan and Ethiopia had forged ahead in coordinated water management; Europe had excelled in the international control and operation of important rivers such as the Rhine and the Danube; Israel demonstrated that it probably led the world in the efficient use of scarce water.

United Nations World Population Conference, 1974: Held in Bucharest in August, 1974, the First World Population Conference was attended by representatives from 136 governments. The purpose of the Conference was to consider population policies and programs needed to promote human welfare and development, and to review basic demographic problems and their relationship to economic and social development. The principal achievement of the Conference was the adoption of a World Population Plan of Action and a number of related resolutions on such matters as the status of women, the observance of national sovereignty, the environment and world food production. A less tangible but equally important result was the emergence of a consensus that national governments should integrate their policies on population questions with their plans for economic and social development.

The Conference accepted that irrespective of population policies devised now, the population of the world was likely to at least double over the next forty years. A growth in world population from the present level of around 3.9 billion to 5.8 billion was accepted as the lowest likely figure by the year 2000. It was generally accepted that a more likely figure was about 6.5 billion. The Conference therefore concluded that, apart from the effect of rising living standards in lowering fertility rates, measures to increase the availability of resources to sustain and improve the conditions of life must be as much a part of population policy as measures designed ultimately to affect growth rates.

Delegates from 132 countries attended a Second World Population Conference held in Mexico City in August, 1984. There was now general agreement that the world's population was no longer increasing so rapidly as to constitute a continuing crisis, although some countries were still having difficulties balancing their rates of population growth and economic development. The latest estimates suggested that world population might stabilize around 10 billion people during the 21st century.

United States of America: See *Acid Rain; Agency for International Development; Air Quality Act, 1967; Alaska National Interest Lands Conservation Act, 1980; Anadromous Fish and Great Lakes Fisheries Act, 1965; Asbestos-in-Schools Rule, 1982; Bhopal Disaster, 1984; Chesapeake Bay Program; Clean Air Acts, 1963 and 1965; Clean Air Amendment Act, 1970; Clean Air Amendment Act, 1977; Clean Water Act, 1977; Clean Water Amendment Act, 1981; Clean Water Restoration Act, 1966; Coastal Barrier Resources Act, 1982; Coastal Zone Management Act, 1972; Community Development Block Grant; Comprehensive Environmental Response, Compensation and Liability Act, 1980; Council on Environmental Quality; Duncan Classification; Emissions Trading Program; Enabling Acts; Endangered Species Act, 1973; Environmental Protection Agency; Environmental Technical Information System; Federal Land Policy Management Act, 1976; Federal Water Pollution Control Act, 1956; Food, Drug and Cosmetic Act, 1938; Forest and Rangeland Renewable Resources Planning Act, 1974; Great Lakes Water Quality Agreement; Habitat Evaluation Procedure; Insecticide, Fungicide, and Rodenticide Act, 1978; International Security and Development Act, 1985; Joint Committee for Environmental Protection, US-USSR; Litter Laws; Love Canal; Marine Protection, Research, and Sanctuaries Act, 1972; Michigan Episode; Model cities Program; Multiple-Use Sustained-Yield Act, 1960; National Ambient Air Quality Standards; National Environmental Policy Regulations, 1979; National Forest Management Act, 1976; National Human Monitoring Program; National Parks; National Stream Quality Accounting Network; National Trails System Act, 1968; National Wildlife Refuge System Administration Act, 1966; National Wildlife Refuges; New Communities Program; New Source Performance Standards; Noise Control Act, 1972; Non-Attainment Areas; Nuclear Regulatory Commission; Official Controls; Official Map; Planning Commission; Plat; Pollutant Standards Index; Prevention of Significant Deterioration; Primitive Areas; Reclamation Reform Act, 1982; Regional Development Commission; Regulatory Impact Analysis; Resource, Conservation, and Recovery Act, 1976; Safe Drinking Water Act, 1974; Scoping; Skyways, Skywalks, and Pedways; Solid Waste Disposal Act, 1965; State and Local Air Monitoring System; Surface Mining Control and Reclamation Act, 1977; Times Beach; Toxic Substances Control Act, 1976; Urban Development Action Grant Program; Urban Renewal; US-France Memorandum of Understanding, 1984; Variance; Water Pollution Control Amendment Act, 1972; Water Quality Act, 1975;*

Water Quality Amendment Act, 1970; Wild and Scenic Rivers Act, 1968; Wilderness Act, 1964; Williamsburg Restoration. See *Individual States*.

Unplanned Growth: City and regional development left entirely to the forces of supply and demand within a relatively unconstrained competitive economic system. In basic terms, this means that the purchaser of land has absolute rights to use the land for any purpose whatever without restraint and without regard to the rights, welfare, or health of others. A prairie psychology (q.v.) prevails while the land becomes littered with abandoned industrial equipment, worked-out quarries, failed enterprises and despoiled landscape; the individual sustains loss and disappointment as an investment in home or land is compromised by unforeseen adjacent developments. More specifically:

- Natural resources are destroyed or sterilized;
- Open spaces are despoiled;
- Agricultural land is rendered forever unproductive;
- People settle without regard to the enormous costs of the public facilities needed to support them;
- Whole areas of cities decay;
- Undesirable suburban spread and ribbon development occur;
- Housing for the lower-income groups is not provided;
- Houses and factories exist side by side throughout city areas;
- Junkyards spring up along highways and in residential districts;
- Privately constructed roads lead to areas which should not be developed;
- Housing is constructed on cheap, flood-prone land;
- The absence of housing codes leads to mixed residential development;
- Incompatible uses occur frequently;
- Pollution control and environment protection is at the owner's sole discretion;
- Transport or communications corridors for future development cannot be provided;
- Heavy traffic takes the shortest and often most unsuitable routes.

See *Environmental Planning, Objectives of; Incompatible Uses*.

Urban and Rural Zones Act, 1969-75, Denmark: The first stage of a major planning reform, an Act splitting Denmark into urban zones, rural zones, and areas for summer cottages. Its purpose was to ensure planned urban development, a suitable supply of land for building and urban development, and at the same time provide protection for the recreational interests of the population and the preservation of scenic values. The second phase of planning reform was the introduction of the National and Regional Planning Act, 1973. Under this Act, it is a requirement that regional planning be a continuous process throughout Denmark.

The third stage in planning reform was the Municipal Planning Act, 1977; under this Act each local authority was required to produce a municipal structure plan which related to the whole municipality, defined its main structure and urban pattern, and provided a framework for the preparation of local plans. See *Denmark*.

Urban Areas: Areas of land or localities where the principal land uses are residential, industrial, business and commercial. Special uses and related open spaces, within which a comprehensive range of public utility services and community facilities are provided, are generally available to the majority of occupants.

Urban Conservation Area: See *Conservation Zones*.

Urban Consolidation: A policy of increasing the number of houses and home units in existing serviced areas, the aim being to reduce urban sprawl and the costs of providing urban services to outer suburbs. It involves the redefinition of existing planning codes and rezoning to provide for medium density housing. It means "floating" medium density housing throughout low density residential areas, not always a popular measure with established residents.

Urban Design: The integration of human-made and natural features into understandable patterns which are composed in skillful or artistic arrangements for city or town-like development and can be appreciated as a whole. An urban design framework is a skeletal structure of urban design components which contribute to one's sense of place and orientation.

Urban Development Action Grant (UDAG) Program, US: A program established in the United States by the Housing and Community Development Act, 1977, to provide federal financial assistance to cities demonstrating community development need. Participation in the program is in fact limited to those cities and urban centers that meet criteria applied by the Department of Housing and Urban Development (HUD); even then assistance is concentrated on the most distressed cities and towns within the eligible group. The criteria embrace such factors as the age of housing, per capita income, population characteristics, unemployment, and evidence of poverty. Cities have considerable discretion in the types of activities undertaken with action grants; assistance has been running at well over $1 billion per annum. See *Community Development Block Grant (CDBG), US.*

Urban Development Corporations (UDCs), UK: Agencies set up by the Secretary of State for the Environment under the Local Government, Planning and Land Act, 1980, with far-reaching powers to secure the regeneration of outworn urban areas. By 1983, two UDCs had been established covering parts of the dockland areas of London and Merseyside. UDCs are required to prepare plans for their respective areas and secure their implementation, particularly through partnership with the private sector. it was considered that these special bodies could achieve the aims of revitalization more readily than established local authorities.

The London Docklands Development Corporation was established in 1981, charged with the task fo rehabilitating some 2,000 hectares (5,000 acres) of derelict dockland in London's working class east end. Of the 50,000 residents, over 90 per cent were housed in council-owned (local government owned) houses, distinctly downmarket in character; some 20 per cent of the labor force was unemployed. The area had little industry, poor public transport and bad roads. Within three years significant progress had been made.

The Merseyside Development Corporation, created also in 1981, has carried out an extensive program of land reclamation, physical construction work, and servicing, within its designated area on both sides of the River Mersey. The designated area includes the huge derelict docklands district of Liverpool and Birkenhead. Major initial projects have included the International Garden Festival, the refurbishment of the Albert Dock Warehouse buildings, and the development of the New Enterprise Workshops in the Brunswick industrial zone.

Urban Forest Programs: A description sometimes used for the intensification of woodland in city and urban areas as a means of bringing back wildlife and improving the appearance of suburbs. Such forests may also be used to create pleasant buffer zones between conflicting land uses, such as freeways and housing; improving awareness of the environment; and providing more scope for recreation. Community gardens, city farms, and the planting of more native flora may be encouraged; derelict land, such as disused tips, may be converted into parkland.

An urban forest program may be achieved through large-scale public and private plantings, involving state agencies, municipal authorities, corporations, and community groups.

Urban Renewal: The rehabilitation of districts, often surrounding the central business district (q.v.). Measures include slum clearance, redevelopment through private and public sector initiatives, enforcement of health and safety standards, encouragement of building maintenance and repairs, and the restoration of buildings which are still structurally sound and have an economic use. Urban renewal, in the fullest sense, aims not only at freeing urban communities of slums and blighted areas, but also at eliminating the causes of slums and blight. US federal assistance has been available for these purposes, though the biggest problem in the United States has been the relocation of displaced persons.

The deterioration of once fine residential districts and the development of slums surrounding the central business district has been a universal characteristics of urban growth; derelict industrial sites and abandoned warehousing have contributed to the decay. Progressive obsolescence and shifts in economic emphasis and opportunitites account for much of the problem. However, the social, economic and political consequences of slums and blighted areas have focused attention upon the need to combat the physical deterioration and its causes. Many American communities have made serious efforts to meet the situation through both private and public redevelopment efforts. In Britain, extensive slum clearance and redevelopment programs have been completed. See *Housing Act, 1949, US; Housing Improvement Grants, UK; Slum Clearance Program, UK; Urban Development Action Grant, US; Urban Development Corporations (UDCs), UK.*

Urban Renewal Act, 1974, Austria: Introduced by the Austrian Government in 1974, an Act providing legal machinery for the reshaping of residential areas where serious defects exist in terms of urban structure. Among the criteria for such defects, the Act lists external diseconomies resulting from certain uses of land, from industrial operations, or from traffic, especially in the form of noise, vibration, smoke, dust, waste gases, noxious odors, or waste water. These defects could be removed under renewal programs by erecting new buildings or improving existing ones. In restructuring such areas, consideration has to be given to the requirements of environmental protection, e.g., by moving industrial plants causing nuisances to other areas, or by surrounding them with plant and tree covered open spaces.

Urban Sociology: A subject of study traditionally concerned with the nature of community in the city, e.g., the relationship between race and community, ethnicity, neighborhood organizations, racial compositions of neighborhoods, and community power structures. New concerns have been with the decline of cities, and social network research.

The classical Chicago school of urban ecology was the first body in the United States to systematically deal with the issue of variations in the nature of community in the urban setting. Essentially, researchers saw the uses of land in the city as being analogous to the competitive process among plants and animals in natural ecological areas. The Chicago school sought to investigate the nature of community life in the distinctive sub-areas of the city. Sometimes these have been defined within a system of concentric circles, or within other types of ecological framework.

The need for a more systematic comparative approach has led to "social areas analysis," concentrating on social rank (economic status), urbanization (family status), and segregation (ethnic status); other variables relating to density and spatial arrangements have been added in more recent years. The use of complex computer programs permits the inclusion of a large number of variables that can be used to plot out the social composition and community characteristics both within and between cities.

Much of the research in urban sociology focuses on the aggregate social characteristics of individuals (e.g., variations in socioeconomic status) and aggregate characteristics of the social contexts in which they live and work. Traditionally, less attention has been given to actual interpersonal relationships, though some important breakthroughs in research have occurred in recent years.

Urban Structure Plan: A plan illustrating the basic land use distribution and communications network (such as main roads and railways) of the urban areas identified in the strategy plan (q.v.). Normally, it would comprise a number of districts for which district structure plans would then be prepared in more detail. An urban structure plan whilst giving a firm urban pattern does not deal with precise boundaries (which are left to development control plans and district structure plans). See *District Structure Plan.*

Urbanization: A process leading to a profound societal change, characterized by the movement of people from rural to urban areas. Environmental problems are likely to be aggravated also by the present large scale transition from rural to urban societies. During the 1950s, the urban population of the world was estimated to be growing by 3.4 per cent per annum, but in developing countries the rate was much higher. Urban regions are at present growing at something like twice the rate of overall population growth. By the year 2000, more than half the population of the world will be living in urban regions rather than rural areas; this proportion will include about 80 per cent of the people in developed countries and 40 per cent in less developed countries. There will be an increase of nearly 2,000 million in the number of city dwellers, i.e., about 450 million in developed and 1,500 million in developing countries.

In 1950, there were only 70 cities in the world with populations of one milion or more; today there are about 85 in the developed countries and 75 in the developing "third world." By 2000 it is estimated that there will be 276 such cities in the third world alone. There are many cities in the world where large numbers of people endure poverty, overcrowding, and pollution, notwithstanding that opportunities for the poor are, or ought to be, better in urban than in rural areas. This is due to an imbalance in the process of change in which many environmental elements become casualties. Certainly more attention will be given to community planning to obviate the worst aspects. However, the success of such planning hinges in substantial part on an allocation of resources not necessarily acceptable to those who control them. See *Table 20.*

Table 20: Urbanization in Selected Countries

Country	Urban Population as Percentage of Total Population 1960	Urban Population as Percentage of Total Population 1980	Number of Cities of over 500,000 Persons 1980
United States of America	67	73	67
United Kingdom	86	91	18
Sweden	73	87	3
West Germany	77	85	12
Canada	69	80	10
Australia	81	89	5
Israel	77	89	1
South Africa	47	50	7
Soviet Union	49	65	50
China	19	25	65
Japan	62	78	9
Malaysia	25	29	1
Philippines	30	36	3
Brazil	46	65	14

Source: World Bank (1980) *World Development Report*, Washington, D.C.

US-France Memorandum of Understanding, 1984: A bilateral arrangement between the United States of America and France to maintain and enhance cooperation between the two countries in respect to environmental affairs. The understanding was reached between the US Environmental Protection Agency (q.v.) and the French Ministry for the Environment and the Quality of Life.

US-USSR Environmental Protection Agreement: See *Joint Committee for Environmental Protection, US-USSR*.

Uthwatt Report, 1942, UK: The report of an expert committee on compensation (q.v.) and betterment (q.v.); the report recommended that the development rights in rural land should be nationalized, and the nation should buy the land when it was needed for development by compulsory purchase (eminent domain). It recommended also a levy on gains in land value attributable to planning schemes. See *Town and Country Planning Act, 1947*.

Utopian Planners: Descriptive of individuals who carried forward the idea of new "model communities." They included Sir Patrick Geddes (1854-1932), Sir Ebenezer Howard (1850-1928), and Lewis Mumford (1895-   ). Early communities in Britain included Robert Owen's New Lanark, Titus Salt's Saltaire, Bournville, Port Sunlight, Letchworth, and Welwyn Garden City. See *Garden City Movement; New Towns Act, 1946.*

UVCE: An unconfined vapor cloud explosion. A potential hazard with the storage of inflammable gases and other combustibles. See *Risk and Hazard Assessment.*

V

Vancouver Plan of Action: A product of Habitat: the United Nations Conference on Human Settlements (q.v.); the Vancouver Plan of Action urges all countries to establish as a matter of urgency a national policy on human settlements, embodying the desired distribution of population together with related economic and social activities over the national territory. A national policy for human settlements and the environment was regarded as an integral part of any national economic and social development policy.

While the Plan envisaged the continuous improvement, renewal and rehabilitation of existing settlements, the process must respect the rights and aspirations of inhabitants, especially the least advantaged, and preserve the cultural and social values embodied in the existing fabric. The Plan recognized that in the development of human settlements the quality of the environment must be preserved, pollution prevented by minimizing the generation of wastes, and unavoidable wastes effectively managed and whenever possible turned into a resource. Transportation policies should favor mass transportation, reduced congestion and reduced pollution by motor vehicles.

In respect to land, the Plan declared that the management of land should be subject to public surveillance or control in the interest of the nation. The increased increment resulting from the rise in land values deriving from changes in the use of land, from public investment or decision, or due to the general growth of the community must be subject to appropriate "recapture" by public bodies (i.e., the community), unless the situation calls for other measures such as new patterns of ownership, or the general acquisition of land by public bodies. Past patterns of ownership rights should be transformed to match the changing needs of society and be collectively beneficial. The supply of usable land should be maintained by appropriate methods, including soil conservation, control of desertification and salination, prevention of pollution, and use of land capability analysis, and increased by long-term programs of land reclamation and preservation.

Variance, US: Any modification or variation of official controls (q.v.) where it is determined by a planning commission (q.v.) or other body that, by reason of exceptional circumstances the strict enforcement of the official controls would cause unnecessary hardship. A "dimensional variance" means a departure from the terms of a zoning regulation pertaining to the height or width of structures and the size of yards and open spaces, granted where such departure would not be contrary to the public interest.

Vermont, US: See *Litter Laws, US*.

Vibration: The rapid to-and-fro movement of the ground or a structure. The extreme sensitivity of human beings to vibration ensures that the problem is taken into account in environmental studies and environmental impact statements. The problem may arise in situations ranging from the proximity of highways and railways to blasting operations in mines and quarries.

Victor Gruen Foundation for Environmental Planning: Founded by the pre-eminent architect Victor Gruen (1903-). an organization which seeks to bring about a greater public understanding of the decisive role which environmental planning (q.v.) should and must have if the ecological and biological balance of this planet is to be assured. The Foundation devotes itself to this task by undertaking research, education, and training directed towards the appreciation of environmental values. These activities have been influenced by the ideas of Le Corbusier (1887-1965). The Foundation is based in Los Angeles, California.

View Plane Technique: A technique of constructing sightlines for the purpose of determining whether proposed buildings will obscure the view of a landscape feature. The unobstructed lines of sight to the nearest part of the landscape element are known as the "view plane;" the view point is the point from which the view plane and sightlines originate. The view plane indicates the maximum building height for any land between the view point and the landscape feature if that view is to be preserved; in fixing building heights, however, this is only one factor to be taken into consideration.

Viewer Catchment: The locations or areas from which a proposed project will be visible, either fully or partially; and the numbers of persons likely to be involved. A full view would encompass all major features of the proposal including low level activity, buildings and stacks; a partial view might encompass only, for example, the upper portions of cooling towers and stacks.

Virginia, US: See *Acid Rain; Litter Laws, US; Williamsburg Restoration*.

Visual Elements: The single physical features, either natural or man-made, which are present within a landscape unit (q.v.), and which contribute to its visual quality (q.v.); may be referred to also as visual components.

Visual Impact Measurement: A measure of the change to landscape character and/or decrease in visual quality (q.v.) resulting from the inclusion of a proposal in a viewer catchment (q.v.). Ratings of high, medium, or low may be applied with particular reference to the number of viewers, proximity of viewers and types of viewing.

Assessing visual intrusion is essentially subjective. Techniques have been developed in the hope of introducing some objectivity. By and large the techniques have not been successful. Certainly the eye is the ultimate test and nothing can replace or disguise the subjective judgments which must be made.

Yet the techniques do provide valuable insights. The attention of the "eye" is directed towards certain matters which it might otherwise overlook. They have a valuable role to play in the evaluation of options (though certainly some techniques are more valuable than others). See *View Plane Technique; Visual Elements; Landscape Analysis, Factors to be Considered; Landscape Evaluation, UK.*

Visual Pollution: Visual squalor in an environmental context including such items as:

- Overhead wirescape;
- Litter and unauthorized tipping;
- Abandoned cars and large items of equipment;
- Derelict factory premises and abandoned industrial equipment;
- Unattractive hoardings and advertisements;
- Overhead highways;
- Parked cars along the length of residential roads;
- Large commercial vehicles parked "at home" in residential streets;
- Unmaintained residential property;
- Accumulations of bottles, garbage, and unwanted items on residential premises (in the yard or on the roof);
- Squatter settlements;
- Some kinds of alternative life-styles.

See *Visual Impact Measurement; Figures 19 and 20.*

Visual Quality: The degree to which visual elements (q.v) or components are present in a landscape or view, and are combined in such a way as to create impressions of harmony, contrast, and diversity. See *Visual Impact Measurement.*

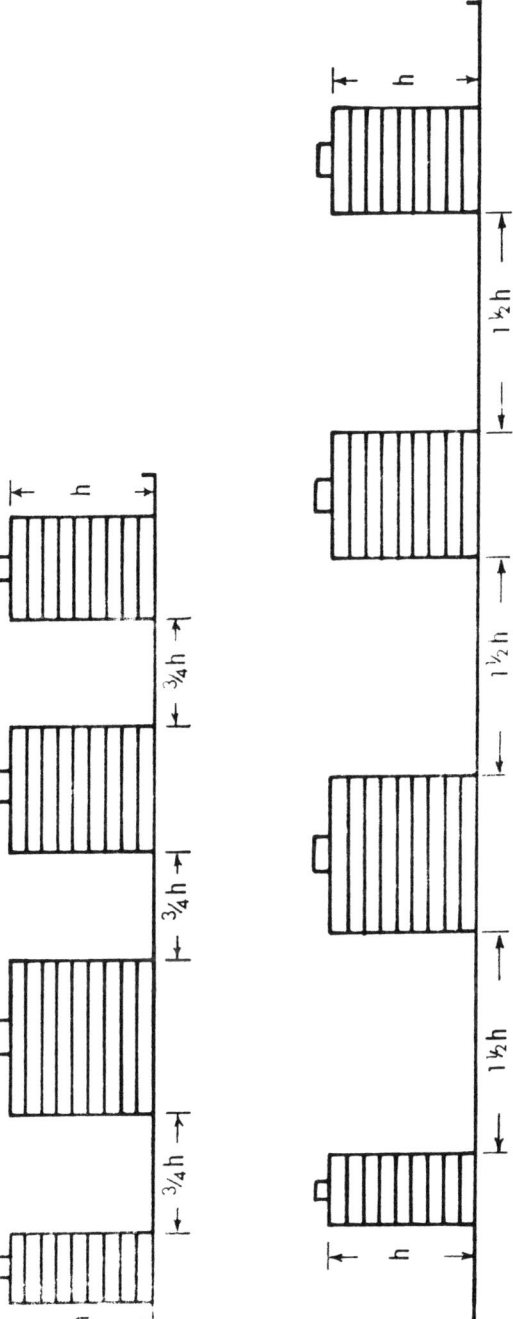

Figure 19: The visual effects of the spacing of high buildings. Top: Nine story buildings separated by a distance equal to three-quarters their height. Bottom: The same buildings if they were separated by a distance equal to one and one-half times their height.

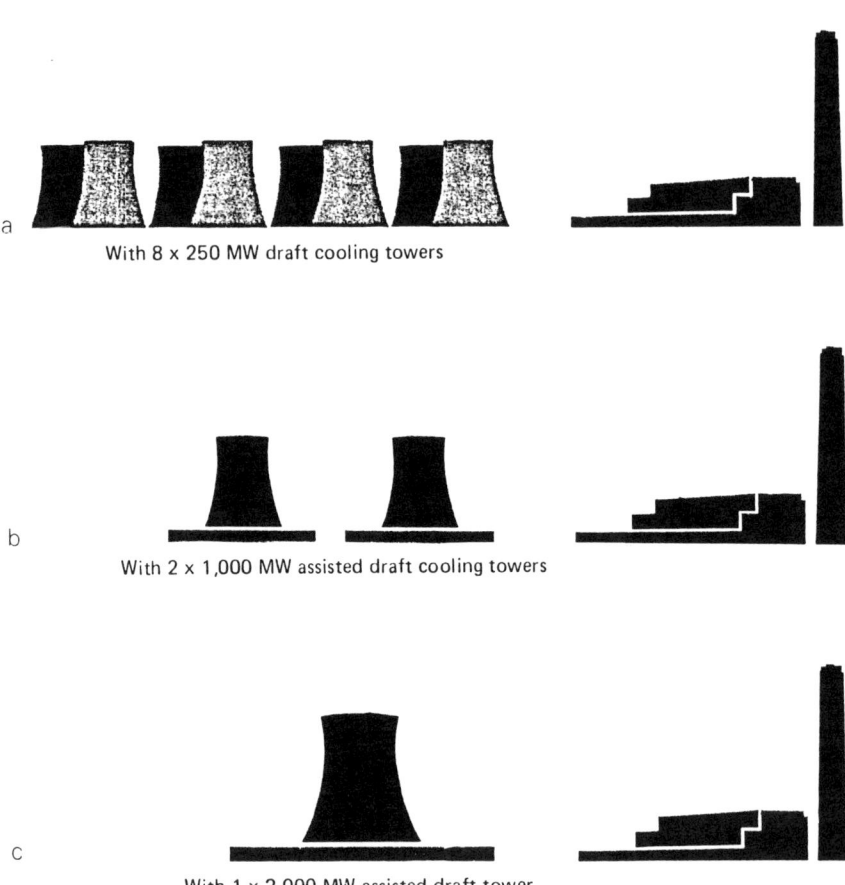

Figure 20: Visual appearance of cooling towers. Assisted draft cooling towers at a 2,000 MW power station, compared with the previous complement of natural draft towers. Source: Central Electricity Generating Board, London, England.

W

Washington, US: See *Litter Laws, US.*

Water Act, 1984, Sweden: An Act providing for the protection and conservation of water in Sweden as a common natural asset; it contains also several provisions of importance relating to nature conservancy. See *National Environment Protection Board, Sweden.*

Water Act, 1973, UK: An Act under which ten Regional Water Authorities were established in Britain to take over and integrate the functions of several statutory bodies. In addition to enforcing of pollution control legislation, the duties of the new authorities were extended to include environmental and aesthetic considerations. Thus in carrying out their functions they are to have regard to the preservation of natural beauty, the conservation of flora and fauna and features of scientific interest, and the protection of buildings of architectural value and public rights of way. The Act recognizes the amenity and recreational value of water space and requires the authorities to provide, whereever practicable, for public access to, and utilization of, the amenity. See *Figure 21.*

Water Classification, New South Wales: The six classifications of water prescribed in the Clean Water Regulations of New South Wales, Australia. These classifications, now covering all water bodies in the state, restrict what may be discharged into them. The following is a summary of them:

> Class S—Specially Protected Waters. No wastes whatsoever can be discharged into Class S waters, and all waters flowing into Class S waters are to be classified as Class P. Special precautions may be taken to control any activity carried out within the direct catchment area of Class S waters. The classification is intended to provide for protection of impounded waters for public water supply, of a stream in the vicinity of a water intake structure, waters in nature reserves and national parks, and waters in places of special scientific interest.

> Class P—Protected Waters. These are waters discharging into waters that are primarily used for domestic water supplies. The discharges permitted into Class P waters are limited to those meeting the standard required for drinking-quality water.

Figure 21: United Kingdom—Boundaries of the water authorities. Source: Department of the Environment, UK.

Class C—Controlled Waters. These are waters in outer catchments or drainage areas which may eventually reach public water supplies. Some discharges may be permitted subject to appropriate treatment for removal of pollutants, and subject to adequate dilution.

Class R—Restricted Waters. This classification is appropriate for waters to be protected for other than domestic water supply. The degree of protection required is that appropriate to provide satisfactory conditions for the proper protection and encouragement of aquatic life and water-associated wildlife.

Class O—Ocean Outfall Waters. These are open ocean waters into which waste discharges are permitted under specified conditions, but generally subject to the removal of gross solids and floating matter.

Class U—Underground Protected Waters. Waste discharges may be approved subject to requirements aimed at protecting the quality of underground waters for particular uses.

See *State Pollution Control Commission, New South Wales.*

Water Pollution: Substances, bacteria or viruses present in such concentrations or numbers as to impair the quality of the water, rendering it less suitable or unsuitable for its intended use and presenting a hazard to people or to their environment. Pollution may be caused by:

- Bacteria, viruses and other organisms that can cause disease, e.g., cholera, typhoid fever, and dysentery;
- Inorganic salts that cannot be removed by any simple conventional treatment process, making the water less suitable for drinking, for irrigation and for many industries;
- Plant nutrients such as potash, phosphates, and nitrates which, while largely inorganic salt, have the added effect of increasing weed growth, promoting algal blooms and producing, by photosynthesis, organic matter which may settle to the bottom of a lake;
- Oily materials that may be inimical to fish life, cause unsightliness, screen the river surface from the air thus reducing reoxygenation, accumulate in troublesome quantities, or have a high oxygen demand;
- Specific toxic agents, ranging from metal salts to complex synthetic chemicals;
- Waste heat that may render the river less suitable for certain purposes;

332   Environmental Planning

- Silt that may enter a river in large quantities causing changes in the character of the river bed;
- Radioactive substances.

Widespread disease is the main result of dirty water. Two-thirds of India's most devastating diseases, e.g., typhoid, infective hepatitis, cholera, dysentery and diarrhea, are water-borne. Of India's 3,119 cities and towns, only 217 have sewage treatment facilities.

The Ganges takes the sewage and industrial wastes of 48 large cities and 66 towns. Fortunately, the Ganges has a remarkable capacity for self-purification, and at many points people can still drink the water with minimal danger. However, the Yamuna which flows through Delhi is considered to be unfit for drinking or even washing all the way from Delhi to Agra.

The U.S. Public Health Service began its studies of water pollution as a result of the outbreaks of water-borne disease that took place around the turn of the century. The Public Health Service Act, 1912, specifically authorized surveys and studies of water pollution, particularly as it affected human health. Those early investigations provided the basis for technical services to local, state, and federal agencies, which remain an essential part of the federal program.

Traditionally, however, the control of pollution in the United States has been a state and local responsibility. For nearly half a century, Congress authorized research, studies, and technical assistance in the field, but did not take any further action concerning water pollution control until 1948. In that year, it passed the first Federal Water Pollution Control Act. Administered by the U.S. Public Health Service, this was experimental legislation, to be reviewed after a trial period of five years, and revised on the basis of that experience. It was extended for an additional three years to June 1956.

In 1956, Congress enacted the first permanent comprehensive Federal Water Pollution Control Act, which was a broad extension and improvement of the 1948 Act. Over the next five years much was accomplished. In 1961, Congress further amended the Water Pollution Control Act to launch a still more vigorous attack on the problem.

The development of state water pollution legislation has reflected increasing recognition of the need to provide for pollution control on a comprehensive basis, to protect all beneficial water uses within each state. Traditionally, state water pollution control legislation developed in connection with the preservation of public health and most early legislation vested authority for enforcement of the water pollution control program in state health departments.

Water Pollution Control Amendment Act, 1972, US: An Act of Congress amending earlier United States water pollution measures. The effects of these 1972 amendments were far-reaching, effectively rewriting the 1956 Act and the subsequent amending and augmenting legislation. The objective of the new Act was to restore and maintain the chemical, physical and biological integrity of the nation's waters. It reaffirmed the primary responsibilities and rights of the states to prevent, reduce, and eliminate pollution; and to plan the development and use of land and water resources. The federal agencies were to cooperate with State and local agencies to develop comprehensive solutions to prevent, reduce, and eliminate pollution, in concert with programs for managing water resources.

The legislation set out a number of policies and goals, namely that: the discharges of pollutants into navigable waters be eliminated by 1985; wherever attainable, interim goals for water quality to provide for the protection and propagation of fish, shellfish, and wildlife, and for recreational purposes in and on the water be achieved by July 1983; the discharge of toxic pollutants in toxic amounts be prohibited; federal financial assistance be provided for the construction of publicly-owned waste treatment works; area-wide waste treatment management processes be developed to ensure adequate control of the sources of pollution in each state; and a major research and demonstration effort be mounted to develop technologies necessary to eliminate the discharge of pollutants into navigable waters and the oceans. See *Clean Water Amendment Act, 1981; Federal Water Pollution Control Act, 1956, US.*

Water Quality Act, 1965, US: An Act to improve water quality approved unanimously by the US Congress in 1965. The Act established the Federal Water Pollution Control Administration under a Commissioner. While the legislation set up the new Administration within the Department of Health, Education and Welfare, a Presidential reorganization order transferred the new Administration to the Department of the Interior on 10 May 1966.

A key provision of the Act provided for the establishment of water quality standards. The states were given the option of setting water quality standards by 30 June 1967, for interstate waters within their borders. These standards had to "enhance the quality" of waters and be accompanied by a plan of implementation, including enforcement, of the proposed criteria. As a first step states were given a deadline of October 1966, to give notice to the Secretary of the Interior that they were willing to set their own standards. All 50 states, the District of Columbia, the Territories of Guam and the Virgin Islands and the Commonwealth of Puerto Rico did so and most of the states held the required public hearings to give interested groups an opportunity to comment on what uses of water should be provided for and what the required water quality should be for these uses.

The standards and implementation plans submitted by the states were subject to approval by the Secretary of the Interior. In order to receive approval, the standards had to be consistent with the objective of enhancing the qualityof these interstate waters for such purposes as public water supplies, fish and wildlife, recreation, agriculture, industry and other legitimate uses. If a state failed to adopt appropriate water quality standards, the Secretary of the Interior could, after consulting the representatives of federal departments, the states, municipalities, industries and other affected parties, prepare suitable water quality regulations for the interstate waters in the state concerned. See *Federal Water Pollution Control Act, 1956, US; Water Pollution Control Amendment Act, 1972, US.*

Water Quality Improvement Act, 1970, US: An Act of the US Congress passed in 1970 which required, inter alia, that a complete investigation and study of all methods of financing the cost of water pollution control be undertaken and submitted to Congress. Responsibility for water pollution control was transferred forthwith from the Department of the Interior to the new Environmental Protection Agency. In addition, the state governments became the final arbiters in the establishment of water quality standards, the Environmental Protection Agency simply acting as adviser. This was a retrograde step, in the opinion of many. See *Clean Water Amendment Act, 1981, US.*

West Germany: See *Convention on Long-Range Trans-Boundary Air Pollution; Federal Environmental Agency; Information and Documentation System for Environmental Planning (UMPLIS); International Commission for the Protection of the Rhine.*

Wetlands: Areas that are inundated by surface or groundwater with a frequency sufficient to support a prevalence of vegetatitve or aquatic life that requires saturated or seasonally saturated soil conditions for growth and reproduction. Wetlands may be swamps, marshes, fens, or bogs, depending on soil wetness and vegetation type. Wetlands may be located in uplands, along floodplains, around lake margins, in glaciated depressions, and in estuarine tidal marshes.

Wetlands perform important hydrologic functions such as to maintain the base flow of streams, to stabilize groundwater tables, and to provide groundwater discharge. They protect water quality by filtering out sediments and maintaining the capacity of watercourses to dilute pollutants from agriculture, urban runoff, and waste discharges. Wetlands also regulate stormwater runoff and minimize downstream flood potential.

Marshes and swamps provide important habitat for wildlife. This habitat makes excellent feeding, nesting, and wintering grounds for waterfowl and non-game bird species. Fur-bearing animals utilizing wetlands include, in the United States, mink, muskrat, and beaver. Human activity has been responsible for the significant reduction in America's wetlands during the past century. In the 1980s, the United States was still losing about 300,000 acres of wetlands each year due to development and drainage activities, despite a range of federal and state programs to protect wetlands. See *Convention on Wetlands of International Importance, Especially as Waterfowl Habitat.*

Wild and Scenic Rivers: Those rivers, or sections of rivers, which, together with their corridors, represent select parts of the environment which have significant value for recreation, or for aesthetic, nature conservation, historical or other purposes. Generally, wild rivers are those that traverse essentially natural areas that have had only minimal disturbance, do not have pondages or other water control structures, and where vehicle access is precluded or strictly limited. The term "wild" does not imply anything about the character of a river; it may have a low gradient with no obstructions or be steep and rocky. The term "corridor" is not an arbitrary distance, such as one kilometer, each side of the river; it simply embraces the immediate environment of the river which has recreational, scenic and nature conservation values associated with the river.

Wild and Scenic Rivers Act, 1968, US: A United States measure for the protection of wild and scenic rivers; the idea was to create a national system of designated rivers, restored as necessary to a condition suitable for fishing and swimming. However, implementation of the Act lagged somewhat. By 1979, only 27 river segments (some 2,200 miles or 3,500 kilometers) were components of the national system. In that year, the US President directed that: all federal agencies shall act to avoid or mitigate adverse effects on rivers identified as candidates for designation by the Heritage Conservation and Recreation Service; all federal land management agencies shall assess whether rivers located on their lands are suitable for inclusion in the wild and scenic rivers system, and take appropriate action; the Secretaries of Agriculture and Interior shall jointly revise their guidelines for evaluating wild and scenic rivers and shorten the time for studying rivers for designation.

In addition to these initiatives, the President affirmed his support for the designation of four river segments named in his previous Environmental Message, and added four new segments. The eight segments were: Gunnison River, Colorado; Encampment River, Colorado; Priest River, Idaho; Illinois River, Oregon; Bruneau River, Idaho; Dolores River, Colorado; Upper Mississippi River, Minnesota; and the Salmon River, Idaho. By 1985, the national wild and scenic rivers system embraced 7,000 miles or 11,200 kilometers of river.

Wilderness: A large tract of primitive country with its land and waters and its native plant and animal communities substantially unmodified by humans and their works. Large size and spaciousness are the essential characteristics of wilderness. Wilderness provides opportunities for primitive and unconfined forms of recreation, for self-education, and for observation by scientists. It enables visitors to experience solitude in areas where survival depends on personal resources. Access should be restricted to people travelling only on foot, ski, or in hand- or sail-powered boats. It is essential to the survival of wilderness and its genetic diversity, and to the preservation of its unique recreational advantages that mechanical forms of transport for visitors should be prohibited.

This concept of wilderness, conserved and made available for hardy and indeed dangerous recreation, was first developed in North America.

Most of the wilderness and primitive areas are west of the Great Plains. The domimant attributes of such areas are: first, that visitors to them have to depend exclusively on their own efforts for survival; and, second, that they preserve as nearly as possible the essential features of the primitive environment. Most of them are high, rugged terrains of peaks and forests, tundra and mountain glades, clear rivers and lakes. Some are vast stretches of arid and semi-arid country, interspersed with varied ecological conditions. Others have special attributes, such as the Boundary Waters canoe area (formerly known as the Superior Roadless area) in Minnesota, which protects the finest lakeland canoeing country in the United States by prohibiting logging on shorelines or the use of airplanes for transport, and which is contiguous with similar regions in Canada's Quetico Provincial park, in Ontario. Hikers and campers may follow thousands of kilometers of trails, on foot or with packhorses, to experience the kind of life their forefathers knew. The wild areas provide similar pursuits but are of smaller size. They are equally valuable as undisturbed examples of primeval ecology important to scientific research and for the determination of proper conservation practices on other land. A national Wilderness Act (q.v.) was passed in 1964.

Wilderness Act, 1964, US: An Act providing Congressional protection for a few named wilderness areas, while providing a procedure whereby lands within the national forests, national parks, and national wildlife refuges, would be reviewed and appropriate areas set aside within a National Wilderness Preservation System. See *Wilderness.*

Williamsburg Restoration, US: A striking example of a heritage conservation program in the restoration of Williamsburg, the capital of colonial Virginia, to much of its early 18th century appearance. From 1699 until 1780, when the capital was moved to Richmond, Williamsburg was the political, social and cultural center of the entire colony. Virginia was at that time the largest and most populous of the original British colonies. Its prosperity was largely dependent on the growth and sale of tobacco and the institution of slavery. It was also the center of protest against British authority. Acquisition and renovation has continued since 1926. The work has been carried on by a non-profit organization, dedicated to the theme "that the future may learn from the past."

The first major edifice to be restored was the Wren Building at the College of William and Mary, the second oldest college in the United States. The design of the building is thought to have originated with Sir Christopher Wren, the English architect. The 68 hectare historic area contains today more than eighty original 18th century structures, many other buildings that have been reconstructed after extensive archaeological and documentary research, and a hundred gardens and greens. The larger buildings, in addition to the College, include the Governor's Palace, the Capital building which housed the Assembly, the Bruton Parish Church, the Courthouse of 1770, the public gaol, the magazine and guardhouse. The principal street, Duke of Gloucester Street, was described by President Franklin D. Roosevelt as "the most historic avenue in all America."

In 1977, a heritage conservation program was introduced by the Secretary of the Interior, and a national register of historic places established. In 1979, the US President foreshadowed the establishment of a more comprehensive federal program to identify and protect significant natural areas as well as historic places with a new register of natural areas to supplement the original register. Federal agencies, state and local governments, Indian tribes and citizens would be encouraged to identify potential heritage areas.

Wisconsin, US: See *Acid Rain; New Communities Program.*

World Bank: A group of three institutions: the International Bank for Reconstruction and Development (IBRD), the International Development Association (IDA) and the International Finance Corporation (IFC). The common objective of these institutions is to help raise standards of living in developing countries by channelling financial resources from developed countries to the developing world. The Bank was established in 1945.

Since 1947, the Bank has been a "specialized agency" of the United Nations. The United Nations recognizes, however, the rather special nature of the Bank, its semi-autonomous role, and that the Bank must be the sole judge of to whom to lend and how much.

In 1970, the Bank created the position of Environmental Adviser, since converted into the Office of Environmental and Health Affairs. With this step, the Bank became the first development assistance agency to screen development projects, on a systematic basis, for their environmental and health implications. At first, such screening tended to occur at the "eleventh hour" as decisions had to be made on projects that had already progressed far along the path of "identification, preparation, appraisal, and negotiation." In time, however, environmental questions came to be handled as a routine and integral part of the analysis of each Bank project. Borrowers were urged to take environmental and health issues into account when identifying projects.

At the same time, lending for traditional projects continues, and is being redirected, to be more responsive to the new strategy of deliberately focusing on the poorest segments of society in the developing countries.

For example, during the eight-year period 1971-78, the Bank provided assistance amounting to almost $600 million for water supply and sewerage facilities in Latin America and the Caribbean, most of these facilities being in urban areas. These projects were expected to benefit about 8 million urban poor. All these measures have been carried out within the context of financially self-sustaining operations.

The Bank also assists city and regional governments in establishing environmental information and planning tools as part of the local urban planning and budget cycle, and in subjecting the Bank's own investments to scrutiny with these tools. The matrix in Figure 22 indicates the type of

338   Environmental Planning

Project Cycle	Environmental Analyses			
	1. Elaboration of regional development pattern	2. Selection of location for given development scheme	3. Design of project for given location	4. Evaluation of compatibility between given development scheme and given location
Identification	Promotion of efficient development pattern (e.g. infrastructure)	Selection of location for target sector (e.g. low-cost housing)	Selection of development scheme for target area (e.g. industry)	Identification of problem areas for env. projects (e.g. drainage channels)
Preparation	Detail design of projects promoting growth pattern (e.g. road alignment)	Detail site selection, site planning	Selection of alternatives, detail project design	
Appraisal				Review of environmental soundness
Supervision			Design and implementation of environmentally sound procedures	
Evaluation				Review of actual environmental performance

Figure 22: Environmental analyses in the World Bank's project cycle.

analysis suitable for each stage in the Bank's project cycle (identification, preparation, appraisal, supervision, and evaluation).

World Conference on Agrarian Reform and Rural Development: A United Nations Conference held in Rome in 1979; it adopted an Action Program for the equitable distribution and efficient use of land, water and other productive resources, with due regard to ecological balance and environmental protection.

World Conservation Strategy (WCS): A concept formulated in 1980 as a result of the cooperative endeavors of the International Union for Conservation of Nature and Natural Resources (IUCN), the World Wildlife Fund (WWF), and the United Nations Environment Program (UNEP) in consultation with over 700 scientists throughout the world. The WCS asserts that sustainable development can be achieved only through the conservation of living resources, identifies priority conservation issues and outlines the main requirements for dealing with them. The Strategy urges every country to prepare and implement a National Conservation Strategy. It should be noted that the WCS is concerned with the conservation of living resources and not with the conservation of non-living resources such as minerals, oil and gas. See *Living Resource Conservation.*

World Data Center on Microorganisms: Registers of microbial genetic resources prepared on behalf of the United Nations; the world data center is in Australia.

World Food Conference: A United Nations Conference held in 1974; it resulted in action to deal with crises in food supply including the establishment of a World Food Council, the International Fund for Agricultural Development, and an information system for food stocks and requirements, fertilizer availability, and crop conditions. An international agricultural research and training program now covers most major food crops and animal production. Environmental considerations are taken into account in all these activities.

World Health Organization (WHO): A permanent World Health Organization, supported initially by 54 countries, which came into being in April 1948 under the auspices of the United Nations. Based in Geneva, the new organization absorbed the international health activities of the United Nations Relief and Rehabilitation Administration which had assisted the health departments of liberated countries with both advice and practical aid, the Paris office of epidemic intelligence and the health organization of the League of Nations. WHO became the sole international health organization with the exception of the Pan American Sanitary bureau. Unfortunately, in 1947 the Soviet Union and associated countries withdrew from active participation in the creation of the new organization.

The World Health Organization operates, like the World Bank, with a high degree of autonomy as a specialized agency of the United Nations. The

Organization is directed and guided by a World Health Assembly which meets annually, and by an executive board representing 18 nations. The Organization itself is headed by a Director-General.

The constitution of WHO defines the organization's objective as "the attainment of all peoples of the highest possible level of health." This is to be regarded "as one of the fundamental rights of every human being without distinction of race, religion, political belief, economic or social condition." Health is defined as "a state of complete physical, mental and social well-being and not merely the absence of disease or infirmity." Thus the responsibilities of governments, according to the constitution, can be fulfilled only by the provision of adequate health and social measures.

The work of WHO in relation to the environment encompasses diverse matters such as the control of communicable diseases; water supply and waste disposal; air and water pollution; standards for biological and chemical substances; nutrition, food hygiene and food standards; occupational health; the effects of radiation; and psychosocial influences.

The WHO-UNEP Environmental Health Criteria Program is currently providing information on the quantitative relationship between exposure to chemical and physical agents and effects; some 38 criteria documents have been published. The carcinogenic risk of some 300 substances has been examined.

World Meteorological Organization (WMO): A United Nations agency that evolved from its predecessor, the International Meteorological Organization originally founded in 1873. WMO has a distinct role to play in grappling with the environmental issues now facing the world as a whole. Its contribution has found expression in the Global Atmospheric Research Program (GARP) and in an operational atmospheric monitoring system known as World Weather Watch (q.v.). In 1979, WMO undertook a World Climate Program to be executed in cooperation with the United Nations Environment Program (q.v.) and several other United Nations agencies. The WMO Background Air Pollution Monitoring Network (BAPMON) has been developed as part of the Global Environment Monitoring System (GEMS) (q.v.).

World Population Conferences, 1974 and 1984: See *United Nations World Population Conference, 1974.*

World Population Plan of Action: A product of the United Nations World Population Conference, 1974 (q.v.). The basic objectives and approach of the World Population Plan of Action were:

- To help coordinate population trends and the trends of economic and social development. The basis for an effective solution of population problems is, above all, socioeconomic transformation... The Plan of Action must be considered as an important component of the system of international strategies and as an instrument of the international community for the promotion of economic development, quality of life, human rights and fundamental freedoms.

The principles of the Plan assumed that:

- The formulation and implementation of population policies is the sovereign right of each nation. This right is to be exercised... without external interference, taking into account universal solidarity in order to improve the quality of life of the peoples of the world.
- Population and development are interrelated: The socio-economic nature of the recommendations contained in the Plan reflect its awareness of the crucial role that development plays in affecting population trends.
- All couples and individuals have the basic right to decide freely and responsibly the number and spacing of their children and to have the information, education and means to do so; the responsibility of couples and individuals in the exercise of this right takes into account their responsibilities towards the community.
- Women have the right to complete integration in the development process particularly by means of an equal access to education and equal participation in social, economic, cultural and political life.
- In formulating national population goals and policies, consideration must be given, together with other economic and social factors, to the supplies and characteristics of natural resources and to the quality of the environment and particularly to all aspects of food supply.
- The growing interdependence among nations makes international action increasingly important to the solution of development and population problems.

On the basis of these principles, the Plan defines its primary aim as:

"To expand and deepen the capacities of countries to deal effectively with their national and subnational population problems...."

The greater part of the Plan is made up of recommendations for action and implementation. These cover population goals and policies, socio-economic policies, promotion of knowledge and policies, the role of national governments and international cooperation, and monitoring, review and appraisal.

The Plan was a remarkable international consensus, and it was also a solid foundation for the development and refinement of a world population policy consistent with the motto adopted for World Population Year: "One World for All."

World Resources Institute: An international policy research center created in late 1982; its research is aimed at providing accurate information about global resources and population, identifying emerging issues, and developing politically and economically workable proposals. Its recommendations in respect to some urban and environmental issues are as follows:

- Slow the rate of migration into the largest cities through policies that decentralize industries and provide employment opportunities in rural hinterlands and that improve total living conditions by providing health, education, and other basic services.

- Take advantage of self-help efforts by legalizing informal urban settlements and encouraging neighborhood improvement initiatives.

- Provide the financial resources for basic environmental services in urban areas by strengthening cities' tax and rate systems, and by sharing national and regional revenue sources with urban governments. This marshaling of resources must be accompanied by a devolution of authority from the central government, a significant upgrading of municipal administrative capacity, and an intensive effort to develop and demonstrate low-cost sanitation and other technologies appropriate to densely populated areas.

- Agree on key "environmental indicators" and standard "geographical information systems" to be used by international organizations and governments.

- Fully develop the Global Environment Monitoring System (GEMS), first proposed in 1972, providing sufficient support to enable the system to track the major changes in the physical and chemical, biochemical and physiological, resource and industrial, demographic and socioeconomic parameters that determine the quality of the human environment.

- Undertake an international scientific research program to develop an understanding of the earth and its environs as a single physical, biological, and geological system and build a predictive capability based on that understanding.

- Spur the development, in the public and private sectors, of computer applications that enable local, regional, and national governments to make use of international environmental databases.

World Weather Watch: An operational atmospheric monitoring system devised by the World Meteorological Organization (WMO) (q.v.). Its evolution has been a long and complex matter, an important recent development being the introduction of a global network of special stations for monitoring atmospheric background pollution.

World Weather Watch has four essential elements:

- A global observing system in which surface stations (some 8,500 of them), merchant ships (some 5,500 of them), ocean weather ships, commercial aircraft and meteorological satellites all fit into a fully coordinated observing system which is constantly in operation;
- A global data processing system, involving world and regional centers, some of them equipped with the most capable electronic computers;
- A complex operational telecommunications system circling the globe which encompasses not only every country in the world but almost all the major communities in each country, providing a storm and flood warning system;
- A global network of special stations (some 59 stations in 25 countries initially) to monitor atmospheric background pollution.

It has been found necessary to study whether long-term changes in atmospheric pollutants both gaseous and particulate, may affect atmospheric processes and induce climatic changes and, if so, the likely scope and nature of such changes. The way in which pollutants are transported and undergo chemical changes is also of crucial importance. In view of the importance of the atmosphere in transferring some pollutants to the oceans, it is clear that atmospheric and oceanic monitoring systems should be closely coordinated.

World Wildlife Fund (WWF): The World Wildlife Fund is an international foundation, based in Switzerland. It was founded in 1961 when many people were becoming seriously alarmed at the rapid decline of wildlife and the destruction of natural habitats. Scientists and conservationists from twelve countries joined together to issue a manifesto from Morges, Switzerland, the headquarters of the International union for the Conservation of Nature and Natural Resources (IUCN) declaring:

> "All over the world today vast numbers of fine and harmless wild animals are losing their lives or their homes as a result of thoughtless and needless destruction... the emergency must be tackled with vigour and efficiency."

Described as a new Noah's Ark, the World Wildlife Fund was launched—with the task of raising money from the public to support action to conserve the wildlife and natural habitat of the planet Earth. The secretariat was established at the office of IUCN in Morges, thus ensuring close liaison between the scientists and the fund-raisers.

The first national organization was set up in the United Kingdom followed by the United States of America, Switzerland, and the Netherlands. National organizations now exist also in Australia, Austria, Belgium, Canada, Denmark, Finland, France, Germany, India, Italy, Japan, Kenya, Luxembourg, Malaysia, New Zealand, Norway, Pakistan, Peru, South Africa, Spain, Sweden, Turkey, and Venezuela—27 countries in all. These national organizations are active forces for conservation in their countries through direct action, fund-raising and support for projects, and through educational and youth activities.

The scope of the World Wildlife Fund is the conservation of species of native wildlife in danger of extinction. It aims to create awareness of the threats to nature, and to convert financial support into action based upon scientific priorities. Between its foundation in 1961 and 1979, the Fund has channeled over $55 million into more than 2,000 projects which have saved animals from extinction all over the world.

X Y Z

Yugoslavia: See *Council for Mutual Economic Assistance (COME-COM)*.

Zoning: The most common development control; conventional zoning operates to reserve certain areas for specific land uses. The zoning ordinance generally divides the community into districts, and specifies for each the permitted uses, height and bulk of structures, minimum lot sizes, and dimensions and density of development. Many refinements to conventional zoning ordinances have been employed to protect environmental characteristics. Some of these are flexible zoning (q.v.), environmental protection districts (q.v.), performance standards (q.v.), development zones and holding zones (q.v.).

Traditionally, zoning is a technical or physical approach to the segregation of incompatible activities, such as housing and industry; it has certainly been effective in preventing generally the worst aspects of mixed development in new areas. However, it can achieve little in the short-term to solve problems in existing areas. Zoning controls future development or redevelopment, while activities already established are granted "existing use" rights which ensure continuity and even some limited expansion. The success of zoning depends upon the regulations being strictly applied, and rezoning applications being agreed to or refused only after full inquiry by a higher level of government. Many examples exist where slack administration, or simple non-enforcement, has allowed residential development to spring up in "industrial zones" and around airports and large factories; further, green belts and buffer zones tend to have limited lives and to be slowly whittled away. There is also the problem where the large corporation selects its preferred site and seeks approval to construct a facility; if the proposal is of great economic and social value, the zonings are then swiftly adapted to the new situation. Planning then becomes reactive, responding as best it can to situations outside of its control. Zoning has some profound implications, for in separating homes from workplaces a transportation problem emerges involving many in the lower-income categories. Zoning has also proved somewhat inflexible in its categories, for once they are written down in schedules they are found not to be suitable for specific cases. One solution suggested has been the introduction of performance standards, instead of zones; access to certain locations would thus depend on meeting criteria relating to aesthetics, noise, pollution, traffic, parking and other standards; however, such a concept is difficult to administer because of the scientific or subjective nature of the criteria. Zoning remains the preferred approach in North America and Europe. It has been applied in the area of national physical planning or macro-scale planning, to confine certain industries to certain areas and to protect major resources such as clay, sand, gravel, limestone, hard-rock, and coal. See *Back-Zoning; Board of Zoning Appeals, Los Angeles; Conservation Zones; Development Zones and Holding Zones; Exclusionary Zoning; Flexible (Average Density of Cluster) Zoning; Interim Zoning; New York Zoning Ordinance, 1916; Residential Zones; Spot Zoning and Rezoning; Subdivision; Taking Issue; Transfer of Development Rights; Zoning Administrator, Los Angeles.*

Zoning Administrator, Los Angeles, US: A member of the Los Angeles City Planning Department, appointed by the Planning Director. An administrator conducts hearings in respect to zoning matters and makes decisions which are final unless appealed to the Board of Zoning Appeals (q.v.). See *Hearing Examiners.*

Zoning Ordinances, US: Laws passed by municipalities relating to the zoning (q.v.) of land. In the United States, these ordinances are made by municipalities without, in most cases, the endorsement of a higher level of government, though subject to challenge in the courts.

The zoning ordinance is the principal legal regulator of urban design; it specifies the uses to which property may be legally put and the intensity of development allowed, stated usually in terms of floor area. A zoning plan may often specify off-street parking requirements or off-street truck loading facilities as a ratio of floor area.

Zoning designations in the United States have become more and more elaborate, with numerous subclassifications to encompass complex variations and combinations. However, zoning ordinances are negative in the sense that they can protect the individual owner and the public from ill-suited developments but they cannot plan school locations, traffic movements, or create beauty, order, and amenity. Further zoning ordinances are a parochial exercise, in the hands of local administrations and may not be linked with any community plan; indeed, they are often adopted in the absence of a city or town plan.

Generally, zoning ordinances or regulations contain:

- A text which lists the types of zones which may be used, and the regulations which may be imposed in each zone; these must be uniform throughout the zone. The text usually makes provision for the granting of dimensional variances, conditional use permits, and for nonconforming use of land and structures. Procedures for the granting of any zoning change are set out. See *Figure 23.*

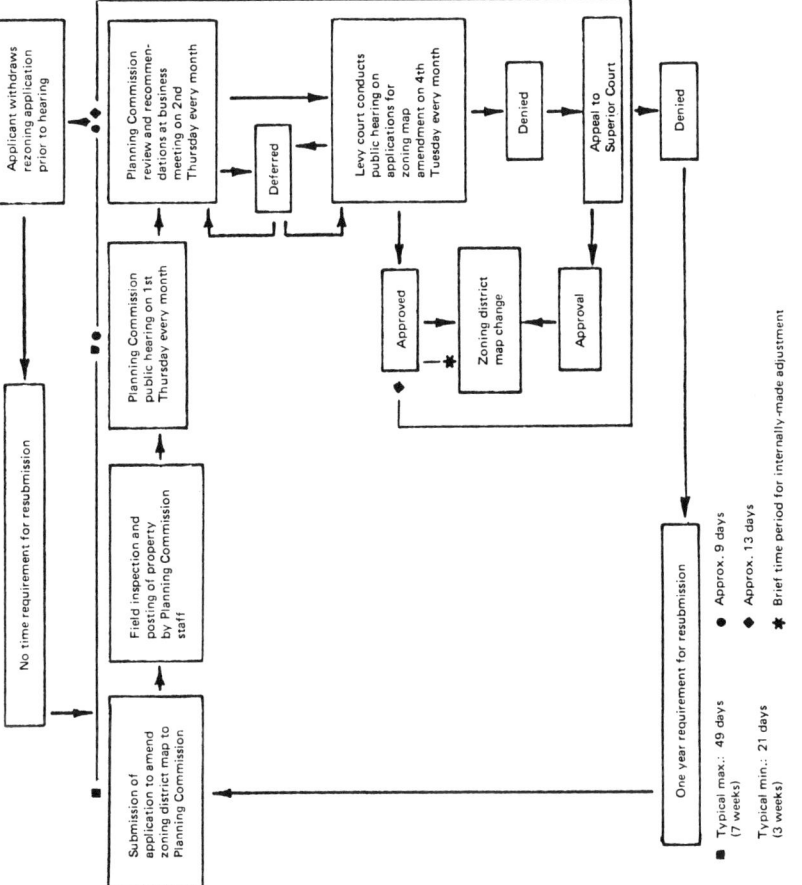

Figure 23: Kent County, Delaware rezoning process.

- A map showing the boundaries of the zoned area and the boundaries of each zone.
- Requirements in respect to the filling or excavation of land, the removal of natural resources, the use of watercourses and flood-prone land; the size, width, height, bulk, and location of structures and buildings; setback lines; minimum or maximum areas or percentages of areas, courts, yards, or other open spaces which are left unoccupied, and minimum distances between buildings or other structures; floor area to ground area ratios; intersections, interchanges, and transportation arteries.
- Requirements in respect to historical and conservation districts.